P9-CBK-600

Active Defense Mechanisms in Plants

NATO ADVANCED STUDY INSTITUTES SERIES

A series of edited volumes comprising multifaceted studies of contemporary scientific issues by some of the best scientific minds in the world, assembled in cooperation with NATO Scientific Affairs Division.

Series A: Life Sciences

Recent Volumes in this Series

Volume 32 – The Blood – Retinal Barriers
edited by Jose G. Cunha-Vaz

Volume 33 – Photoreception and Sensory Transduction in Aneural Organisms
edited by Francesco Lenci and Giuliano Colombetti

Volume 34 – Lasers in Biology and Medicine
edited by F. Hillenkamp, R. Pratesi, and C. A. Sacchi

Volume 35 – Environmental Physiology of Fishes
edited by M. A. Ali

Volume 36 – The Prostaglandin System: Endoperoxides, Prostacyclin, and Thromboxanes
edited by F. Berti and G. P. Velo

Volume 37 – Active Defense Mechanisms in Plants
edited by R. K. S. Wood

Volume 38 – Cell Growth
edited by Claudio Nicolini

Volume 39 – Drug Development, Regulatory Assessment, and Postmarketing Surveillance
edited by William M. Wardell and Giampaolo Velo

Volume 40 – Chromosome Damage and Repair
edited by Erling Seeberg and Kjell Kleppe

Volume 41 – Protein Biosynthesis in Eukaryotes
edited by R. Pérez-Bercoff

This series is published by an international board of publishers in conjunction with NATO Scientific Affairs Division

A Life Sciences
B Physics
Plenum Publishing Corporation
London and New York

C Mathematical and Physical Sciences
D. Riedel Publishing Company
Dordrecht, Boston, and London

D Behavioral and Social Sciences
E Applied Sciences
Sijthoff & Noordhoff International Publishers
Alphen aan den Rijn, The Netherlands, and Germantown, U.S.A.

Active Defense Mechanisms in Plants

Edited by
R.K.S. Wood
Imperial College of Science and Technology
London, England

PLENUM PRESS • NEW YORK AND LONDON
Published in cooperation with NATO Scientific Affairs Division

n?

Library of Congress Cataloging in Publication Data

NATO Advanced Study Institute on Active Defense Mechanisms in Plants (1980 : Cape Sounion, Greece)
Active defense mechanisms in plants.

(NATO advanced study institutes series. Series A, Life sciences ; v. 37)
"Proceedings of a NATO Advanced Study Institute on Active Defense Mechanisms in Plants, held April 21-May 3, 1980, in Cape Sounion, Greece"—Verso of t.p.
Bibliography: p.
Includes index.
1. Plants—Disease and pest resistance—Congresses. I. Wood, R. K. S. II. Series.
SB750.N36 1980 632'.3 81-15706
ISBN 0-306-40814-7 AACR2

Proceedings of a NATO Advanced Study Institute on Active Defense Mechanisms in Plants, held April 21-May 3, 1980, in Cape Sounion, Greece

A Division of Plenum Publishing Corporation
233 Spring Street, New York, N.Y. 10013

Printed in the United States of America

PREFACE

A NATO Advanced Study Institute on "Active Defence Mechanisms in Plants" was held at Cape Sounion, Greece, 21 April - 3 May 1981. It succeeded a similar Institute held at Porte Conte, Sardinia in 1975 on "Specificity in Plant Diseases."

What are active defence mechanisms in the context of plant disease in which a plant, the host, may be damaged by a pathogen? Defence mechanisms comprise properties of the host that decrease this damage. The mechanisms are passive when they are independent of the pathogen. They are active when they follow changes in the host caused by the pathogen. Thus for a fungal pathogen, cell walls of a higher plant which are lignified before infection would be a passive defence mechanism if they decreased damage by impeding growth of the fungus. Cell walls known to become lignified as a response to the pathogen would be an active defence mechanism if it were established that this response decreased damage.

The papers and discussions at this Advanced Study Institute were about active defence mechanisms in higher plants, mainly economically important crop plants, against fungi, bacteria and viruses as pathogens.

Taking the microorganisms first it is a truism but one that bears repeating that although plants almost always grow in close association with a wide variety of fungi and bacteria, often of types that can be pathogens, they rarely become diseased, at least not sufficiently so as to attract notice. A particular plant is a host to only a few of the many microorganisms to which it is exposed during its life; these are its pathogens. It is highly resistant to the rest. A plant may also exhibit considerable resistance even to its own pathogens in the sense that they may cause only limited damage. Furthermore, some forms of the plant may be highly resistant to certain forms of the pathogens. A main purpose of the Institute was to present and discuss the evidence that these types of resistance to fungi and bacteria depend upon active defence mechanisms.

The relations between viruses and their host plants are very different from those between the other pathogens and their hosts so much so that there are some who think that the concept of active defence as outlined above hardly applies to viruses. But there are many who think otherwise so that there was little difficulty in deciding that viruses as pathogens of higher plants should be an important part of the Institute. This conformed with an earlier agreement with Professor G. Loebenstein that there should be an international meeting on active defence which would include viruses. I am much indebted to Professor Loebenstein for his interest in and advice about the Institute both in the early planning and later.

This book comprises the main lectures upon which the meetings and discussions were based and many shorter papers related to the subjects of the lectures. I am most grateful to the lecturers and to the other speakers as I am to those who accepted invitations to be Chairmen of the many sessions and leaders of the long and fruitful discussions. The discussions are not included in the book but a record of them has been kept to which reference may be made by those who attended the Institute.

The success of an Advanced Study Institute depends very much on the advice and help of innumerable colleagues and assistants. I thank each of them and I am sorry that I can name only a few. First and foremost I and indeed all other participants are greatly indebted to Dr Eris Tjamos who was responsible for all the many and complex arrangements in Greece. Dr Tjamos was, in effect, a co-Director of the Institute and such success as it achieved depended very much on his splendid efforts for many months before and then during the Institute. I am also greatly in debt to Mrs June Cheston for her secretarial work before the Institute, for her editorial work with manuscripts and for preparing the excellent typescripts from which this book was copied.

I thank my colleagues Dr Simon Archer and Dr Robert Coutts for much help before and during the Congress and in particular Dr Archer for agreeing to be my understudy and Dr Coutts for organizing a session on the use of protoplasts and tissue cultures in the study of active defence.

It is a pleasure to record the warmth of our reception in Greece by Professor S.D. Demetriades, Professor S.G. Georgopoulos and other members of the Hellenic Phytopathological Society, and to thank Professor Georgopoulos for the stimulating address with which he opened the Institute.

The financing of the Institute was helped greatly by the many bodies, especially a number of Universities and the National Science Foundation in the U.S.A., which paid substantial parts of the expenses of many of the participants.

Lastly and on behalf of all who participated in, profited from and enjoyed what is one of the best types of scientific meeting, I thank most warmly the Scientific Affairs Division of NATO for sponsoring the Institute and for the very generous grant which covered its costs and met a large part of the expenses of participants, and Dr Mario di Lullo of the Division for his advice about the Institute and his interest in it at all stages.

R.K.S.W.

CONTENTS

GENERAL REVIEW OF ACTIVE DEFENCE MECHANISMS IN PLANTS AGAINST PATHOGENS

R. HEITEFUSS

Institut für Pflanzenpathologie und Pflanzenschutz
der Georg-August-Universität, Grisebachstrasse 6
3400 Göttingen-Weende, Federal Republic of Germany

INTRODUCTION

The honour of presenting the introductory review-paper concerning the overall topic of this two weeks' symposium is certainly more than balanced by the burden and by the challenge to the invited speaker. These were admittedly my feelings when I was preparing this lecture, to be given to a distinguished audience of experts in their own and related areas of research, who of course are much more intimately familiar with progress and problems in their respective fields. Thus, to relieve this burden, I decided that it should not be my task to tell you about your special subjects, but rather to set the stage for the presentations and discussions we are going to have during the following two weeks. This will require a certain amount of generalisation, with all the difficulties and disadvantages that the generalist may encounter in discussion with the specialist. But it also gives the opportunity of a broad view of an ever enlarging scene, and does not exclude the bringing of one or the other aspect or detail into higher magnification or better focus, or the aiming of the spotlight on to special subjects which definitely will need much more discussion during our symposium, while permitting the selection of such special subjects which reflect my personal interest and experience.

Some people say that every good lecture should start either with a new joke or with an appropriate definition of the subject you are going to talk about. I shall confine myself to the latter. What do we understand as active defence or resistance in plants against pathogens ? Although such definitions have been given before by many plant pathologists, I will try one which may be general enough to suit specific needs.

Active resistance in plants to microorganisms comprises a series of interconnected processes which, following recognition, are induced in the host cell through its continuing irritation by structures or products of the parasite and which results in exclusion, inhibition or elimination of the potential pathogen.

You will readily recognize in this definition and know from your own experience, that it is not just one factor which determines active resistance, but rather a multitude of interconnected reactions within a dynamic system. I am following, for example, Kuć (44) and others who describe resistance as a multicomponent system. I think, that this definition would also fit into Bateman's (8) recently proposed 'Multiple component hypothesis of pathogenesis and parasitism', in which he presumably would include resistance as the induction of an unfavourable environment within a given plant to a potential parasite or pathogen.

However these general definitions do not relieve us from the attempt to dissect the system into its single but interconnected components. This can only be done by keeping in mind that sequences in time, space and location, as well as qualitative and quantitative aspects require consideration. In other or simpler words, when, where, which and to what extent do resistance determining events take place during the host-parasite encounter ? And only in consequence of such analysis for single and specific host-parasite combinations may we then attempt to integrate the key factors into groups of common elements within the multicomponent system of active defence in plants in general.

Using a sequentual approach, it might be possible to divide mechanisms of active defence into two distinct phases or periods (11), *a determinative phase*, during which recognition of a plant pathogen by the host cell may occur, and *an expressive phase*, during which plant cells undergo biochemical and structural changes resulting in resistance or incompatibility in contrast to susceptibility or compatibility.

I shall discuss selected aspects of these different phases with examples taken primarily from fungal diseases and also to a limited extent from bacterial and viral diseases. Finally, I shall also discuss some aspects of acquired resistance or induced protection in plants against microbial parasites, thereby pointing out how knowledge obtained by basic research on host-parasite interactions and active resistance may in the future be applied in plant protection.

RECOGNITION PHENOMENA

This topic touches very much on the theme of the last NATO

Symposium on 'Specificity in plant diseases' (1975) and has also been treated extensively in recent reviews (14, 66). It is connected with the problem of varietal specificity in gene-for-gene systems and raises the question of recognition and interaction between parasite genes for avirulence and the host genes for resistance. Based mainly on analogies and results from other related fields, the most likely candidates for these recognition events in plants are carbohydrate containing proteins, more generally called lectins and formerly called phytohaemagglutinins, which not only are surface-localised in plant cell walls and membranes but are also cytoplasmic constituents. In parasites the recognition molecules or signals may be polysaccharides or glycoproteins localized at the surface of cell walls or secreted into inter- or intracellular surrounding media within the host.

The question however remains, how do these primary events of recognition trigger or elicit the resistance responses that follow ? I will try to give some possible answers to this question in later sections related to more complicated reaction sequences.

Here it should be indicated that lectins directly might be responsible for resistance by functioning as agglutinins as in the original meaning. Sequeira and Graham (65) reported that avirulent strains of *Pseudomonas solanacearum* were strongly agglutinated *in vitro* by lectins isolated from potatoes which reacted weakly or not at all with virulent bacteria. The first indication that variety specific responses might also be explained on the basis of differential interactions between host lectins and different races of *P. phaseolicola* was recently obtained by El-Banoby and Rudolph (24). Possibly extracellular polysaccharides, as produced by all plant pathogenic bacteria, may be involved in these differential interactions, as will be reported in more detail by Rudolph at this meeting. These and the following examples of recognition phenoma however indicate, that the border lines between 'preformed' and 'induced' determinants of resistance in the strict sense are not as rigid in nature as in a clearcut definition. The host lectins responsible for recognition are apparently present before infection or contact with the parasite. Their interaction with specific signals from the parasite may however start the process of induced, active defence.

INDUCED INHIBITION OF PENETRATION INTO THE HOST CELL

In our discussion of the sequence in time of events in active resistance we have now to return to an earlier phase which usually takes place before specific recognition within the host tissue as indicated above. I refer to the process of active penetration by a parasite through the cell wall into the host cell, in the case of leaf parasites such as powdery mildews into the epidermal cell only. This confinement of the parasite to one cell layer has definite advantages in histological analysis by modern techniques of light and electron microscopy. Recent investigations have yielded extensive information especially for *Erysiphe graminis* on coleoptiles and leaves of wheat or barley (1, 4, 39, 47, 64, and others).

The epidermal cells respond to attempted or successful penetration by the parasite or to mechanical probing by microneedles with appositions to and modifications of the cell wall. These phenomena have been known in many plants since the time of de Bary and described as the halo response, the forming of cytoplasmic aggregates and the formation of papillae or cell wall appositions. Relevant to our topic today is the question whether papilla formation is a form of active defence. There has been considerable debate about this since both successful and unsuccessful or prevented penetrations may be accompanied by papillae. Recent, very detailed investigations by Zeyen and Bushnell (80) using time-lapse microcinematography with barley coleoptiles and transmission electron microscopy with leaf epidermal cells indicated that papilla deposition can be completed within 30 minutes to three hours after the appearance of the cytoplasmic aggregate. They suggested that papilla deposition can be divided into four sequential stages; 1 the deposition of osmiophilic, lipid materials; 2 the deposition and partial compaction of non-osmiophilic, amorphous material, probably insoluble polysaccharides; 3 the compaction of this amorphous material and 4 the incorporation of osmiophilic material into the host cell wall and into the amorphous material.

At maturity, the papillae are hardened, electron opaque wall appositions. Zeyen and Bushnell (80) postulate that the effectiveness of papillae as defence mechanism against epidermis penetrating fungi may be related to the ability of the epidermal cell to complete fully deposition and compaction of papillae before there is appreciable development of the fungal penetration peg. Other circumstantial evidence supports this conclusion (3). Challenge appressoria of *Erysiphe graminis* fail to penetrate pre-formed papillae of a compatible barley cultivar (4). Centrifugally enhanced formation of papillae or mechanisms linked to this process have the potential to prevent fungal ingress (78). A binary system for analysing primary infection and host response in populations of powdery mildew was developed by Johnson, Bushnell and Zeyen (39); it allowed evaluation of 'critical event points' in the sequence of events and reactions responsible for the expression of compatibility or incompatibility. Much more could be stated and discussed here on the chemical and functional nature of papillae or wall appositions in relation to active defence of plants against powdery mildew and other fungi and even to non-pathogenic bacteria (1, 2, 12, 47, 61, 62, 67, 73). But we have to turn our spotlight on to other aspects which need discussion.

ENCAPSULATION OF FUNGAL STRUCTURES WITHIN THE INVADED HOST CELL

To illustrate this possible form of active resistance I shall mainly use as examples the haustoria forming rusts and mildews. Apparently several lines of defence have been breached by the time the fungus has developed its haustoria within host cells. Such fully developed haustoria may also be found during an early phase of pathogenesis in a resistant, incompatible host. In this case the system has not yet arrived at the 'switching point' according to Heath (32). The haustorium usually is surrounded by a 'zone of apposition','extrahaustorial matrix' or 'haustorial sheath', using here the three different terms assigned to the same structure which is the immediate environment in which the haustorium exists (9) or to the host-parasite interface. Although there is still some controversy, most cytologists assume it to be a product of the host (48). Our studies with ^{3}H orotic acid labelled haustoria of *Uromyces phaseoli* within *Phaseolus vulgaris* strongly support this conclusion (52).

The function of the haustorial sheath may be to provide a means of transport from host to parasite for metabolites required by the fungus. But it may also function by preventing such uptake, as indicated by microradioautographic studies (53, 54), in which less radioactive lysin was taken up from the resistant host into the haustoria of the parasite. Manocha (50) comes to a similar conclusion from his studies with stem rust and wheat with the temperature sensitive gene *Sr*6. Observations of Coffey (17) with *Melampsora lini* on flax indicated an encasement of haustoria especially in intermediate types of resistance. So the haustoria may be encapsulated by a material apparently deposited by the host cell thereby rendering them non-functional. Again, these phenomana cannot be regarded as isolated events. They can be very different even in closely related host-parasite combinations and in cultivar-race interactions as discussed recently in considerable detail by Littlefield and Heath (48). This is especially evident in the next topic.

HYPERSENSITIVITY AND ACTIVE DEFENCE

Even from the first description and definition of hypersensitivity by Stakman (71), who noted that cells of wheat cultivars resistant to *Puccinia graminis tritici* were rapidly killed at and around the site of penetration, this phenomenon has been re-investigated and interpreted. There has recently been much argument as to whether hypersensitivity is the cause or the consequence of the resistance reaction (33, 41). As Ingram (38) and Littlefield and Heath (48) have pointed out, it may be either, according to circumstances, the host-parasite combination and the specific genes of avirulence and resistance involved. The most critical and thorough investigations with regard to this question have been carried out by Rohringer and Samborski and their co-workers for wheat and *Puccinia graminis tritici* (28, 29, 30, 62, 63, 70). About the technique used and the results obtained, I assume Dr Rohringer will report at this symposium in considerable detail. I hope that I interpret his conclusions correctly. In some host-parasite combinations the invaded host cell obviously shows necrosis and consequently inhibits fungal development. In other incompatible combinations the fungus is affected first and cell death occurs later and non-specifically. In a third group of interactions host cell necrosis and death is not always accompanied by necrosis of the haustorium. This essentially confirms earlier results (10) which showed no clear correlation between hypersensitive cell death and resistance to *Puccinia graminis tritici*.

The other classical example of hypersensitivity and active defence is the resistance of potato to *Phytophthora infestans*. The classical studies by Müller and Börger (55), were carried further by Tomiyama and co-workers (42, 43) and more recently by Hohl and Stössel (34), Hohl and Suter (35) and Shimony and Friend (69).

In contrast to the view of Király *et al.* (41), the hypersensitive death of host cells preceded inhibition of fungal growth by several hours in incompatible combinations. So at least in this host-parasite combination the cytological sequence of events is clear. What, however, remains to be answered are the following questions :

What are the factors inducing or eliciting the hypersensitive response; can the hypersensitive response be regarded as an energy requiring process of active defence, or is it merely an induced disorganisation and breakdown of cell constituents and compartments; what are the chemical means of defence in this case and are they related to the hypersensitive response ?

This leads naturally to the next topic.

ELICITORS OF PHYTOALEXIN SYNTHESIS AND THE ROLE OF PHYTOALEXINS IN ACTIVE DEFENCE

I feel like "Carrying owls to Athens" in discussing this topic in an audience including the most active researchers in this rapidly expanding field. Therefore, I will restrict myself to a few remarks, since this subject has been covered in several recent reviews (5, 23, 27) and will certainly be taken up in much more detail during this symposium.

1 Phytoalexin production can be induced by very low concentrations of 'elicitors' isolated from cell walls or from filtrates from cultures of several pathogens or non-pathogens .

2 The elicitor of *Phytophthora megasperma* has been characterised

as a glucan oligosaccharide; elicitors from other sources also seem to be glucans or glycoproteins.

3 Most of the available evidence indicates that elicitors have no race specificity.

4 Specific elicitors of glyceollin accumulation have been recently reported from a *Pseudomonas glycinea* - soybean system (11).

5 Partially purified glycoproteins isolated from incompatible races of *Phytophthora megasperma* protected soybean seedlings against infection by compatible races; glycoproteins from compatible races were ineffective (77). These glycoproteins were, however, poor elicitors of glyceollin production. It is claimed that these race specific glycoproteins trigger the defence reaction in incompatible but not in compatible cultivars.

These few statements are only an introduction to the discussion which will follow and in which the following questions may be elucidated or answered.

1 Is the elicitation of phytoalexin synthesis always accompanied by and causally related to induction of resistance ?

2 How are the hypersensitive response and phytoalexin production interrelated ?

3 What are the receptors within the host cell which receive the message of the elicitor or the race-specific 'triggering' substances?

4 Does the elicitor or triggering substance start a chain of events involving the action of m-RNA and the induction of protein synthesis and are these related to phytoalexin synthesis and to resistance ? What are the enzymes and pathways of phytoalexin synthesis and how are they regulated ?

5 What are the mechanisms suppressing the hypersensitive response and phytoalexin production in compatible host-parasite combinations ?

Before turning to this last question I will leave my role as a generalist for a few remarks on the first question citing a specific example from investigations of co-workers and myself (36, 37).

We isolated a so-called elicitor from cell walls of germinated spores of *Uromyces phaseoli* which after injection into bean leaves induced resistance against bean rust but also against other pathogens. The elicitor was only effective when injected at least two days before inoculation. Production of the phytoalexin phaseollin was detectable by TLC and bioassay two days after injection of the elicitor and continued to increase. However, the fungus was able to develop in the presence of the accumulating phaseollin if infection started simultaneously with elicitation of phytoalexin synthesis. Our conclusions from these results may be of general interest.

The parasite may be sensitive to inhibition by a phytoalexin only during very early stages of development and much higher concentrations of phytoalexins are required for inhibition at later stages in which the parasite may be more or less insensitive.

The localization of the phytoalexin within the tissue is such that direct contact with the parasite, as is necessary for its inhibition, is not possible.

Although phytoalexin synthesis is elicited, resistance in the potential host is mainly due to other factors or to additional determinants.

But let us now come back to question 5 which may lead to the next topic which I have to deal with briefly. At first sight it seems to be diametrically in contrast to the general subject of this lecture. It is the question as to whether it is possible to prevent or suppress the mechanisms of active defence or more specifically, hypersensitivity and the synthesis of phytoalexins.

SUPPRESSION OF HYPERSENSITIVITY AND PHYTOALEXIN SYNTHESIS

I will restrict myself again to one example, *Phytophthora infestans* on potatoes although the phenomena have been reported for other host-parasite combinations (56, 69). The hypersensitive reaction and phytoalexin synthesis in potatoes can be elicited non-specifically by components of cell walls of *Phytophthora infestans* (76). Pre-inoculation with a compatible race inhibits hypersensitivity and phytoalexin accumulation (72, 75) and water soluble, high molecular weight constituents from zoospores, and germination fluids of compatible races of *Phytophthora infestans* were also effective (19, 20). Recently the hypersensitivity inhibiting factors (HIF) were isolated from zoospores and mycelia of the fungus and characterized as glucans containing β-1, 3 and β-1, 6 linkages and 17 - 23 glucose units. Interactions between these glucans from compatible races and cell membranes may protect plants from damage or reactions caused by elicitors of hypersensitivity and phytoalexin production (21, 22, 25, 51). I assume that Dr Kuć and Dr Friend will report at this symposium more about these recent results.

They are also of general interest in another connection. Do they mean that in this and possibly other host-parasite combinations specificity is associated with the compatible interaction and not with incompatibility (25) ? In other words, would active defence be specifically inhibited ? Or is this induced susceptibility ?

ACQUIRED RESISTANCE - INDUCED PROTECTION

Much of what has been said about different aspects of active defence in the preceding sections may apply also to acquired resistance, induced protection or cross-protection known for a long time in phytopathology especially through the extensive and pioneering investigations of the virologists, but also receiving more general attention recently (49). Here I would like to restrict the subject to the type of protection against subsequent inoculation as induced by pre-inoculation of plants with non-pathogens or pathogens and as observed with fungi, bacteria and viruses. This subject will be much more extensively treated during this meeting, so only a few general remarks are appropriate here.

Induced protection has much in common with immunisation in mammalian systems in which we differentiate between passive and active immunisation. Can we also use this distinction in plant systems ? Passive protection would mean that the protecting organism reacts directly with the pathogen by physical or chemical means. Active protection would mean that physiological or structural changes are induced in the plant which render it resistant or enable it to respond with reactions leading to resistance. Obviously active protection seems to be much more important, not only as localized but also as systemic protection. And apparently in most cases it also may be regarded as a multicomponent system as, for example, the systemic protection in cucurbits against *Colletotrichum lagenarium*. One mechanism may be restriction of penetration into the host, a second may be responsible for agglutination of hyphae in penetrated tissue and a third the production and accumulation of phytoalexins around the site of infection (45, 59).

Studies of induced protection in tobacco plants against viruses have shown that resistance was correlated with the appearance of at least four proteins not present in healthy plants (40, 74). The formation of these proteins also occurred after the injection of polyacrylic acid which similarly induced protection (26). The physiological role of these newly formed proteins remains to be elucidated. However, similar additional proteins were also observed in leaves resistant to *Uromyces phaseoli* (18, 79), and in cucumber leaves with induced resistance to *Colletotrichum lagenarium* (6).

The observation that such acquired systemic protection was also effective in the field (16) supports the hope that in the future this principle may be developed to practical application in disease control. And this will be my final topic.

INDUCTION OF SYSTEMIC PROTECTION AS MEANS OF ACTIVE DEFENCE AGAINST PLANT DISEASES

The idea of protecting plants against diseases by means of activating their natural resistance mechanisms has been discussed repeatedly (44, 46 and others). Grossmann (29) coined the term of "conferred resistance" to include both the determinants of passive and active resistance in this "artificially" induced protection against plant pathogens. Relevant to our symposium is only the activation of latent structural or biochemical mechanisms of active defence against the potential pathogen in accordance with the definition given earlier in this lecture. Important with respect to the practical consequences is the observation that many plants, independently of their genetically determined susceptibility or resistance, have the potential for this acquired, non-specific protection which needs activation by biological, biochemical or chemical means.

I have discussed the biological means in the proceeding section. That the elicitors of phytoalexin production may be potential biochemical candidates for the induction of protection has been pointed out by Albersheim and associates (7). At this meeting Schönbeck will report about bacterial extracts which were not fungicidal, but even when applied under field conditions protected plants against several diseases. That chemical means can also be effective was recently shown by Cartwright *et al.* (15). A systemic fungicide, dichlor-dimethyl-cyclopropane carboxylic acid, increases the capacity of rice to synthesize phytoalexins (momilactones A and B = diterpenes) in response to infection by *Pyricularia oryzae* and thereby seems to protect the plant against the rice blast.

Another case, admittedly less clear cut, may be the systemic fungicide triadimephon (Bayleton) which is one of the most successful fungicides introduced in recent years especially for control of *Erysiphe graminis*, *Puccinia striiformis* and other diseases mainly in cereals. This fungicide does not inhibit conidial germination and appressoria formation but only haustoria formation of mildew at recommended field application rates (57). In addition it apparently stimulates defence reactions within the host. Mildew haustoria may be encapsulated by host material, possibly callose, and thereby rendered non-functional,as shown recently with special staining techniques in barley leaves (79).

Admittedly there are only very few examples, in which the principle of "conferred resistance" or induced protection has been developed to practical importance in plant protection. However it may be a future way by which many of the difficulties resulting from the development of resistance or tolerance against fungicides with specific mechanisms of action can be avoided.

At the end of this general review of active defence mechanisms in plants against pathogens, I will not try to summarize my incomplete remarks on the different aspects of structural and biochemical mechanisms of resistance. Instead, I now hand over the subject to you to complete the review in more detail during the next two weeks, to close the gaps which I have left, and to secure by your presentations and discussions the success of this symposium.

REFERENCES

1. AIST, J.R. (1976). Papillae and related wound plugs of plant cells. *Annual Review of Phytopathology* 14, 145 - 163.

2. AIST, J.R. (1977). Mechanically induced wall appositions of plant cells can prevent penetration by a parasitic fungus. *Science* 197, 568 - 571.

3. AIST, J.R. & ISRAEL, H.W. (1977). Effects of heat-shock inhibition of papilla formation on compatible host penetration by two obligate parasites. *Physiological Plant Pathology* 10, 13 - 20.

4. AIST, J.R., KUNOH, H. & ISRAEL, H.W. (1979). Challenge appressoria of *Erysiphe graminis* fail to breach preformed papillae of a compatible barley cultivar. *Phytopathology* 69, 1245 - 1250.

5. ALBERSHEIM, P. & VALENT, B.S. (1978). Host-pathogen interactions in plants. *Journal of Cell Biology* 78, 627 - 643.

6. ANDEBRHAN, T., COUTTS, R.H.A., WAGIH, E.E. & WOOD, R.K.S. (1980). Induced resistance and changes in the soluble protein fraction of cucumber leaves locally infected with *Colletotrichum lagenarium* or tobacco necrosis virus. *Phytopathologische Zeitschrift* 98, 47 - 52.

7. AYERS, J.R., VALENT, B.S., EBEL, J. & ALBERSHEIM, P. (1976). Host-pathogen interactions. XI. Composition and structure of wall-released elicitor fractions. *Plant Physiology* 57, 766 - 774.

8. BATEMAN, D.F. (1978). The dynamic nature of disease. In : *Plant Disease* Vol. III. Ed. by J.G. Horsfall and E.B. Cowling, pp. 53 - 83, Academic Press, New York.

9. BRACKER, C.E. & LITTLEFIELD, L.J. (1973). Structural concepts of host-pathogen interfaces. In : *Fungal pathogenicity and the plant's response*. Ed. by R.J.W. Byrde and C.V. Cutting, pp. 159 - 318, Academic Press, London.

10. BROWN, J.F., SHIPTON, W.A. & WHITE, N.H. (1966). The relationship between hypersensitive tissue and resistance in wheat seedlings infected with *Puccinia graminis tritici*. *Annals of Applied Biology* 58, 279 - 290.

11. BRUEGGER, B.B. & KEEN, N.T. (1979). Specific elicitors of glyceollin accumulation in the *Pseudomonas glycinae*-soybean host-parasite system. *Physiological Plant Pathology* 15, 43 - 51.

12. BUSHNELL, W.R. & BERGQUIST, S.E. (1975). Aggregation of host cytoplasm and the formation of papillae and haustoria in powdery mildew of barley. *Phytopathology* 60, 1848 - 1849.

13. BUSHNELL, W.R. & ZEYEN, R.J. (1976). Light and electron microscope studies of cytoplasmic aggregates formed in barley cells in response to *Erysiphe graminis*. *Canadian Journal of Botany* 54, 1547 - 1655.

14. CALLOW, J.A. (1977). Recognition, resistance and the role of plant lectins in host-parasite interactions. *Advances in Botanical Research* 4, 1 - 49.

15. CARTWRIGHT, D., LANGCAKE, P., PRYCE, R.J. & LEWORTHY, D.P. (1977). Chemical activation of host defence mechanisms as a basis for crop protection. *Nature* 267, 511 - 513.

16. CARUSO, F.L. & KUĆ, J. (1977). Protection of watermelon and muskmelon against *Colletotrichum lagenarium* by *Colletotrichum lagenarium*. *Phytopathology* 67, 1285 - 1289.

17. COFFEY, M.D. (1976). Flax rust resistance involving the K gene : an ultrastructural survey. *Canadian Journal of Botany* 54, 511 - 513.

18. DEEPEN, A. (1979). Elektrophoretische Untersuchungen der löslichen Blattproteine einer resistenten Bohnensorte (*Phaseolus vulgaris*) nach Inokulation mit Bohnenrost (*Uromyces phaseoli*). Vergleich von inokulierten und nicht inokulierten Blattbereichen. *Diplomarbeit, Landwirtschaftliche Fakultät Göttingen.*

19. DOKE, N. (1975). Prevention of the hypersensitive reaction of potato cells to infection with an incompatible race of *Phytophthora infestans* by constituents of the zoospores. *Physiological Plant Pathology* 7, 1 - 7.

20. DOKE, N. & TOMIYAMA, K. (1977). Effect of high molecular substances released from zoospores of *Phytophthora infestans* on hypersensitive response of potato tubers. *Phytopatholo-*

gische Zeitschrift 90, 236 - 242.

21. DOKE, N., GARAS, N.A. & KUĆ, J. (1979). Partial characterization and aspects of the mode of action of the hypersensitivity-inhibiting factor (HIF) isolated from *Phytophthora infestans*. *Physiological Plant Pathology* 15, 127 - 140.

22. DOKE, N., GARAS, N.A. & KUĆ, J. (1980). Effect on host hypersensitivity of suppressors released during the germination of *Phytophthora infestans* cystospores. *Phytopathology* 70, 35 - 39.

23. DRYSDALE, R.B. (1978). The elicitation of phytoalexin production. *Annals of Applied Biology* 89, 340 - 344.

24. EL-BANOBY, F.E. & RUDOLPH, K. (1980). Agglutination of *Pseudomonas phaseolicola* by bean leaf extracts. *Phytopathologische Zeitschrift* 98, 91 - 95.

25. GARAS, N.A., DOKE, N. & KUĆ, J. (1979). Suppression of the hypersensitive reaction in potato tubers by mycelial components from *Phytophthora infestans*. *Physiological Plant Pathology* 15, 117 - 126.

26. GIANINAZZI, S. & KASSANIS, B. (1974). Virus resistance induced in plants by polyacrylic acid. *Journal of General Virology* 23, 1 - 9.

27. GRISEBACH, H. & EBEL, J. (1978). Phytoalexine, chemische Abwehrstoffe höherer Pflanzen ? *Angewandte Chemie* 90, 668 - 681.

28. GROSSMANN, F. (1968). Conferred resistance in the host. *World Review of Pest Control* 7, 176 - 183.

29. HARDER, D.E., ROHRINGER, R., SAMBORSKI, D.J., KIM, W.K. & CHONG, J. (1978). Electron microscopy of susceptible and resistant near-isogenic(*sr*6/*Sr*6) lines of wheat infected by *Puccinia graminis tritici*. I. The host-pathogen interface in the compatible (*sr*6/*P*6) interaction. *Canadian Journal of Botany* 56, 2955 - 2966.

30. HARDER, D.E., ROHRINGER, R., SAMBORSKI, D.J., RIMMER, S.R., KIM, W.K. & CHONG, H. (1979). Electron microscopy of susceptible and resistant near-isogenic (*sr*6/*Sr*6) lines of wheat infected by *Puccinia graminis tritici*. II. Expression of incompatibility in mesophyll and epidermal cells and the effect of temperature on host-parasite interactions in these cells. *Canadian Journal of Botany*

57, 2617 - 2625.

31. HARDER, D.E., SAMBORSKI, D.J., ROHRINGER, R., RIMMER, S.R., KIM, W.K. & CHONG, J. (1979). Electron microscopy of susceptible and resistant near-isogenic (*sr6/Sr6*) lines of wheat infected by *Puccinia graminis tritici*. III. Ultrastructure of incompatible interactions. *Canadian Journal of Botany* 57, 2626 - 2634.

32. HEATH, M.C. (1974). Light and electron microscope studies of the interactions of host and non-host plants with cowpea rust *Uromyces phaseoli* var. *vignae*. *Physiological Plant Pathology* 4, 403 - 414.

33. HEATH, M.C. (1976). Hypersensitivity, the cause or the consequence of rust resistance ? *Phytopathology* 66, 935 - 936.

34. HOHL, H.R. & STÖSSEL, P. (1976). Host-parasite interfaces in a resistant and a susceptible cultivar of *Solanum tuberosum* inoculated with *Phytophthora infestans* : leaf tissue. *Canadian Journal of Botany* 54, 900 - 912.

35. HOHL, H.R. & SUTER, E. (1976). Host-parasite interfaces in a resistant and a susceptible cultivar of *Solanum tuberosum* inoculated with *Phytophthora infestans* : leaf tissue. *Canadian Journal of Botany* 54, 1956 - 1970.

36. HOPPE, H.H., HÜMME, B. & HEITEFUSS, R. (1980). Elicitor induced accumulation of phytoalexins in healthy and rust infected leaves of *Phaseolus vulgaris*. *Phytopathologische Zeitschrift* 97, 85 - 88.

37. HÜMME, B., HOPPE, H.H. & HEITEFUSS, R. (1978). Resistenzinduzierende Faktoren isoliert aus Zellwänden der Uredosporenkeimschläuche des Bohnenrostes (*Uromyces phaseoli*). *Phytopathologische Zeitschrift* 92, 281 - 284.

38. INGRAM, D.S. (1978). Cell death and resistance of biotrophs. *Annals of Applied Biology* 89, 201 - 294.

39. JOHNSON, E.B., BUSHNELL, W.R. & ZEYEN, R.J. (1979). Binary pathways for analysis of primary infection and host response in populations of powdery mildew fungi. *Canadian Journal of Botany* 57, 497 - 511.

40. KASSANIS, B., GIANINAZZI, S. & WHITE, R.F. (1974). A possible explanation of the resistance of virus-infected tobacco plants to second infection. *Journal of General Virology* 23, 11 - 16.

41. KIRÁLY, Z., BARNA, B. & ERSEK, T. (1972). Hypersensitivity as a consequence, not a cause, of plant resistance to infection. *Nature* 239, 456 - 458.

42. KITAZAWA, K. & TOMIYAMA, K. (1969). Microscope observations of infection of potato cells by compatible and incompatible races of *Phytophthora infestans*. *Phytopathologische Zeitschrift* 66, 317 - 324.

43. KITAZAWA, K., INAGAKI, H. & TOMIYAMA, K. (1973). Cinephotomicrographic observations on the dynamic responses of protoplasm of a potato plant cell to infection by *Phytophthora infestans*. *Phytopathologische Zeitschrift* 76, 80 - 86.

44. KUĆ, J. (1968). Biochemical control of disease resistance in plants. *World Review of Pest Control* 7, 42 - 55.

45. KUĆ, J., SHOCKLEY, G. & KEARNEY, K. (1975). Protection of cucumber against *Colletotrichum lagenarium* by *Colletotrichum lagenarium*. *Physiological Plant Pathology* 7, 195 - 199.

46. KUĆ, J. (1977). Plant protection by the activation of latent mechanisms for resistance. *Netherlands Journal of Plant Pathology* 83, Supplement 1, 463 - 471.

47. KUNOH, H. & ISHIZAKI, H. (1975). Silicon levels near penetration sites of fungi on wheat, barley, cucumber and morning glory leaves. *Physiological Plant Pathology* 5, 283 - 287.

48. LITTLEFIELD, L.A. & HEATH, M.C. (1979). Ultrastructure of rust fungi. Academic Press, New York.

49. LOEBENSTEIN, G. (1972). Localization and induced resistance in virus-infected plants. *Annual Review of Phytopathology* 10, 177 - 206.

50. MANOCHA, M.S. (1975). Autoradiography and fine structure of host-parasite interface in temperature-sensitive combinations of wheat stem rust. *Phytopathologische Zeitschrift* 83, 207 - 215.

51. MARCAN, H., JARVIS, M.C. & FRIEND, J. (1979). Effect of methyl glycosides and oligosaccharides on cell death and browning of potato tuber discs induced by mycelial components of *Phytophthora infestans*. *Physiological Plant Pathology* 14, 1 - 9.

52. MENDGEN, K. & HEITEFUSS, R. (1975). Micro-autoradiographic

studies on host-parasite interactions. I. The infection of *Phaseolus vulgaris* with tritium labelled uredospores of *Uromyces phaseoli*. *Archives of Microbiology* 105, 193 - 199.

53. MENDGEN, K. (1977). Reduced lysine uptake by bean rust haustoria in a resistant reaction. *Naturwissenschaften* 64, 438.

54. MENDGEN, K. (1979). Microautoradiographic studies on host-parasite interactions. II. The exchange of ^{3}H-lysine between *Uromyces phaseoli* and *Phaseolus vulgaris*. *Archives of Microbiology* 123, 129 - 135.

55. MÜLLER, K.O. (1959). Hypersensitivity. In : *Plant Pathology* I. Ed. by J.G. Horsfall and A.E. Dimond, pp. 469 - 519. Academic Press, New York.

56. OKU, H., SHIRAISHI, T. & OUCHI, S. (1977). Suppression of induction of phytoalexin, pisatin. *Naturwissenschaften* 64, 643.

57. PAUL, V. & SCHEINPFLUG, H. (1979). Untersuchungen zur Wirkung von [R] Bayleton auf die Pathogenese des Gerstenmehltaus. *Pflanzenschutz-Nachrichten Bayer* 32, 83 - 92.

58. POLITIS, D.J. & GOODMAN, R.N. (1978). Localized cell wall appositions : Incompatibility response of tobacco leaf cells to *Pseudomonas pisi*. *Cytology and Histology* 68, 309 - 316.

59. RICHMOND, S., KUĆ, J. & ELLISTON, J.E. (1979). Penetration of cucumber leaves by *Colletotrichum lagenarium* is reduced in plants systemically protected by previous infection with the pathogen. *Physiological Plant Pathology* 14, 329 - 338.

60. RIDE, J.P. (1978). The role of cell wall alterations in resistance to fungi. *Annals of Applied Biology* 89, 302 - 306.

61. RIDE, J.P. & PEARCE, R.B. (1979). Lignification and papilla formation at sites of attempted penetration of wheat leaves by non-pathogenic fungi. *Physiological Plant Pathology* 15, 79 - 92.

62. ROHRINGER, R., KIM, W.K. & SAMBORSKI, D.J. (1979). Histological study of interactions between avirulent races of stem rust and wheat containing resistance genes *Sr*5, *sr*8, or *Sr*22. *Canadian Journal of Botany* 57, 323 - 331.

63. SAMBORSKI, D.J., KIM, W.K., ROHRINGER, R., HOWES, N.K. & BAKER, R.J. (1977). Histological studies on host-cell necrosis conditioned by the *Sr6* gene for resistance in wheat to stem rust. *Canadian Journal of Botany* 55, 1445 - 1452.

64. SARGENT, C. & GAY, J.L. (1977). Barley epidermal apoplast structure and modification by powdery mildew contact. *Physiological Plant Pathology* 11, 195 - 205.

65. SEQUEIRA, L. & GRAHAM, T.L. (1977). Agglutination of avirulent strains of *Pseudomonas solanacearum* by potato lectin. *Physiological Plant Pathology* 10, 43 - 50.

66. SEQUEIRA, L. (1978). Lectins and their role in host-pathogen specificity. *Annual Review of Phytopathology* 16, 453 - 481.

67. SHERWOOD, R.T. & VANCE, C.P. (1976). Histochemistry of papillae formed in reed canarygrass leaves in response to non-infecting pathogenic fungi. *Phytopathology* 66, 503 - 510.

68. SHIMONY, C. & FRIEND, J. (1975). Ultrastructure of the interaction between *Phytophthora infestans* and leaves of two cultivars of potato (*Solanum tuberosum* L.) Orion & Majestic. *New Phytologist* 74, 59 - 65.

69. SHIRAISHI, T., OKU, H., YAMASHITA, M. & OUCHI, S. (1978). Elicitor and suppressor of pisatin induction in spore germination fluid of pea pathogen, *Mycosphaerella pinodes*. *Annals of the Phytopathological Society of Japan* 44, 659 - 665.

70. SKIPP, R.A. & SAMBORSKI, D.K. (1974). The effect of the *Sr6* gene for host resistance on histological events during the development of stem rust in near-isogenic wheat lines. *Canadian Journal of Botany* 52, 1107 - 1115.

71. STAKMAN, E.C. (1915). Relations between *Puccinia graminis* and plants highly resistant to its attack. *Journal of Agricultural Research* 4, 193 - 199.

72. TOMIYAMA, K. (1966). Double infection by an incompatible race of *Phytophthora infestans* of a potato plant cell which has previously been infected by a compatible race. *Annals of the Phytopathological Society of Japan* 32, 181 - 185.

73. VANCE, C.P. & SHERWOOD, R.T. (1976). Cycloheximide treatments implicate papilla formation in resistance of reed canarygrass to fungi. *Phytopathology* 66, 498 - 502.

74. VAN LOON, L.C. & VAN KAMMEN, A. (1970). Polyacrylamide disc electrophoresis of the soluble leaf proteins from *Nicotiana tabacum* var. 'Samsun' and 'Samsun NN'. II. Changes in protein constitution after infection with tobacco mosaic virus. *Virology* 40, 199 - 211.

75. VARNS, J.L. & KUĆ, J. (1971). Suppression of rishitin and phytuberin accumulation and hypersensitive response on potato by compatible races of *Phytophthora infestans*. *Phytopathology* 61, 178 - 181.

76. VARNS, J.L., CURRIER, W.W. & KUĆ, J. (1971). Specificity of rishitin and phytuberin accumulation by potato. *Phytopathology* 61, 968 - 971.

77. WADE, M. & ALBERSHEIM, P. (1979). Race-specific molecules that protect soybeans from *Phytophthora megasperma* var. *sojae*. *Proceedings of the National Academy of Sciences, U.S.A.* 76, 4433 - 4437.

78. WATERMAN, M.A., AIST, J.R. & ISRAEL, H.W. (1978). Centrifugation studies help to clarify the role of papilla formation in compatible barley powdery mildew interactions. *Phytopathology* 68, 797 - 802.

79. WOLF, G. & FRIČ, F. (1981). A rapid staining method for *Erysiphe graminis* f. sp. *hordei* in and on whole barley leaves with the protein specific dye Coomassie Brilliant Blue. *Phytopathology*, in press.

80. ZEYEN, R.J. & BUSHNELL, W.R. (1979). Papilla response of barley epidermal cells caused by *Erysiphe graminis* : rate and method of deposition determined by microcinematography and transmission electron microscopy. *Canadian Journal of Botany* 57, 898 - 913.

A STRUCTURAL VIEW OF ACTIVE DEFENCE

D.S. INGRAM

Botany School
Downing Street
Cambridge CB2 3EA, U.K.

INTRODUCTION

The following poem illustrates the all too frequent approach of research scientists to the problems confronting them, including the problems associated with the active defence of plants to potential pathogens.

The Blind Men and the Elephant

It was six men of Hindostan
To learning much inclined,
Who went to see the elephant-
(Though all of them were blind),
That each by observation
Might satisfy his mind.

The first approached the elephant
And happening to fall
Against its broad and sturdy side,
At once began to bawl,
"God bless me! but the elephant
Is very like a wall!"

The second, feeling of the tusk,
Cried "Ho! what have we here
So very round and smooth and sharp?
To me 'tis mighty clear
This wonder of an elephant
Is very like a spear! "

The third approached the animal,
And happening to take
The squirming trunk within his hands,
Thus boldly up and spake;
"I see," quoth he, "the elephant
Is very like a snake!"

The fourth reached out his eager hand
And felt about the knee;
"What most this mighty beast is like
Is mighty plain," quoth he;
"'Tis clear enough the elephant
Is very like a tree! "

The fifth who chanced to touch the ear
Said, "E'en the blindest man
Can tell what this resembles most;
Deny the fact who can,
This marvel of an elephant
Is very like a fan! "

The sixth no sooner had begun
About the beast to grope
Than, seizing on the swinging tail
That fell within his scope,
"I see," quoth he, "the elephant
Is very like a rope! "

And so these men of Hindostan
Disputed loud and long,
Each in his own opinion
Exceeding stiff and strong,
Though *each* was *partly* in the right,
They *all* were in the wrong.

J.G. Saze

1816 - 1887

The resulting confusion and argument is the inevitable consequence of studies which focus on only a part of the whole.

A proper understanding of active defence depends upon the *integrated*, not *isolated*, research efforts of the scientists studying the three components of the phenomenon, as depicted below.

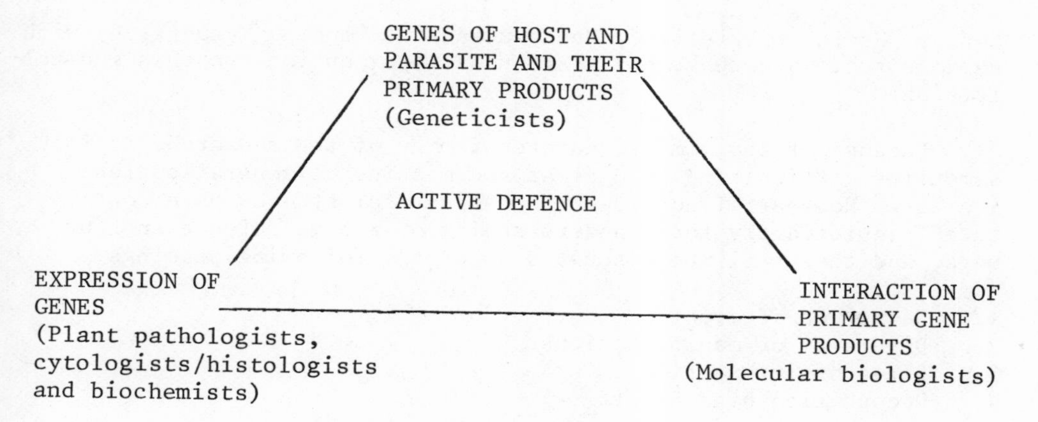

Without such integration there exists the danger of falling into the same trap as the wise, but blind, men faced with an elephant and describing an incomplete, and quite probably erroneous, concept of the whole.

Cytological/histological (structural) studies, like genetical studies, occupy a special position in this scheme, for they form the foundation upon which the molecular-biological and biochemical superstructure is built. Precise investigation and description of the structure of both compatible and incompatible interactions should, therefore, be indispensable first steps in the elucidation of the mechanisms of active defence. In addition, through application of the developing technologies of cytochemistry, microbeam analysis and microautoradiography, structural studies should be used to bridge the gap between mere description of interactions and molecular-biological and biochemical definition of the underlying processes.

In the following discussion I shall first assess the contribution of structural studies to our understanding of active defence so far; then I shall briefly outline the potential of microscopical analytical technology for the future development of such research.

THE PRESENT POSITION

There has been an enormous number of light and electron microscopic studies of active defence in recent years. Unfortunately most of these have been isolated, in that only one or two gene-gene interactions have been examined for each host-parasite combination, and there have been few comprehensive studies where the interaction between a *series* of genes for avirulence/virulence and for resistance/susceptibility have been compared. The investigations of Tomiyama and associates (45) with potato and *Phytophthora infestans*, of Maclean and Tommerup (28) with *Bremia lactucae* and lettuce and of

Coffey, Heath, Littlefield, Mendgen and the Winnipeg group (26) with various rusts and their hosts are notable exceptions to this generalisation.

Because of the limited nature of most of the research, it is sometimes difficult to make clear and meaningful generalisations from it. However, I consider that structural studies have contributed significantly to an understanding of active defence in four ways, and these will be discussed under the following headings.

1. Sequences of events
2. Diversity of manifestations
3. Sites of interaction
4. Recognition of non-self

These topics are all interlinked, but there is some advantage in considering each one separately.

Sequence of events

Structural studies have played a large part in defining the order in which events happen in a wide range of incompatible host-parasite interactions. The importance of such research is that it helps to distinguish between events which are components of active defence and events which merely result from it. The point can be illustrated by reference to the relationship between rapid cell death (the 'hypersensitive response') and resistance to biotrophic fungi, especially the rusts (Uredinales).

The connection between the rapid death of host cells and resistance to biotrophs was first noted by Ward (47) in Cambridge at the turn of the century, and by his pupil, Dorothea Marryat (30). They studied *Bromus* spp. and *Puccinia dispersa* (= *P. recondita*) and wheat bred by Biffen and *Puccinia glumarum* (= *P. striiformis*), respectively. Subsequently Stakman (44) at Minnesota described a similar phenomenon in wheat resistant to *Puccinia graminis*, and coined the term 'hypersensitivity' to describe it. He outlined the sequence of events in hypersensitivity as invasion by the fungus, death of host cells upon or before contact with the fungus, followed by cessation of growth of the fungus.

For many years following the work of Stakman a 'traditional view' grew up that since rapid cell death was normally associated with extreme resistance to biotrophs it was quite likely to be the cause of it (34). This was first challeneged by Brown and associates during the late 1960s (3, 35) as a result of their research with various races of *P. graminis* and a range of resistant genotypes of wheat. The findings were complex, but may be summarised as follows. Three categories of host-parasite interaction could be recognised on the basis of parasite growth and host necrosis. In the highly susceptible interactions the rust colonies became very large and host necrosis was negligible. In the extremely resistant interactions the opposite picture obtained, with very small rust colonies being greatly exceeded by the area of necrotic tissue. Both these observations are consistent with the traditional view of rapid cell death as a cause of resistance. In the third category, however, involving hosts of intermediate to extreme resistance, rust colonies of moderate diameter were considerably larger than the area of host necrosis, an observation which is at variance with the traditional view.

The work was important, for it provoked a re-examination of the role of cell death in resistance to biotrophs. Later, others challenged the traditional view, notably Király *et al.* (22), and suggested that cell death in a resistant host *results* from death or debilitation of the pathogen. However, this and several other light microscopical studies which have examined the same point have led to somewhat inconclusive results (23, 31, 39, 43). In this context, however, the light microscope is probably too imprecise an instrument, and the electron microscope may provide more valuable insights.

In most cases, for a wide variety of hosts and pathogens (26, 27, 42), the sequence of events revealed by the electron microscope has corresponded to the one proposed by Stakman in 1915, that a host cell, or cells, dies soon after contact with the fungus, followed by the cessation of growth of the fungus (Fig. 1). Recently, however, the studies of Harder *et al.* (14, 15) with wheat carrying the temperature sensitive *Sr6* gene and a race of *P. graminis* with the *P6* gene for avirulence, have shown that the relationship between the limitation of the fungus and host cell death in that interaction is variable. At 19°C, which conditions incompatibility, host cell necrosis was not always accompanied by fungal necrosis, and vice versa, and in neither case did it appear that the death of one partner in the interaction was responsible for the death of the other.

Many more electron microscopical studies of many more gene for gene interactions in as wide a range of hosts and parasites as possible are necessary. However, the overall picture which I believe may be slowly emerging is one of diversity where, in different interactions, cell death may : (a) precede the cessation of fungal growth and possibly be an integral *component* of the resistance process; (b) follow the cessation of fungal growth and be a *consequence* of the operation of the resistance mechanism, triggered by substances emanating from the inhibited fungus itself; or (c) be an event which is *associated* with resistance, perhaps initiated by general stress accruing from resistance, but having no role in resistance itself. These three possibilities are not mutually exclusive, and each could apply in different gene-gene interactions, even within a single host species-parasite species combination.

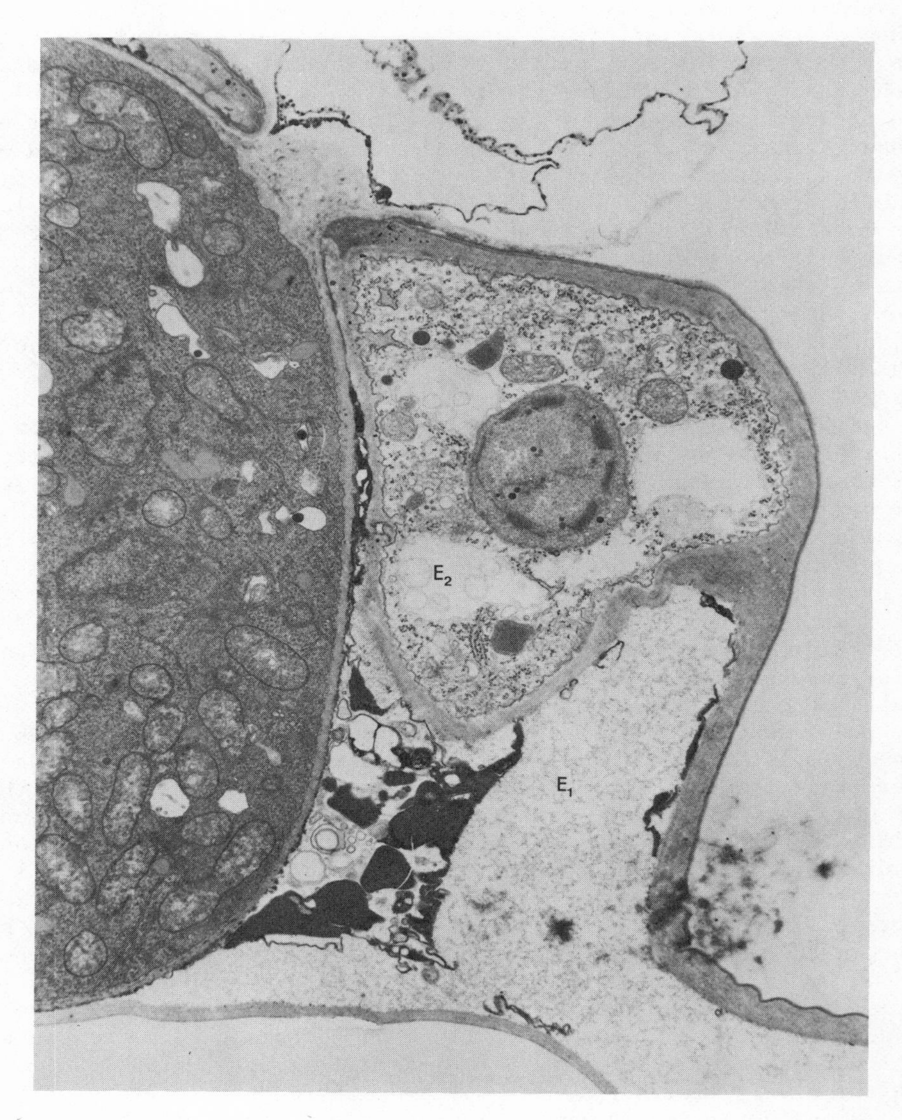

Figure 1 Electron micrograph showing a primary vesicle (PV) of an incompatible race of *Bremia lactucae* (race W5) in an epidermal cell (E_1) of the lettuce cultivar Avondefiance. Note the complete disruption of the membranes and cytoplasm of this cell and the apparent normality of the cytoplasm of the fungus and of the adjacent, non-penetrated cell (E_2). (Reprinted by permission from *Nature*, Vol. 249, No. 5453, pp. 186 - 187. Copyright (c) 1974 Macmillan Journals Ltd.).

Structural studies have not, so far, thrown any clear light on the nature of the triggering of rapid cell death during resistance, or on its role in restricting fungal growth in those cases where it appears to be an integral part of the resistance mechanism. The many possibilities have been discussed in detail by Ingram (21).

It is a sobering thought that more than sixty years after the publication of Stakman's paper in 1915 we still do not know for certain whether rapid host cell death really does have a role to play in the resistance of any host to any biotrophic fungus, let alone what that role might be (48). Real progress will only be made when : (a) many more *genetically defined* host-parasite combinations have been examined; (b) there has been closer integration of structural and metabolic studies; and (c) 'rapid cell death (hypersensitivity)' and 'limitation of fungal growth' have been better defined in both structural and biochemical terms.

Diversity of manifestations

The clear theme which emerges from this discussion of the sequence of events in relation to cell death and resistance to biotrophic fungi is one of probable diversity. Diversity in the manifestation of active defence is also apparent in other structural aspects of the cell death phenomenon. For example, research with *Melampsora lini* and flax (26), with *Uromyces phaseoli vignae* and cowpea (26), with *U. phaseoli* and French bean (33) and with *B. lactucae* and lettuce (28) has shown that even in a single host species-parasite species combination there are marked differences both in the speed of cell death and the stage of fungal development at which host cell death and fungal limitation occur. It also suggests that there may be many different types of cell death associated with resistance to biotrophs. This is often manifest even at the macroscopic level by the fact that different cell death responses often involve different coloration of the necrotic cells.

In discussing diversity it is important to emphasise that rapid cell death is a far from universal phenomenon in resistance to biotrophic fungi, and has dominated structural studies simply because it can easily be seen with the naked eye. Stakman himself, however (44),singled out the wheat cultivars Emmer and Einkorn as being distinctive in that, unlike other cultivars, their resistance to *P. graminis* did not involve extensive necrosis of host cells. Similarly Mendgen (33) has shown that at least one type of resistance of French bean to *U. phaseoli* does not involve cellular necrosis, while Rohringer *et al.* (37) have shown that although cell death is associated with resistance of wheat carrying the *Sr6* gene to *P. graminis*, it is not associated with resistance of wheat carrying the *Sr5* or *Sr8* genes.

In addition to necrosis, many other cellular changes in plants resisting attack by biotrophic fungi have been reported. For example, Littlefield and Heath (26) recently listed the following from studies of various hosts responding to rust fungi : cytoplasmic changes e.g. increases in the size and/or number of organelles, the appearance or disappearance of crystal-containing microbodies, changes in the amount and distribution of endoplasmic reticulum, rapid loss of starch grains and changes in the form and structure of the invaginated host plasma membrane; the accumulation of electron-opaque, granular or fibrillar material in extra-haustorial matrixes and elsewhere; and the encasement of haustoria by electron-lucent material. While some of these changes may represent early stages in cellular necrosis, others certainly do not. All coincide with resistance, but as with cell death the problem is of distinguishing between those which have a role in resistance and those which merely result from or are associated with the process. The important point to be noted, however, is that this wide diversity in the cellular features of resistance does not only represent differences between different hosts responding to different pathogens, but is also apparent in the expression of different resistance genes in a single plant species interacting with a single pathogen species and in different plant species responding to the same pathogen species. These observations, together with those relating to cell death discussed above, may be interpreted as evidence of a wide diversity of different basic mechanisms of active defence.

Close cytological examination of the diversity of cellular changes in a series of four host cultivars and in three non-host cultivars reacting to one race of *U. phasoli vignae* (16) has led Heath to suggest that there are many stages during the infection process whereby interaction between plant and parasite may result in the expression of active defence (Fig. 2). Heath refers to these stages as 'switching points', where the outcome of the interaction determines the subsequent progress of the infection. Thus susceptibility depends upon what she calls the 'correct' response at every stage, while resistance requires an 'incorrect' response at only *one* stage. An incorrect response can be interpreted as the triggering of an active defence mechanism or the failure to induce active susceptibility.

The switching point hypothesis draws attention to the fact that resistance and susceptibility may have evolved many times during the history of a relationship between a plant and a biotrophic fungus, and that each such interaction is superimposed, one upon another. I like to envisage this as meaning that susceptibility is a situation where the fungus has 'grown right through' the evolutionary history of its relationship with a host.

Unfortunately Heath was not able to show that each of the switching points in the *U. phaseoli vignae*-cowpea interaction re-

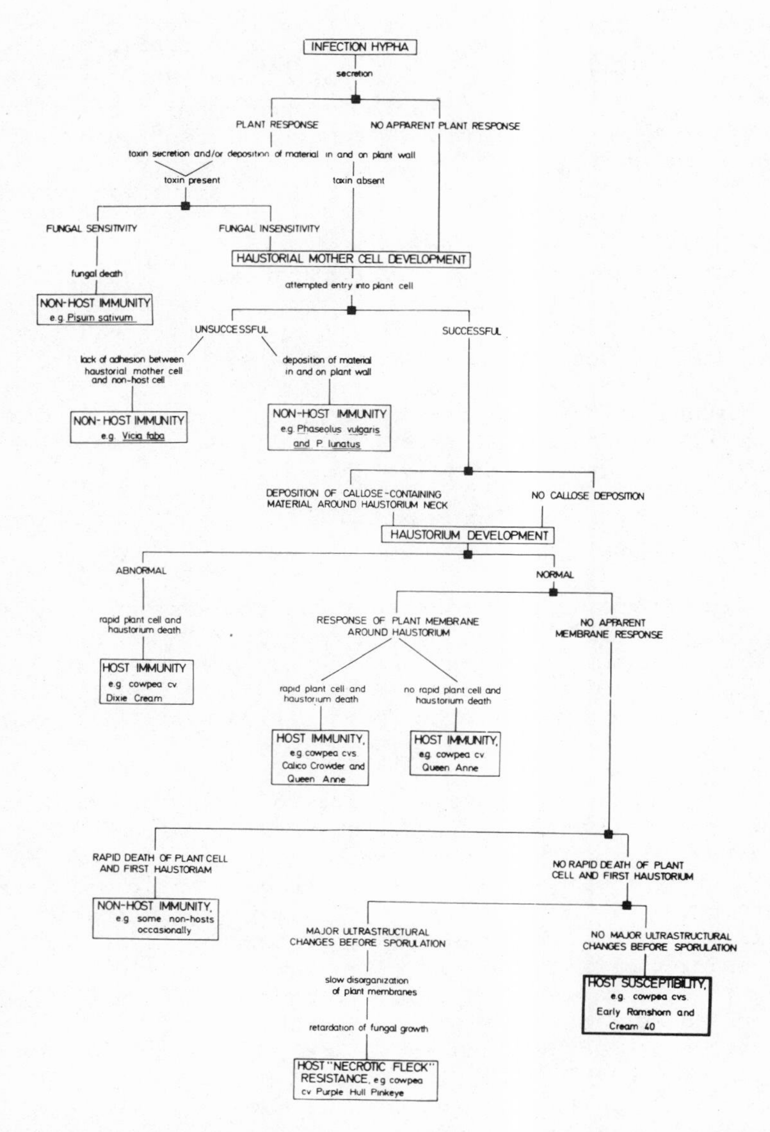

Figure 2 Proposed scheme (based on ultrastructural observations) of events during the infection process of cowpea rust (*Uromyces phaseoli* var. *vignae*) which could lead to the observed resistance or susceptibility of host and non-host plants. It is proposed that the responses induced at each "switching point" (■) determine the subsequent progress of infection. (Redrawn with permission from Heath, M.C. (1974), *Physiological Plant Pathology* Vol. 4, pp. 403 - 414. Copyright by Academic Press Inc. (London) Ltd.).

presented the involvement of different major genes, but suggested that one would expect this to be so from, for example, the light microscopical work of Ellingboe and colleagues (8), who have shown that different genes for resistance to wheat and barley mildew in their respective hosts are manifested at different points and in different ways during the infection process, and from the studies of Littlefield (25), who has shown a similar situation in the resistance of flax to *M. lini*.

A final aspect of diversity which has been revealed in a number of cytological studies (Tommerup and Ingram, unpublished) is that the cells of a leaf or other organ may be a heterogeneous population so far as disease resistance is concerned, with individual cells reacting in different ways to a potential pathogen. This may in part be a positional effect (14) resulting from physiological factors such as growth substance concentration (11), or it may point to genetic differences between the different cells of a leaf (2).

Sites of interaction

A third important role of structural studies has been in pinpointing the precise site within a tissue or cell of a host where the critical interaction occurs leading to the triggering (or not) of active defence or active susceptibility mechanisms. This can best be illustrated by reference to recent work with phytopathogenic bacteria.

For example, in a careful and detailed study of the interaction of avirulent and compatible strains of *Pseudomonas solanacearum* with tobacco Sequeira *et al.* (40) have shown that the site of the initial interaction with avirulent strains is the *cell wall*, with attachment of the bacterial cell to the host wall being the initial event (Fig. 3a). This is followed by envelopment of the bacterial cell by fibrillar and granular material extruded from the cell wall and bounded by the outer cell layer (Fig. 3b). Concurrently with this envelopment changes occur in the cytoplasm of the host cell: the plasma membrane becomes separated and convoluted, and numerous membrane-bound vesicles occur in the space between the plasma membrane and the wall (Fig. 3c). This is followed by host cell collapse, the so-called 'hypersensitive response'.

In contrast, virulent strains of *P. solanacearum* do not become attached to the host wall, but remain free to multiply in the intercellular fluid, causing no visible changes in host cells in the short term. This lack of attachment may be due to the production by the virulent strains of an extracellular polysaccharide which prevents binding.

Thus the initial site of interaction between the incompatible

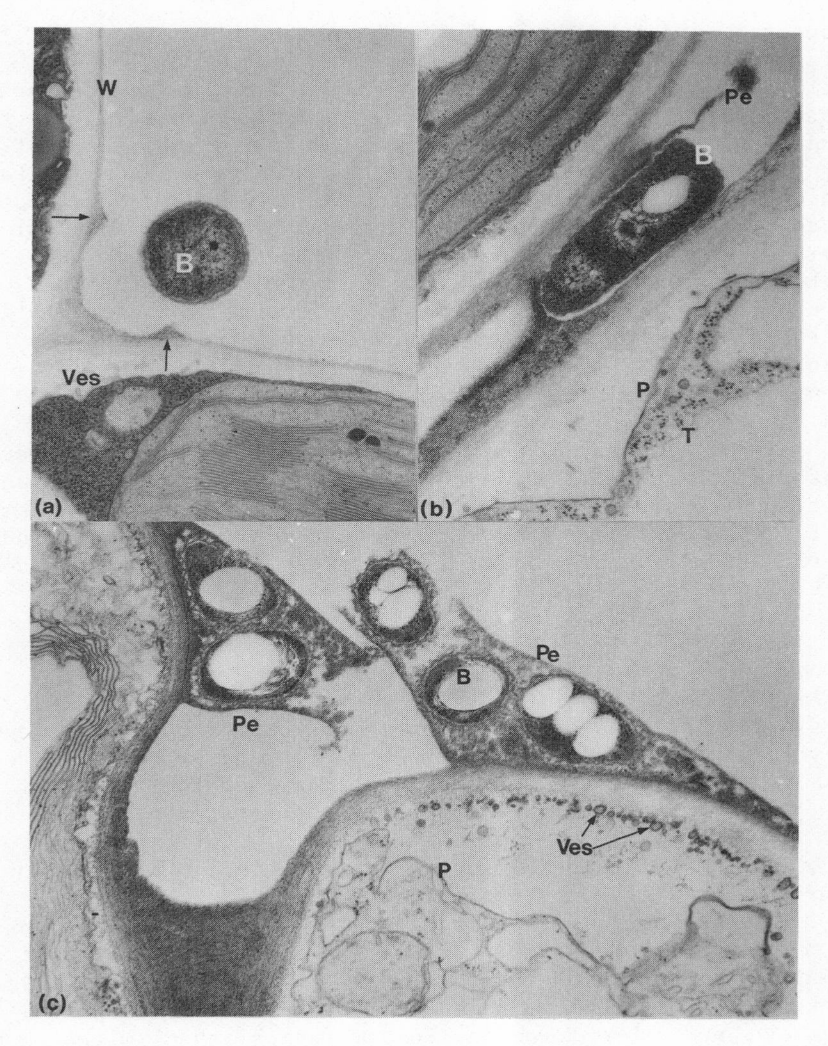

Figure 3 Electron micrographs showing the interaction of the cells of tobacco and incompatible (B1) *Pseudomonas solanacearum* bacteria. In (a) and (b) granular and fibrillar material progressively surround the bacterium (B) as it rests on the host cell wall (W) (a, arrows; b, 3 hours after infiltration); note containment of the granular material by the host cell wall pellicle (Pe). In (c) note further the extreme separation of the host plasmalemma (P) and the accumulation of vesicles (Ves) at and beyond the point of contact with the bacteria. (Reproduced with permission from Sequeira, L., Gaard, G. and De Zoeten, G.A. (1977), *Physiological Plant Pathology* Vol. 10, pp. 43 - 50. Copyright by Academic Press Inc. (London) Ltd.).

bacterium and tobacco is the cell wall. This attachment is not the trigger for active defence, however, for heat-killed avirulent bacteria become attached but do not induce a resistance response in the host; the attachment, presumably, allows transfer of the resistance-inducing factor from the living bacterial cell to the host cell. Many previous efforts to isolate the resistance-inducing factor, which must pass through the host wall, may have failed because of a lack of understanding of the phenomenon of attachment, which may not only allow transfer but may also be a necessary prerequisite for the activity of the substance which is transferred. Interactions that are essentially similar to the *P. solanacearum*-tobacco interaction, although different in detail, have been described for the interaction between *Xanthomonas malvacearum* and cotton (5) and for *Pseudomonas phaseolicola* and *Phaseolus vulgaris* (36).

In all these cases structural studies have provided critical information on the sites of interaction(s) between hosts and parasites which could significantly affect the further development of efforts to elucidate the mechanisms involved. Similar kinds of information have been provided for many other host-parasite interactions.

Recognition of non-self

It is of considerable importance to a proper understanding of host-parasite specificity to know whether plant cells possess an inherent and basic mechanism for the early detection of all alien organisms, or whether recognition mechanisms have evolved separately after the establishment of each interaction. The evidence for or against the existence of such a basic mechanism for recognition of non-self is slight, but comes, primarily, from structural studies of the interaction between biotrophic fungi and various non-hosts. There have been few such studies, examples being the light and electron microscopic studies of Heath (16, 17) with *U. phaseoli*, and the light microscopic studies of Maclean and Tommerup (28) with *B. lactucae* and of Tommerup and Ingram (unpublished) with *Peronospora farinosa*.

In these cases the structural evidence suggests that the cells of some non-hosts may be penetrated and infection structures formed within them, with the cytological response of the plant cell being little different from that in compatible interactions. Eventually both fungal structures and the invaded plant cells die, but sometimes considerable periods of time elapse before this occurs. That some rusts may make as much mycelial growth in the tissues of a non-host as they do on artificial membranes suggests that in some instances non-host resistance may be due not to any specific resistance mechanism but to lack of some stimulatory substance emanating from the plant. It is therefore possible that in some interactions mechanisms leading to resistance and susceptibility are superimposed not upon a basic recognition of non-self mechanism but upon the presence or absence of some kind of nutritional (in the widest sense) compatibility. However, the evidence is limited and this suggestion may not be widely applicable; only further research will provide the answer.

THE FUTURE DEVELOPMENT OF STRUCTURAL STUDIES

It may be concluded from the discussion above that every gene for gene interaction must be regarded as probably having unique features so far as resistance and susceptibility are concerned, and that a full investigation of the underlying processes must be preceded by careful structural studies to determine the precise sites of initial interactions and the exact sequences of events which follow. Such studies will only be possible if experiments are made using hosts and parasites that are strictly defined genetically, and if the most sophisticated microanalytical procedures are employed. Such microanalytical techniques include the following (12)

1. Selective staining
2. Enzyme cytochemistry
3. Immuno-cytochemistry (using antibodies or lectins as probes)
4. Element localisation
5. Autoradiography
6. Freeze etching

None of these has yet contributed significantly to an understanding of the mechanisms of active defence, yet all have the potential to do so, providing that their limitations are recognised.

Selective staining (38), which depends upon the performance *in situ* of specific and defined chemical reactions at the level of organelles and other cellular components in order to specify their composition, is probably the least useful of the techniques listed for studies of active defence. Procedures are available for characterising polysaccharide deposits, nucleic acids, lipids and proteins in plant cells, although there are important problems of preservation, resolution, specificity and interpretation. Nevertheless, I can envisage a role for such procedures, perhaps combined with selective enzymatic extraction, in for example the characterisation of encasements and other deposits around haustoria (20), or of the granular and fibrillar material which accumulates around bacterial cells following their attachment to the walls of resistant hosts.

Enzyme cytochemistry aims to localise the products of specific enzyme activity at a precise site in the cell by means of an electron dense precipitate. In some cases the reaction product is insoluble, but usually it is soluble and a second, capture reaction is necessary to precipitate the reaction product *in situ*. Inevitably, therefore, there are problems of specificity and localisation, but procedures have now been devised for identifying a wide range of important enzymes in plant cells (41). Applications of enzyme cytochemistry in plant pathology have been few, although notable exceptions are localisation of lipase activity associated with the penetration of lettuce cuticle by *B. lactucae* (6) and the analysis of peroxidase activity in French bean tissue infected by *U. phaseoli* (32). In studies of active defence the technique might be used for : pinpointing the loss of organelle integrity associated with cell death, through an examination of the distribution of such enzymes as acid hydrolases; and, as more is learnt of the biosynthetic pathways involved, in locating both in space and time the appearance of enzymes associated with the accumulation of phytoalexins and related compounds.

Through immuno-cytochemistry (24) it is possible to localise proteins and glycoproteins at plant surfaces using antibodies or lectins as probes. For example, Heslop-Harrison *et al.* (19) have used a complex of gold and the lectin concanavalin-A to locate glycoprotein determinants on the surfaces of stigmas, while Burgess and Linstead (4) have used a similar complex in studies of lectin binding sites on the surfaces of tobacco protoplast membranes. This technique is in its infancy, but it may well be adaptable for use with other lectins which are specific for structures in fungal or host walls or membranes. If so, the possibilities for the future study of recognition phenomena in active defence appear legion.

Energy dispersive X-ray microanalysis (EDX) is the best developed of the many microbeam analytical techniques for element localisation now available for use with plant tissues (13, 46). X-rays produced when an electron beam of sufficient energy strikes any material within the plant cell are detected in the scanning or transmission electron microscope using an X-ray spectrometer. It is possible to detect approximately 10^{-19}g of an element in a static probe analysis with an ultrathin section, with an analytical spatial resolution in the range 20 - 30 nm. Most elements of biological interest can be detected, although analysis becomes more difficult with light elements and sodium is the practical lower limit. Heath (18) has used EDX to partially characterise the electron opaque deposits formed by a non-host (French bean) in response to *U. phaseoli vignae*, and many other similar applications are also possible in studies of active defence. It may also be possible to use EDX to analyse changes in the distribution of ions between host or parasite cytoplasm and organelles as a precise indication of the loss of membrane integrity associated with cell death, providing that the problems of fixation of soluble substances can be overcome (see below).

In micro-autoradiography (9) radioactive isotopes introduced into specimens are located by placing a thin layer of photographic emulsion over the section to produce a latent image which is seen as grains of silver. Statistical methods are available to allow quantification of the distribution of silver grains between cells and organelles, but there are serious problems of dealing with soluble substances (see below) and of knowing the chemical form of the compound in which the label is located (1). The technique could be used extensively in studies of active defence to detect the time during an incompatible interaction at which uptake of labelled molecules from the host into the fungus or secretion from the fungus into the host does or does not commence. For example, Manocha (29) followed the incorporation of radioactive leucine into haustoria of *P. graminis* from host cells and its relation to the development of a haustorial sheath. He concluded that in incompatible interactions the sheath may act as a barrier to nutrient uptake. In addition it may be possible, through the use of labelled precursors, to use micro-autoradiography to locate with precision the cellular or subcellular sites of phytoalexin synthesis and accumulation.

Freeze etching (10), the last of the techniques listed, is not analytical in the same way as those discussed above. It is included here, however, because it could provide interesting insights into cellular changes occurring during active defence processes. The cell is stabilised physically by freezing and is then processed for examination by electron microscopy without the need to remove the frozen water. The size and spatial relationships between organelles is thus maintained and dehydration is avoided. Because the tissue is fractured, not sectioned, unique views of the interior of cells are provided, particularly of the internal architecture of membranes. Quite small conformational changes in membranes can be detected, and herein lies its possible value for active defence studies, both in the characterisation of recognition events and of the disorganisation of membranes during cell death.

In conclusion it is important to emphasise that all the techniques referred to above have their limitations, especially the problem of specificity, the problem of fixation of soluble substances and, arising from these two, the problem of interpretation. These limitations cannot be overemphasised, but they can be surmounted (a) by development and application of the techniques of cryofixation (7) and (b) through experience of their wider application in plant pathology. The future prospects for structural studies of active defence are, therefore, truly exciting and challenging.

REFERENCES

1. ANDREWS, J.A. (1975). Distribution of label from ^{3}H-glucose and ^{3}H-leucine in lettuce cotyledons during the early stages of infection with *Bremia lactucae*. *Canadian Journal*

of Botany 53, 1103-1115.

2. BRETTELL, R.I.S. & INGRAM, D.S. (1979). Tissue culture in the production of novel disease-resistant crop plants. *Biological Reviews* 54, 329 - 345.

3. BROWN, J.F., SHIPTON, W.A. & WHITE, N.H. (1966). The relationship between hypersensitive tissue and resistance in wheat seedlings infected with *Puccinia graminis tritici*. *Annals of Applied Biology* 58, 279 - 290.

4. BURGESS, J. & LINSTEAD, P.J. (1977). Membrane mobility and the concanavalin A binding system of the plasmalemma of higher plant protoplasts. *Planta* 136, 253 - 259.

5. CASON, E.T., RICHARDSON, P.E., ESSENBERG, M.K., BRINKERHOFF, L.A., JOHNSON, W.M. & VENERE, R.J. (1978). Ultrastructural cell wall alterations in immune cotton leaves inoculated with *Xanthomonas malvacearum*. *Phytopathology* 68, 1015 - 1021.

6. DUDDRIDGE, J.A. & SARGENT, J.A. (1978). A cytochemical study of lipolytic activity in *Bremia lactucae* Regel during germination of the conidium and penetration of the host. *Physiological Plant Pathology* 12, 289 - 296.

7. ECHLIN, P., RALPH, B. & WEIBEL, E.R. (1978). *Low temperature biological microscopy and microanalysis*. Blackwell Scientific Publications, Oxford.

8. ELLINGBOE, A.H. (1972). Genetics and physiology of primary infection by *Erysiphe graminis*. *Phytopathology* 62, 401 - 406.

9. EVANS, L.V. & CALLOW, M.E. (1978). Autoradiography. In : *Electron Microscopy and Cytochemistry of Plant Cells*, Ed. by J.L. Hall, pp. 235 - 277. Elsevier/North-Holland Biomedical Press, Amsterdam, New York and Oxford.

10. FINERAN, B.A. (1978). Freeze etching. In : *Electron Microscopy and Cytochemistry of Plant Cells*, Ed. by J.L. Hall, pp. 279 - 341. Elsevier/North-Holland Biomedical Press, Amsterdam, New York and Oxford.

11. HABERLACH, G.T., BUDDE, A.D., SEQUEIRA, L. & HELGESON, J. (1978). Modification of disease resistance of tobacco callus tissues by cytokinins. *Plant Physiology Lancaster* 62, 522 - 525.

12. HALL, J.L. (1978). *Electron Microscopy and Cytochemistry of*

Plant Cells. Elsevier/North-Holland Biomedical Press, Amsterdam, New York and Oxford.

13. HALL, T.A. (1979). Biological X-ray microanalysis. *Journal of Microscopy* 117, 145 - 163.

14. HARDER, D.E., ROHRINGER, R., SAMBORSKI, D.J., RIMMER, S.R., KIM, W.K. & CHONG, J. (1979). Electron microscopy of susceptible and resistant near-isogenic (*Sr6/Sr6*) lines of wheat infected by *Puccinia graminis tritici*. II. Expression of incompatibility in mesophyll and epidermal cells and the effect of temperature on host-parasite interactions in these cells. *Canadian Journal of Botany* 57, 2617 - 2625.

15. HARDER, D.E., SAMBORSKI, D.J., ROHRINGER, R., RIMMER, S.R., KIM, W.K. & CHONG, J. (1979). Electron microscopy of susceptible and resistant near-isogenic (*Sr6/Sr6*) lines of wheat infected by *Puccinia graminis tritici*. III. Ultrastructure of incompatible interactions. *Canadian Journal of Botany* 57, 2626 - 2634.

16. HEATH, M.C. (1974). Light and electron microscope studies of the interactions of host and non-host plants with cowpea rust, *Uromyces phaseoli* var. *vignae*. *Physiological Plant Pathology* 4, 403 - 414.

17. HEATH, M.C. (1977). A comparative study of non-host interactions with rust fungi. *Physiological Plant Pathology* 10, 73 - 88.

18. HEATH, M.C. (1979). Partial characterization of the electron-opaque deposits formed in the non-host plant, French bean, after cowpea rust infection. *Physiological Plant Pathology* 15, 141 - 148.

19. HESLOP-HARRISON, J. (1978). Recognition and response in the pollen-stigma interaction. In : *Cell-cell recognition*, Ed. by A. Curtis, pp. 121 - 137. University Press, Cambridge.

20. HICKEY, E.L. & COFFEY, M.D. (1978). A cytochemical investigation of the host-parasite interface in *Pisum sativum* infected by the downy mildew fungus *Peronospora pisi*. *Protoplasma* 97, 201 - 220.

21. INGRAM, D.S. (1978). Cell death and resistance to biotrophs. *Annals of Applied Biology* 89, 291 - 295.

22. KIRÁLY, Z., BARNA, B. & ERSEK, T. (1972). Hypersensitivity as

a consequence, not a cause, of plant resistance to infection. *Nature, London* 239, 456 - 458.

23. KITAZAWA, K. & TOMIYAMA, K. (1969). Microscopic observations of infection of potato cells by compatible and incompatible races of *Phytophthora infestans*. *Phytopathologische Zeitschrift* 66, 317 - 324.

24. KNOX, R.B. & CLARKE, A.E. (1978). Localization of proteins and glycoproteins by binding to labelled antibodies and lectins. In : *Electron Microscopy and Cytochemistry of Plant Cells*, Ed. by J.L. Hall, pp. 149 - 185. Elseveir/North-Holland Biomedical Press, Amsterdam, New York and Oxford.

25. LITTLEFIELD, L.J. (1973). Histological evidence for diverse mechanisms of resistance to flax rust, *Melampsora lini* (Ehrenb.) Lev. *Physiological Plant Pathology* 3, 241 - 247.

26. LITTLEFIELD, L.J. & HEATH, M.C. (1979). *Ultrastructure of rust fungi*. Academic Press, New York.

27. MACLEAN, D.J., SARGENT, J.A., TOMMERUP, I.C. & INGRAM, D.S. (1974). Hypersensitivity as the primary event in resistance to fungal parasites. *Nature, London* 249, 186 - 187.

28. MACLEAN, D.J. & TOMMERUP, I.C. (1979). Histology and physiology of compatibility and incompatibility between lettuce and the downy mildew fungus *Bremia lactucae* Regel. *Physiological Plant Pathology* 14, 291 - 312.

29. MANOCHA, M.S. (1975). Autoradiography and fine structure of host-parasite interface in temperature-sensitive combinations of wheat stem rust. *Phytopathologische Zeitschrift* 82, 207 - 215.

30. MARRYAT, D. (1907). Notes on the infection and histology of two wheats immune to the attacks of *Puccinia glumaraum*, yellow rust. *Journal of Agricultural Science, Cambridge* 2, 129 - 138.

31. MAYAMA, S., REHFELD, D.W. & DALY, J.M. (1975). The effect of detachment on the development of rust disease and the hypersensitive response of wheat leaves inoculated with *Puccinia graminis tritici*. *Phytopathology* 65, 1139 - 1142.

32. MENDGEN, K. (1975). Ultrastructural differentiation of different peroxidase activities during the bean rust infection process. *Physiological Plant Pathology* 6, 275 - 282.

33. MENDGEN, K. (1978). Der infectionsverlauf von *Uromyces phaseoli* bei anfälligen und resistenten bohnensorten. *Phytopathologische Zeitschrift* 93, 295 - 313.

34. MÜLLER, K.O. (1959). Hypersensitivity. In : *Plant Pathology* Vol. 1, Ed. by J.G. Horsfall & A.E. Dimond, pp. 469 - 519. Academic Press, New York.

35. OGLE, H. & BROWN, J.F. (1971). Quantitative studies of the post-penetration phase of infection by *Puccinia graminis tritici*. *Annals of Applied Biology* 67, 309 - 319.

36. ROEBUCK, P., SEXTON, R. & MANSFIELD, J.W. (1978). Ultrastructural observations on the development of the hypersensitive reaction in leaves of *Phaseolus vulgaris* cv. Red Mexican inoculated with *Pseudomonas phaseolicola* (race 1). *Physiological Plant Pathology* 12, 151 - 157.

37. ROHRINGER, R., KIM, W.K. & SAMBORSKI, D.J. (1979). A histological study of interactions between avirulent races of stem rust and wheat containing resistance genes *Sr5*, *Sr6*, *Sr8* or *Sr22*. *Canadian Journal of Botany* 57, 324 - 331.

38. ROLAND, J.C. (1978). General preparation and staining of thin sections. In : *Electron Microscopy and Cytochemistry of Plant Cells*, Ed. by J.L. Hall, pp. 1 - 62. Elsevier/North-Holland Biomedical Press, Amsterdam, New York and Oxford.

39. SAMBORSKI, D.J., KIM, W.K., ROHRINGER, R., HOWES, N.K. & BAKER, R.J. (1977). Histological studies on host cell necrosis conditioned by the *Sr6* gene for resistance in wheat to stem rust. *Canadian Journal of Botany* 55, 1445 - 1452.

40. SEQUEIRA, L., GAARD, G. & DE ZOETEN, G.A. (1977). Interaction of bacteria and host cell walls : its relation to mechanisms of induced resistance. *Physiological Plant Pathology* 10, 43 - 50.

41. SEXTON, R. & HALL, J.L. (1978). Enzyme cytochemistry. In : *Electron Microscopy and Cytochemistry of Plant Cells*, Ed. by J.L. Hall, pp. 63 - 147. Elsevier/North-Holland Biomedical Press, Amsterdam, New York and Oxford.

42. SHIMONY, C. & FRIEND, J. (1975). Ultrastructure of the interaction between *Phytophthora infestans* and leaves of two cultivars of potato (*Solanum tuberosum* L.), Orion and Majestic. *New Phytologist* 74, 59 - 65.

43. SKIPP, R.A. & DEVERALL, B.J. (1972). Relationships between fungal growth and host changes visible by light microscopy

during infection of bean hypocotyls (*Phaseolus vulgaris*) susceptible and resistant to physiological races of *Colletotrichum lindemuthianum*. *Physiological Plant Pathology* 2, 357 - 374.

44. STAKMAN, E.C. (1915). Relations between *Puccinia graminis* and plants highly resistant to its attack. *Journal of Agricultural Research* 4, 193 - 199.

45. TOMIYAMA, K. (1971). Cytological and biochemical studies of the hypersensitive reactions of potato cells to *Phytophthora infestans*. In : *Morphological and Biochemical Events in Plant-Parasite Interaction*, Ed. by S. Akai & S. Ouchi, pp. 387 - 401. Phytopathological Society of Japan, Tokyo.

46. VAN STEVENINCK, R.F.M. & VAN STEVENINCK, M.E. (1978). Ion localization. In : *Electron Microscopy and Cytochemistry of Plant Cells*, Ed. by J.L. Hall, pp. 187 - 234. Elsevier/North-Holland Biomedical Press, Amsterdam, New York and Oxford.

47. WARD, H.M. (1905). Recent researches on the parasitism of fungi. *Annals of Botany* 19, 1 - 54.

48. WILLIAMS, P.H. (1976). Cytochemical aspects of specificity in plant pathogen interactions. In : *Biochemistry and Cytology of Plant-Parasite Interaction*, Ed. by S. Tomiyama, J.M. Daly, I. Uritani, H. Oku & S. Ouchi, pp. 70 - 77. Elsevier Scientific Publishing Co., Amsterdam, Oxford and New York.

PHYSIOLOGICAL AND BIOCHEMICAL EVENTS ASSOCIATED WITH THE EXPRESSION OF RESISTANCE TO DISEASE

J.A. BAILEY

Long Ashton Research Station
Long Ashton
Bristol, BS18 9AF, U.K.

INTRODUCTION

The mechanisms by which a plant cell may recognise potentially pathogenic microorganisms are discussed by Professor Keen earlier in this Symposium. It is the requirement of this paper to consider the processes which result from recognition and which lead to resistance, a phenomenon usually associated with inhibition of pathogen growth within infected tissues. Inhibition of fungal growth within plant cells is characteristic of the resistance shown by cultivars of a host plant species to different physiological races of host-specific pathogens. This specialization is shown by many obligate fungi, e.g. rusts and mildews, but also by some facultative pathogens, e.g. *Phytophthora infestans, Fulvia fulva* and *Colletotrichum lindemuthianum*. An important feature common to these pathogens, but also to many fungi which do not show race-specific reactions, is the ability of pathogens to grow biotrophically in susceptible cultivars, while in resistant cultivars, the inhibition of pathogen growth is often associated with the premature death of the infected cells. The significance, if any, of the relationship between infected cell death and resistance of plants to disease has been the subject of several recent reviews (21, 39, 60). In one, Ingram (39) pointed out that although visible expressions of resistance may be similar, the processes responsible for these effects could be very different. In many host-pathogen interactions, detailed knowledge of the events leading to resistance are often lacking and hence any generalisations from one situation to another must be considered very carefully. For this reason, much of my paper will be concerned with experiments pertinent to the mechanisms of resistance shown by *Phaseolus vulgaris* to the host-specific pathogen *Colletotrichum lindemuthianum*.

EXPRESSION OF SUSCEPTIBILITY AND RESISTANCE IN *PHASEOLUS VULGARIS* INFECTED WITH *COLLETOTRICHUM LINDEMUTHIANUM*

Colletotrichum lindemuthianum (Sacc. and Magn.) Briosi and Cav. is a facultative pathogen existing as several physiological races which can be differentiated by their reaction with various cultivars of *Phaseolus vulgaris*. It is a seed-borne pathogen able to invade the lower regions of seedling hypocotyls. Most studies comparing mechanisms of resistance and susceptibility have therefore been carried out with hypocotyls (9, 43, 57), although pods have also been used and the possibility of using leaves was reported recently (68). Spores of *C. lindemuthianum* behave similarly when placed on young hypocotyls of resistant or susceptible cultivars. Within 24 to 48 hours they germinate to produce very short germ-tubes, at the end of which appressoria are formed. Differences between resistant and susceptible hypocotyls only become evident when infection hyphae from the appressoria penetrate epidermal cells. Infection is normally direct and does not occur through stomata. On excised hypocotyls of susceptible cultivars, infection by a virulent race causes no observable host response and hyphae continue to grow between the cell wall and the plasma membrane of the epidermal cells and then the parenchymatous cells without causing any deleterious effects either to themselves or to the infected cells. At temperatures below 20°C this harmonious biotrophic relationship can continue for several days, e.g. for more than five days on excised hypocotyls at 16°C (12). This leads to extensive colonization of hypocotyl tissue. After this period of biotrophy, the infected cells die, the tissue collapses and a brown lesion is produced. In some circumstances, especially in excised hypocotyls which are incubated at lower temperatures, considerable necrotrophic fungal growth occurs and rotting may become extensive (5, 12). However, in entire seedlings (54, 55, 57) or on excised seedlings with intact cotyledons (Rowell and Bailey, unpublished data), lesion expansion does not occur and the lesions become limited. Lesion limitation may also result when excised hypocotyls are transferred to between 22 and 25°C after a biotrophic relationship has been established by previous incubation at 16°C (5).

On resistant hypocotyls grown in light, the initially infected cell and perhaps one or two adjacent cells die soon after infection and appear as scattered flecks on the surface. Growth of the avirulent race is restricted within the initially infected cell. This process is often referred to as hypersensitivity (9, 38, 51). However, on hypocotyls grown under predominantly dark conditions many more cells are killed. For example, groups of between five and eight dead cells resulted from each infection in hypocotyls grown in darkness except for brief interludes of light (54). Using hypocotyls grown in total darkness and inoculated under a green safe light many more host cells were killed, so much so that groups of dead cells coalesced to form a lesion. Considerably more fungal growth occurred in the dark-grown than in light-grown hypocotyls (Rowell and Bailey, unpublished data).

Resistance of bean to *C. lindemuthianum* can therefore be seen as restriction of fungal development within a few cells (hypersensitivity) or in many cells (lesion limitation) and can result from infection with either avirulent or virulent races. The cause of this fungal growth inhibition will be considered in the next section.

MECHANISMS OF RESISTANCE

Inhibition of fungal growth occurs within dead host cells (6). *C. lindemuthianum* is a facultative pathogen and thus cell death alone, which could lead to starvation and hence cessation of growth of obligate pathogens, may not be sufficient to explain the restriction of growth of intracellular hyphae of this fungus. Alternatively, it has been shown that resistant tissues contain several highly toxic isoflavonoid phytoalexins (5, 9). Phaseollin and phaseollinisoflavan are usually present in the largest amounts, but phaseollidin and kievitone also occur. Other isoflavonoids which have been isolated from *P. vulgaris* inoculated with other fungi (74) may also be produced but they have not yet been identified from tissue infected with *C. lindemuthianum*.

An early hypothesis suggested that races of a pathogen may differ in their sensitivities to phytoalexins and that this may explain their differential behaviour on cultivars of crop plants (18). Using several races of *C. lindemuthianum* no data was obtained to support this view (13). All the races tested were completely inhibited by concentrations of phaseollin which differed only slightly, i.e. between 10 and 20 μg per ml. Similarly it is possible that the races may differ in their ability to metabolize and ultimately detoxify phytoalexins. Little quantitative information is available, but it has been reported that two races of *C. lindemuthianum* metabolized phaseollin, phaseollidin, phaseollinisoflavan and kievitone extensively. No antifungal products were detected after these phytoalexins had been incubated with the fungus for 24 hours (5). Cultivar resistance cannot, therefore, be explained by differences in the direct interactions between the different fungal races and the phytoalexins.

The timing and extent of phytoalexin accumulation in infected tissue appear more important. Following inoculation with a virulent race, phytoalexins did not accumulate during the infection process nor during the period of biotrophic colonization. However when the infected tissues died and the lesions formed, these phytoalexins were produced. When spreading lesions formed, the amounts of phytoalexin which accumulated were very small, less than 5 μg per g infected tissue. When the lesions became limited the amounts of phytoalexin produced were much greater, several hundred μg per g lesion tissue (5, 9). In limited lesions most phytoalexin was present at the edge of the lesions where one can assume the concentration was consequentially even greater (54). In contrast, following inoculation with an avirulent race, phytoalexin formation occurred very soon after infection and was closely associated with early death of the initially infected cells and inhibition of the pathogen. During the next two to three days the phytoalexins continued to accumulate until concentrations calculated to be between 3,000 and 4,000 μg per g infected tissue were attained (9).

Any total assessment of the contribution of antifungal host metabolites to the observed inhibition of hyphal growth in resistant tissues requires that they be produced at the right time, in the right place and in sufficient concentration to be effective (22, 64). Evidence of the type briefly referred to above suggests that when fungal growth has stopped there is much more phytoalexin present than would be needed to explain this inhibition. Nevertheless, we must also ask whether phytoalexins are present at the time when fungal growth inhibition takes place. Studies on the effect of higher temperature on the process of infection by virulent races have recently allowed this question to be considered. At 16°C a virulent race grew biotrophically for several days. However, hypersensitive flecks or limited lesions were produced when inoculated hypocotyls were transferred to 25°C after infection had taken place. Investigations showed that on transfer to 25°C the established biotrophic relationship was soon upset; the infected cells and two to three adjacent cells died very quickly, turned brown and phytoalexins accumulated. The normally extensive fungal growth was prevented (12). Maintaining infected hypcotyls at 16°C for 72 hours prior to transfer allowed sufficient hyphal growth to occur such that further hyphal development could be measured. Thus, when hypocotyls were transferred to 25°C the pattern of hyphal growth could be compared with the onset of host cell death and of phytoalexin accumulation. The results which were obtained are summarized in Fig. 1 and show that within 13.5 hours after transfer to 25°C the infected cells were granular and contained slight visible pigmentation, suggesting that the process of cell death had originated several hours earlier. Phytoalexins were not present at this time and were first detected 20.7 hours after transfer. Measurements of hyphal growth showed that the maximum rate of growth, which also indicates when growth first became inhibited, occurred 2.3 hours after the phytoalexins formed. The continued reduction in hyphal growth was accompanied by the continued accumulation of phytoalexins. This study suggests that phytoalexins are indeed present when intracellular growth is first inhibited. Unfortunately, at the present time, techniques are not available for measuring the cellular concentration of phytoalexins at these times.

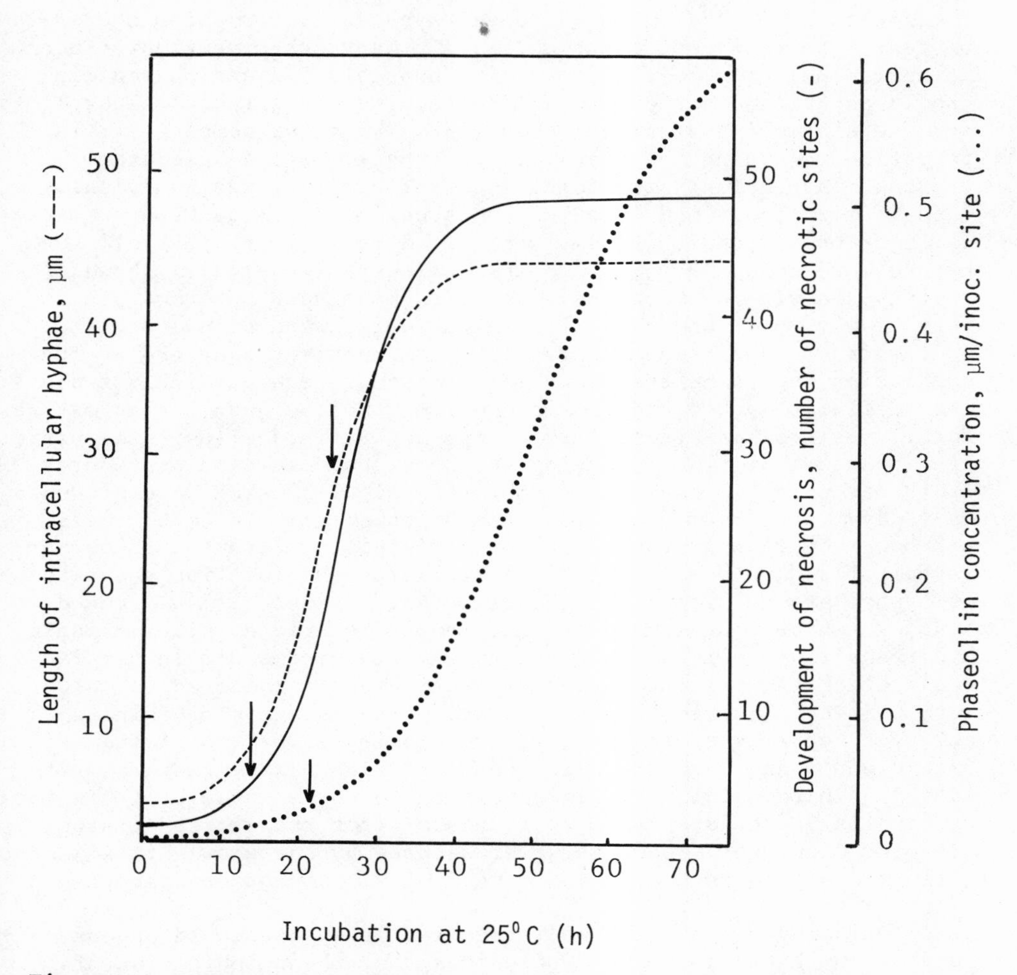

Figure 1. Temporal relationships between necrosis, accumulation of phytoalexins and inhibition of fungal growth

Hypocotyls of *P. vulgaris*, cv. Kievitsboon Koekoek were inoculated with *C. lindemuthianum*, incubated at 16°C for 72 hours and transferred to 25°C. Infected cell death, phytoalexin concentration and length of intracellular hyphae were measured. Arrows indicate time when necrosis had started (——), when phaseollin was present (····) and when rate of hyphal growth began to decline (----)

Another critical approach necessary in order that firm conclusions may be drawn is to compare the physiological and biochemical effects of phytoalexins on fungal hyphae *in vitro* with the state of fungus in resistant plant cells. In many experiments, the bean phytoalexins, especially phaseollin, phaseollidin and phaseollinisoflavan, killed various fungal cells, including spores, sporelings and the tips of mycelial cultures (13). Previous comparative information regarding the viability of fungal cells in resistant tissues, particularly in isolated necrotic cells, was equivocal. Electronmicroscopy had revealed gross cellular damage within intracellular hyphae and this was interpreted to indicate that all such hyphae were dead (43, 48). However, alternative studies showed that hyphae within hypersensitive cells continued to grow, albeit very slowly (61), and that mycelial colonies were produced when excised resistant tissues were placed on nutrient agar (28). The conclusion was therefore drawn that although severely inhibited, intracellular hyphae had remained alive. Neither conclusion may be totally reliable. Electronmicroscopy may fail to illustrate sufficient hyphae. Similarly, during the described excision procedures, it is not possible to know whether the colonies which were produced many days later had originated from hyphae within infected cells, nor how many viable propagules were originally present. These uncertainties have been resolved by examining the location and extent of hyphal growth within excised resistant tissues (11). Excised tissues, which contained avirulent hyphae restricted within single hypersensitive cells or virulent hyphae restricted within single cells after the infected hypocotyls had been transferred to 25°C, were placed on nutrient agar and, at intervals, growth of intracellular hyphae was examined microscopically after the tissue had been cleared and then stained in aniline blue. The great majority of these intracellular hyphae were seen to resume growth within the originally infected cell 10 to 20 hours after the excised tissue was placed on agar. Soon afterwards these hyphae grew into adjacent cells and within 50 hours had emerged on to the agar surface.

This proven viability of intracellular hyphae would appear to conflict with the previous conclusions that their inhibition of growth had been caused by the presence of large quantities of fungicidal phytoalexins within infected cells. More detailed studies of the effects of isoflavonoid phytoalexins on fungal cells, however, have shown that under some conditions, including preexposure to sub-lethal amounts of these compounds, hyphae can survive their potentially fungicidal effects (8, 13, 60, see also 72). In many experiments such surviving, i.e. adapted, hyphae also resumed growth and appeared to grow quite normally. However, in the present context, it is extremely important to realize that, in those experiments, growth only occurred when the amount of phytoalexin available to the fungus was reduced by, for example, metabolism in advance of mycelia on agar media or adsorption on to mycelial inocula in liquid media. While the concentrations of phytoalexins were maintained, as they were in resistant tissues, growth of surviving hyphae remained inhibited. It can therefore be envisaged that during resistance the increase in phytoalexin concentration would lead to adaptation and hence survival of intracellular hyphae, but, because the concentrations of phytoalexins in the tissues remain high, growth would remain inhibited.

On the basis of considerable, but largely circumstantial evidence, it can now be concluded that accumulated phytoalexins inhibit growth of *C. lindemuthianum* within resistant tissues of *P. vulgaris*. Accumulation of phytoalexins and inhibition of pathogen growth are very late, but not necessarily the final consequences of the earlier recognition phenomena between the host and its pathogen.

MECHANISMS OF PHYTOALEXIN ACCUMULATION

The mechanisms responsible for initiating the biosynthesis and subsequent accumulation of phytoalexins in plant tissues have been the subject of considerable research. The initial conclusion that phytoalexins were of host origin was based on the demonstration that phytoalexins were produced in tissues after treatment with various chemicals, as well as after infection with a range of fungi and bacteria (18, 20). Since that time many more chemicals and also short-wave irradiation (59) and nematodes (40) have also been found to be effective. Other important discoveries were that fungal culture filtrates and extracts of mycelial walls could cause phytoalexins to be formed. These materials were originally termed inducers (19), but this term has now been superseded by elicitors (41), and a distinction has also been drawn between biotic elicitors, i.e. from microorganisms, and abiotic elicitors, i.e. non-biological (75). The involvement of biotic elicitors in mechanisms of disease resistance has been reviewed previously (1, 71, 73) and forms part of Keen's contribution to this Institute. This section will concentrate on how these elicitors cause their effects.

Phytoalexin accumulation in response to microbial infection

Studies on processes of disease resistance, as exemplified by the classical investigations of Müller (50) using *P. vulgaris* and *Sclerotinia fructicola* and by subsequent studies on the interactions of *P. vulgaris* with *Colletotrichum lindemuthianum* and *Solanum tuberosum* with *Phytophthora infestans* (68), have often illustrated a close association between death of infected tissues and the accumulation of phytoalexins. When *P. vulgaris* was inoculated with virulent or avirulent races of *C. lindemuthianum*, the production of phytoalexins coincided very closely with the occurrence of cell death. Phytoalexins were not produced during periods of biotrophic growth, but when cells died phytoalexins soon accumulated. In two situations, i.e. when resistance was induced at higher temperatures (Fig. 1) (11), or when lesions were limited (51), death of host cells occurred several hours before phytoalexins could be detected. These phytoalexins were also concentrated within dead lesion tissue. They were not present in surrounding healthy cells (9, 58). Similarly, a virulent isolate of the rust, *Uromyces phaseoli*, grew biotrophically, produced large uredosori but did not cause any phytoalexins to form. In contrast, a partially avirulent race caused both death of infected cells and the accumulation of phytoalexins (10). Infection of leaves with differential races of the bacterial pathogen *Pseudomonas phaseolicola* also showed that much greater quantities of phytoalexin were produced during a resistant reaction when necrosis of the inoculated tissues became extensive than during a susceptible reaction when much more bacterial growth occurred but very little cell death took place (65).

A series of experiments using viruses may be of particular significance. Dark brown lesions formed on leaves of *P. vulgaris* between 36 and 60 hours after inoculation with tobacco necrosis virus. Isoflavonoid phytoalexins were produced only when necrosis had occurred and their concentrations increased as the lesions expanded (10). The phytoalexins were again located only within the lesion tissues. Other plants reacted in similar ways. Virus-induced lesions on other legumes, e.g. soybean, pea and cowpea produced glyceollin, pisatin and kievitone, respectively (42, 47) These effects were not restricted to legumes. Local lesions on leaves of the tobacco species *Nicotiana tabacum, N. clevelandii* (7) and *N. megalosiphon* (Bailey, unpublished data) produced the sesquiterpene capsidiol, a phytoalexin originally isolated from pepper, *Capsicum annuum*. Another antifungal sesquiterpene, glutinosone, was isolated from lesions on *N. glutinosa* (15). As previously, these compounds were present at high concentrations, were only present within the lesions and were produced when the lesions formed. Leaves of *N. glutinosa* inoculated with tobacco mosaic virus produced systemic mosaic symptoms but, in the absence of necrosis, did not produce phytoalexins.

An examination of symptom development not only in *P. vulgaris*, but also in other legumes and solanaceous plants, thus reveals a close association between the occurrence of host cell death and the accumulation of phytoalexins (3). This is particularly evident when the extent of cell death is limited, but phytoalexins were also produced, although in small amounts, when spreading lesions were formed. Biotic elicitors could be operating in some of these interactions but not in those involving viruses. Viruses, *per se*, could not produce elicitors.

Phytoalexin accumulation in response to abiotic elicitors

The possible importance of cell death or injury as features common to all mechanisms of phytoalexin elicitation can be further assessed by examining the activities of several abiotic elicitors.

Short wavelength ultraviolet light (254 nm) is extremely cytotoxic and when beans (Bailey, unpublished data), pea (30), soybeans (14, 59), grapes (44) or potato (16) were exposed to ultraviolet light and incubated in darkness, phytoalexins accumulated 10 to 20 hours later. In some of this work, i.e. with bean, soybean and grape, the detection of phytoalexins was closely associated with visible symptoms of cell death i.e. browning of the exposed surfaces. However, the reports of experiments using peas or potatoes concluded that cell death had not occurred. This conclusion seems to have been based solely on a lack of cellular browning and ignores the possibility that cell injury or death may not necessarily be associated with increased cellular pigmentation. Two other observations reported in this work would suggest, however, that damage had occurred. The production of pisatin by uv-treated tissue was prevented when incubation occurred in light, a phenomenon known to prevent uv-induced damage. Similarly, the uv-treated potato tissue was flaccid. Again this would indicate that cellular injury, although perhaps not death, had occurred.

A great many chemicals including salts of heavy metals, respiratory inhibitors and antimetabolites can cause phytoalexins to form in a wide range of plants. Only in a few cases have the effects of these compounds on the cellular integrity of treated tissue been reported concomitantly with reports on phytoalexin accumulation. This is somewhat surprising in view of the suspected toxicity of many of these materials. Detailed studies have, however, been carried out on the effects of mercuric salts on the accumulation of phytoalexins in both bean and potato (16, 33, 52, 70). Small amounts of phytoalexins accumulated within 24 hours after solutions (5×10^{-5} M) of $HgCl_2$ were placed on the inner surface of bean pod cavities. Plasmolysis and uptake of neutral red were unaffected indicating that gross cellular injury had not occurred, although some cells of the $HgCl_2$ treated tissue gave poor reactions with fluorescein diacetate (52). Greater concentrations of $HgCl_2$ were clearly toxic and produced more phytoalexin (50). As part of a series of experiments in which attempts were made to separate the effects of abiotic elicitors on plant cells and the subsequent formation of phytoalexins, bean cotyledons were immersed in different concentrations of $HgCl_2$ for periods varying from 30 seconds to two hours (33). After thorough washing, to remove unabsorbed $HgCl_2$, the cotyledons were incubated in air for up to five days. Phytoalexins accumulated to greatest amounts when only a few cell layers were killed. Phytoalexins did not occur in cotyledons which were undamaged, i.e. those which had not been treated or had been treated with concentrations of $HgCl_2$ below 5×10^{-4} M, or in those which had been extensively damaged by exposure to higher concentrations of $HgCl_2$ for longer periods of time. Similarly, when discs of potato tuber tissue were immersed in mercuric chloride for increasing periods of time, rishitin production was greatest when the killed tissue remained associated with living tissue. Rishitin accumulated only in the dead tissues (70).

Chloroform vapour is extremely phytotoxic and completely killed bean cotyledons within 20 minutes. Treatment, for a few minutes only, however, caused death of a few superficial cells and during subsequent incubation phytoalexins again accumulated (6). Chloroform is volatile and is thus unlikely to remain within treated tissues. Hence, these treatments represent a simple demonstration that death of bean cells can lead to the synthesis and accumulation of phytoalexins.

Both $HgCl_2$ and $CHCl_3$ pass through cells very quickly and as a result can kill large quantities of tissue. In contrast, although lipophilic Triton surfacants, e.g. Triton X-15 or X-35, are also phytotoxic, they do not penetrate deeply into treated tissue. Irrespective of concentration or time of exposure these surfactants killed between one and four cell layers of bean cotyledons. Phytoalexins formed 20 hours after treatment and continued to accumulate during the next 25 hours. The amount of phytoalexin which accumulated was clearly correlated with the toxicity of the surfactants used. Hydrophilic surfactants were not toxic and did not cause phytoalexin production. An interesting difference in the responses of cotyledons treated with Triton X-35 was that, unlike most diseased tissues and those treated with $HgCl_2$ or $CHCl_3$ where phaseollin predominated, the major phytoalexins were kievitone and licoisoflavone A (34).

The results obtained by treating plants, particularly legumes, but also potato, with abiotic elicitors has demonstrated that in some circumstances death of plant cells can lead to the accumulation of phytoalexins. It must also be emphasised that death of treated cells occurred many hours before phytoalexins were produced, indicating that death cannot be due to any phytotoxic effect of the phytoalexins themselves.

On several occasions it has been concluded (e.g. 29), that phytoalexin formation is not related to tissue death. This is certainly true if one only compares the concentrations of phytoalexins with the extent of cell death. In all the experiments referred to here and also those described later in this paper, extensive cell death did not cause greatest amounts of phytoalexin to form. Greatest accumulation occurred when only a relatively small amount of death had taken place. It was essential that living cells were present and this indicates that the interactions between living and dead plant tissues might be important.

Constitutive elicitors of phytoalexin biosynthesis

In 1975, Rahe and Arnold (56) reported that phytoalexins accumulated in localized tissue in bean hypocotyls which had been killed by contact with solid carbon dioxide. Freezing was immediate and thawing took only a few seconds. On removal of the CO_2, the injured tissue appeared water soaked within a few minutes and within six to 12 hours became light tan in colour. Phaseollin was also detected about six hours after thawing and its concentration increased steadily during the next 12 hours. Phaseollin was not present in undamaged hypocotyls nor in hypocotyls which were killed completely by immersion in liquid nitrogen.

Liquid nitrogen and refrigeration at $-20^{o}C$ have also been used to produce partially killed tissues (35). When hypocotyls were placed at $-20^{o}C$ for 10 minutes, the outer cell layers became frozen while the inner tissue remained alive. Phytoalexins accumulated 20 to 40 hours after treatment and the hypocotyls turned brown. Similarly, partial freezing was also achieved by placing hypocotyl segments on a glass surface cooled with liquid nitrogen. The base of the segment froze immediately and a portion 5 to 8 mm long was frozen within a few minutes. On thawing, a junction was clearly visible between the living and dead tissue and after incubation this area was observed as a narrow band of brown pigmentation. Phytoalexins were also produced during incubation and were concentrated in the dead tissues adjacent to the living tissues. Phytoalexins did not accumulate in either the living or dead tissues away from the junction between them. In all these experiments, phytoalexins did not accumulate in hypocotyls which had been completely frozen.

Additional experiments were used to assess the distribution of phytoalexin between living and dead tissues. Unfrozen hypocotyls, and hypocotyls which had been completely frozen and then thawed, were cut longitudinally. Dead and living hypocotyl halves were incubated with their cut surfaces in close contact. Within 24 hours, phytoalexins had formed and analyses showed that they accumulated in both the living and dead tissues, but that more had accumulated in the dead tissue which also contained more brown pigments (35). Contact between these living and dead tissues for as little as three hours was sufficient to cause subsequent accumulation of phytoalexins in the living halves. Increasing the period of contact up to 24 hours increased the accumulation of phytoalexins very little. After separation, incubated dead tissue either contained no phytoalexins, if contact was for less than 24 hours, or did not accumulate any more phytoalexins, if contact was for 24 or more hours, indicating that synthesis of these compounds only occurred in the living cells.

These results confirm that death of bean cells can initiate the accumulation of phytoalexins when in association with living cells and suggest that when cells die materials are released which stimulate phytoalexin synthesis in adjacent living cells. The identification of the active component(s) has not been established. Nevertheless elicitor activity was obtained using extracts obtained from bean hypocotyls (35, 37). Extracts were placed in the central cavity of bean hypocotyls, adsorbed on filter paper and placed between cut surfaces of bean hypcotyls or added to suspensions of cultured bean cells. Greater quantities of phytoalexins, particularly phaseollin, accumulated in these tissues than in those treated with water or with a series of simple nutrient solutions. Extracts of autoclaved bean tissue caused more than 500 μg phaseollin per g tissue to accumulate in cell suspensions, while other synthetic nutrient solutions elicited less than 5 μg per g. Ion-exchange chromatography and gel filtration indicated that the active components of these extracts were neither strongly acidic nor basic and had a molecular weight less than 5,000.

Elicitor activity was obtained in extracts from both unfrozen and frozen hypocotyls. The elicitor thus appears to be a constituent of healthy bean cells and for this reason was termed a constitutive elicitor. Further evidence for constitutive elicitors in other plants has been presented by Albersheim at this Institute. Aqueous extracts of frozen soybean hypcotyls and hydrolysates of soybean cell walls elicited glyceollin accumulation in soybean cotyledons. It is not yet clear whether the active components from bean, soybean hypcotyls or soybean cell walls are structurally related.

Factors analagous to constitutive elicitors may function to enhance lignin formation in radish, *Raphanus sativus*. Homogenates of healthy roots, as well as of those infected with *Peronospora parasitica*, caused lignin to accumulate when infiltrated into radish leaves. This "lignin-inducing factor" may thus also be a product of plant cells. It was also water soluble and of low molecular weight (3).

PHYTOALEXIN ACCUMULATION : A CONSEQUENCE OF INJURY TO PLANT CELLS

Plant cell death, as illustrated earlier, is not a unique, easily-defined event. It is a complex, very variable and often biochemically active process. It is triggered by an initial metabolic lesion, perhaps malfunction of cell membranes. Total cessation of metabolic activity in individual or groups of cells may eventually result. It must be realised, however, that such inert tissue could continue to influence the behaviour of surrounding healthy cells.

Under some circumstances, e.g. treatment with chloroform or partial freezing, death of bean cells led to considerable synthetic activity, which was evident as the accumulation of many secondary metabolites, particularly brown pigments and isoflavonoid phytoalexins. These effects were most noticeable when localised death of cells occurred within predominantly living tissue. Death of an entire tissue did not cause the formation of phytoalexins and only small amounts were produced when the great majority of cells were killed. An appropriate interaction between living and dying cells seems to be essential for maximum accumulation of these metabolites. Similar relationships have been described during infection of tissue by pathogens where it has also been established that the death of individual or small groups of cell preceded the accumulation of phytoalexins. Evidence has also been presented to show that healthy bean tissues contain an elicitor of phytoalexin production.

On the basis of these results it is suggested that in undamaged cells the constitutive elicitor is inactive, perhaps bound to or compartmentalized within the cells. When cells are injured e.g. by maceration, application of toxic chemicals, exposure to uv irradiation, or freezing, the constitutive elicitor is released. Its interaction with metabolically active tissue initiates the *de novo* synthesis of many secondary metabolites, including antifungal isoflavonoids i.e. phytoalexins. The phytoalexins diffuse from the living cells into the dead cells where they accumulate to high concentrations. An illustration of these proposals, which pertain specifically to bean, is shown in Fig. 2.

This hypothesis may, I believe, be relevant to other plant species and some appropriate findings with other legumes, potatoes, tomatoes and vines have been referred to earlier. The necessity for cell injury is clear but the sites of phytoalexin accumulation may differ. It has already been shown with pea, for example, that, although both cellular injury and constitutive elicitor activity were demonstrated, pisatin accumulated in the absence of dead cells (33, see also 53). It must be remembered that the structures of phytoalexins and other secondary metabolites produced by plant species may be quite different and hence the control mechanisms for their synthesis and accumulation may also differ considerably.

This leaves me with a further important question. Can cell injury be invoked to explain the action of biotic elicitors e.g. glucans and glycoproteins, which have been isolated from fungal mycelia (1)?. It has been suggested that in healthy cells phytoalexins do not accumulate because of a balance between their synthesis and degradation, and that biotic elicitors cause phytoalexins to accumulate by directly enhancing synthesis whilst abiotic elicitors do so by preventing degradation (75). The basis of these conclusions has recently been challenged (49), but it is of equal significance that some biotic elicitors, like abiotic elicitors, are phytotoxic. Glycoproteins caused leakage of cellular constituents (23, 24) and the glucan elicitor from *Phytophthora megasperma* was shown to prevent growth of suspension cells (27). In addition, it is now established that both glucan and glycoprotein elicitors

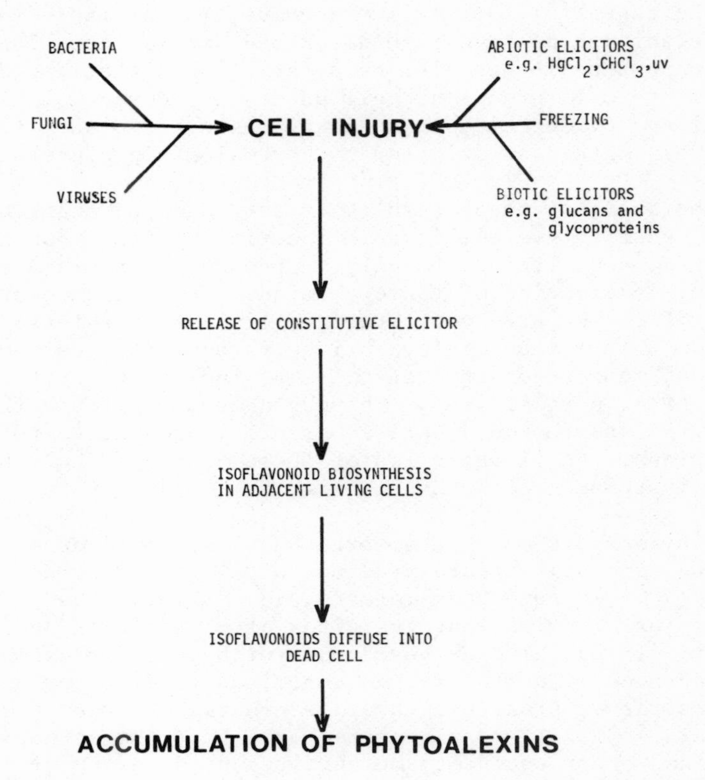

Figure 2 Schematic illustration of the mechanism of phytoalexin accumulation in *Phaseolus vulgaris*

can elicit the formation of different phytoalexins, e.g. isoflavonoids or sesquiterpenes, in different host plants (17, 24). As first discussed by Stoessl (66), it is extremely improbable that the same elicitor could directly activate different biosynthetic pathways and he suggested that phytoalexin biosynthesis must begin from a primary disturbance of cellular metabolism. It seems likely that as for abiotic elicitors this primary disturbance is cellular injury.

The complementary elicitor-suppressor hypothesis (26), in which it is proposed that a virulent race of *Phytophthora* produces a suppressor which prevents the potentially toxic action of the elicitor and hence prevents phytoalexin production, is also consistent with these suggestions.

Hypersensitivity

Hypersensitivity is a response of plant cells to infection by avirulent races which involves both the early death of infected cells and the inhibition of the fungus therein (38, 39, 45, 51, 69). A detailed interpretation of the processes which may be involved in a hypersensitive reaction of bean to avirulent races of *C. lindemuthianum* is shown in Fig. 3. Spores germinate on the cell surface to produce appressoria. At this time there appears to be little host response. However, when cell walls are penetrated by infection hyphae the cell is injured and the constitutive elicitor is released (or activated). Phytoalexin synthesis begins in infected cells, but, as the elicitor affects adjacent cells, phytoalexins are also produced there. Soon after infection, the initially infected cells begin to die (46) and thus phytoalexins accumulate and further fungal development is prevented. Phytoalexin production by living cells and their accumulation in the dead cell continues and their concentrations thus become extremely high (9). Such a mechanism is entirely consistent with the traditional view of hypersensitivity (38, 45).

The failure of phytoalexins to accumulate during biotrophic growth of virulent races of *C. lindemuthianum* would thus be a direct consequence of the continued viability of infected cells and hence the continued inactivity of the constitutive elicitor. However, when cells which had supported biotrophic fungal growth eventually die, the processes outlined above again take place, and if phytoalexins accumulate, fungal development would be restricted. If, on the other hand, the affected cells fail to synthesize sufficient phytoalexins, perhaps because they too are killed, the fungus would continue to grow through the tissue and a spreading lesion would form. It has already been suggested that *Botrytis fabae* causes large lesions to form in leaves of *Vicia faba* because it kills cells so quickly that very little phytoalexin is produced (36).

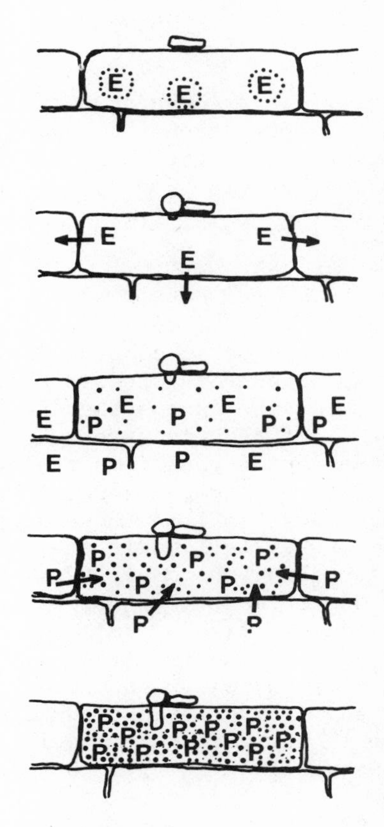

Figure 3 Diagrammatic representation of the involvement of constitutive elicitors in the hypersensitive response of *P. vulgaris* to *C. lindemuthianum*

A CONCEPT OF HOST-CELL COMPATIBILITY

This paper has concentrated on the chain of events which link recognition of host and parasite with subsequent accumulation of phytoalexins and inhibition of pathogen growth. It has been proposed that cell death (or injury) is the event which is directly responsible for activating the required secondary metabolism. Cell death may also be very closely connected with the recognition phenomenon itself, for the differences between infection by avirulent and virulent races of a pathogen seem to depend on when infected cells die. In a previous review, Deverall (22) has discussed the possible existence of compatibility between host and parasite. Compatibility was used by him to indicate a harmonious relationship in which both the host and the parasite remained viable, and incompatibility to indicate a relationship which was deleterious to both host and parasite. I would like to suggest that compatibility and incompatibility should refer only to viability or death of the infected host cells (see 67). Hyphae of virulent races of *C. lindemuthianum* can infect and grow in bean tissues without causing host cell death. These host cells would thus be considered to show cellular compatibility towards the infection hyphae, while interactions with avirulent races which lead to early death of host cells would be considered to illustrate cellular incompatibility.

Compatibility and incompatibility are thus mutually exclusive responses of host cells. The absence or occurrence of death of the initially infected cells appear to be the fundamental genetically controlled events which to a large extent determine the differentiation of cultivars as being susceptible or resistant to a particular race. Thus a period of host cell compatibility enables virulent races to grow extensively in the tissues and this forms the basis for the subsequent development of lesions. However, lesion development itself, i.e. death of infected tissues must be considered to be an example of host cell incompatibility. The ultimate symptoms can, as described earlier, be evident as spreading lesions, limited lesions or as scattered necrotic cells, i.e. ranging from susceptibility to resistance. Conversely, the initially-infected host cells are incompatible with avirulent races and death of these cells is usually equated with production of phytoalexins and restriction of fungal development, i.e. resistance. In dark-grown tissues, however, fungal development may not be prevented and eventually lesions, i.e. symptoms of susceptibility may be produced. It is therefore essential that host cell compatibility and incompatibility are considered to be quite separate from susceptibility or resistance. However, hypotheses concerning the genetical relationships between a host cultivar and its pathogen have usually been based on symptom expression, not on how such symptoms were produced. It has been emphasized that the gene for gene hypothesis relies on grouping apparently related symptoms (21), but details of the processes which lead to their development seem rarely available. The demonstrated variability of the expression of resistance and susceptibility of beans to virulent and avirulent races of *C. lindemuthianum* illustrates quite clearly some of the possible dangers of only using symptoms to establish a gene for gene relationship between host and pathogen. Perhaps a hypothesis based on gene for gene control of host cell compatibility would be more appropriate.

The present discussion has concentrated on interactions between bean and *C. lindemuthianum*. However, this concept of host cell compatibility may also be appropriate to other interactions between host cultivars and specific pathogens. *Phytophthora infestans* is another facultative pathogen which can grow in tissues of some potato cultivars without causing death of infected cells. In other cultivars, death of cells occurs soon after infection (69). Recent studies suggest that host cell viability (compatibility) is maintained by race-specific metabolites produced by the fungus (25, 26).

Interactions of plant cultivars with organisms usually regarded as obligate pathogens can also be interpreted in these terms. For example, *Puccinia graminis* f. sp. *tritici* exists in leaves of susceptible wheat cultivars, not only as intercellular hyphae which do not penetrate cells and which therefore are unlikely to be directly involved in host cell compatibility, but also as haustoria which do penetrate host cells and which are believed to act as the major source of fungal nutrients. Viable host cells which contain haustoria would thus be considered compatible. Many intracellular haustoria are produced as the intercellular hyphae colonize the tissue and produce uredosori. No host cells are killed. Cultivars which contain genes for resistance have been designated resistant predominantly on the basis that uredosori were not produced. Studies on the mechanisms which prevent sporulation have shown, for example, that haustoria produced in cultivars containing the Sr6 gene were associated with death of infected cells, i.e. incompatibility (62, 63). On many occasions the haustoria also appeared to be dead, but immediate detrimental effects on the growth of the intercellular hyphae were not observed (47). Growth of intercellular hyphae resulted in the production of many haustoria, which were usually seen in dead cells. After a few days, however, hyphal growth slowed and uredosori were not produced. These findings suggest that different cultivars show distinct compatible and incompatible reactions to *P. graminis*, but that mechanisms must exist which enable the intercellular hyphae in resistant cultivars to overcome, at least in part, the potentially harmful effects of host cell death. Haustoria within the dying cells may continue to provide some of the nutrients needed by the intercellular hyphae, although their progressive malfunction would eventually explain the cessation of fungal development.

The growth of *P. graminis* in wheat leaves can be very variable (31, 32). The possible merit of a concept based on reactions of host cells is that it might be used to explain such complex host-parasite interactions. Compatibility and incompatibility are used here to refer only to the state of host cells and not to that of the infecting fungus. Incompatibility would not necessarily be equated with inhibition or death of the fungus, i.e. resistance, and similarly compatibility does not of necessity ensure continued fungal viability or growth. An infected host cell could be envisaged as being alive, i.e. compatible, but the pathogen may be unable to grow in it (see 32). Secondary, non cultivar-specific, mechanisms of preventing fungal growth in living host cells or of enabling growth to occur in dead cells could clearly be superimposed on fundamental compatible or incompatible responses.

CONCLUSIONS

Resistance of cultivars of *Phaseolus vulgaris* to *Colletotrichum lindemuthianum* is expressed as an inhibition of intracellular hyphal growth within dead host cells, and may be evident as lesion limitation or hypersensitivity. Isoflavonoid phytoalexins accumulate in these cells at high concentrations and appear to be responsible for the restriction of fungal growth therein.

Phytoalexins are also produced by *P. vulgaris* after infection with other microorganisms or virus, or after treatment with various phytotoxic agents. Dead tissue caused phytoalexins to form when it was in contact with living tissue and extracts of bean tissues were also effective. It is proposed that in all these situations, *de novo* synthesis of isoflavonoids is initiated in living cells by a metabolite which is released from cells when they are injured or killed. The synthesized isoflavonoids diffuse from the living tissues into the dead cells where they accumulate. The accumulation of phytoalexins thus represents a late stage in the responses of bean to *C. lindemuthianum*, and one which, because it is dependent on complex synthetic processes, may vary considerably.

Differentiation between virulent and avirulent races occurs very much earlier, probably during the initial contact between host and pathogen. Evidence suggests that the mechanisms which maintain cell viability after infection by a virulent race or cause immediate death of cells infected with an avirulent race, are of the fundamental processes which determine the cultivar-specific responses.

REFERENCES

1. ALBERSHEIM, P. & VALENT, B.S. (1978). Plants, when exposed

to oligosaccharides of fungal origin, defend themselves by accumulating antibiotics. *Journal of Cell Biology* 78, 627 - 643.

2. ASADA, Y., OHGUCHI, T. & MATSUMOTO, I. (1979). Induction of lignification in response to fungal infection. In : *Recognition and specificity in plant host-parasite interactions*, Ed. by J.M. Daly and I. Uritani pp. 99 - 112, Japan Scientific Societies Press, Tokyo.

3. BAILEY, J.A. (1973). Phaseollin accumulation in *Phaseolus vulgaris* following infection by fungi, bacteria and a virus. In : *Fungal Pathogenicity and the Plant's Response*, Ed. by R.J.W. Byrde and C.V. Cutting, pp. 337 - 353. Proceedings of the Third Long Ashton Symposium 1971. Academic Press.

4. BAILEY, J.A. (1973). Production of antifungal compounds in cowpea *(Vigna sinensis)* and pea *(Pisum sativum)* after virus infection. *Journal of General Microbiology* 75, 119 - 123.

5. BAILEY, J.A. (1974). The relationship between symptom expression and phytoalexin concentration in hypocotyls of *Phaseolus vulgaris* infected with *Colletotrichum lindemuthianum*. *Physiological Plant Pathology* 4, 477 - 488.

6. BAILEY, J.A. & BERTHIER, M. (1980). Phytoalexin accumulation in chloroform-treated cotyledons of *Phaseolus vulgaris*. *Phytochemistry* in press.

7. BAILEY, J.A. BURDEN, R.S. & VINCENT, G.G. (1975). Capsidiol : an antifungal compound produced in *Nicotiana tabacum* and *Nicotiana clevelandii* following infection with tobacco necrosis virus. *Phytochemistry* 14, 597.

8. BAILEY, J.A., CARTER, G.A. & SKIPP, R.A. (1976). The use and interpretation of bioassays for fungitoxicity of phytoalexins in agar media. *Physiological Plant Pathology* 8, 189 - 194.

9. BAILEY, J.A. & DEVERALL, B.J. (1971). Formation and activity of phaseollin in the interaction between bean hypocotyls *(Phaseolus vulgaris)* and physiological races of *Colletotrichum lindemuthianum*. *Physiological Plant Pathology* 1, 435 - 449.

10. BAILEY, J.A., & INGHAM, J.L. (1971). Phaseollin accumulation in bean *(Phaseolus vulgaris)* in response to infection by tobacco necrosis virus and the rust *Uromyces appendiculatus*.

Physiological Plant Pathology 1, 451 - 456.

11. BAILEY, J.A. & ROWELL, P.M. (1980). Viability of *Colletotrichum lindemuthianum* in hypersensitive cells of *Phaseolus vulgaris*. *Physiological Plant Pathology* in press.

12. BAILEY, J.A., ROWELL, P.M. & ARNOLD, G.M. (1980). The temporal relationship between host cell death, phytoalexin accumulation and fungal inhibition during hypersensitive reactions of *Phaseolus vulgaris* to *Colletotrichum lindemuthianum*. *Physiological Plant Pathology* in press.

13. BAILEY, J.A. & SKIPP, R.A. (1978). Toxicity of phytoalexins. *Annals of Applied Biology* 89, 354 - 358.

14. BRIDGE, M.A. & KLARMAN, W.L. (1973). Soybean phytoalexin, hydroxyphaseollin, induced by ultraviolet irradiation. *Phytopathology* 63, 606 - 609.

15. BURDEN, R.S., BAILEY, J.A. & VINCENT, G.G. (1975). Glutinosone, a new antifungal sesquiterpene from *Nicotiana glutinosa* infected with tobacco mosaic virus. *Phytochemistry* 14, 221 - 223.

16. CHEEMA, A.S. & HAARD, N.F. (1978). Induction of rishitin and lubimin in potato tuber discs by non-specific elicitors and the influence of storage conditions. *Physiological Plant Pathology* 13, 233 - 240.

17. CLINE, L., WADE, M. & ALBERSHEIM, P. (1978). Host-pathogen interactions. XV. Fungal glucans which elicit phytoalexin accumulation in soybean also elicit the accumulation of phytoalexins in other plants. *Plant Physiology* 62, 918 - 921

18. CRUICKSHANK, I.A.M. (1963). Phytoalexins. *Annual Review of Phytopathology* 1, 351 - 374.

19. CRUICKSHANK, I.A.M. & PERRIN, D.R. (1968). The isolation and partial characterisation of Monilicolin A a polypeptide with phaseollin-inducing activity from *Monilia fructicola*. *Life Sciences* 7, 449 - 458.

20. CRUICKSHANK, I.A.M. & PERRIN, D.R. (1971). Studies on phytoalexins. XI. The induction, antimicrobial spectrum and chemical assay of phaseollin. *Phytopathologische Zeitschrift* 70, 209 - 229.

21. DALY, J.M. (1975). Specific interactions involving hormonal and other changes. In : *Specificity in Plant Diseases*.

Ed. by R.K.S. Wood and A. Graniti, pp. 151 - 167. Plenum Press.

22. DEVERALL, B.J. (1977). *Defence Mechanisms of Plants.* Cambridge Monographs in Experimental Biology, Cambridge University Press.

23. DE WIT, P.J.G.M. & ROSEBOOM, P.H.M. (1980). Isolation, partial characterization and specificity of glycoprotein elicitors from culture filtrates, mycelium and cell walls of *Cladosporium fulvum* (syn. *Fulvia fulva*). *Physiological Plant Pathology* 16, 391 - 408.

24. DOKE, N., SAKAI, S. & TOMIYAMA, K. (1979). Hypersensitive reactivity of various host and non-host plant leaves to cell wall components and soluble glucan isolated from *Phytophthora infestans. Annals of the Phytopathological Society of Japan* 45, 386 - 393.

25. DOKE, N. & TOMIYAMA, K. (1977). Effect of high molecular substances released from zoospores of *Phytophthora infestans* on hypersensitive response of potato tubers. *Phytopathologische Zeitschrift* 90, 236 - 242.

26. DOKE, N. & TOMIYAMA, K. (1980). Suppression of the hypersensitive response of potato tuber protoplasts to hyphal wall components by water soluble glucans isolated from *Phytophthora infestans. Physiological Plant Pathology* 16, 177 - 186.

27. EBEL, J., AYERS, A.R. & ALBERSHEIM, P. (1976). Host-Pathogen Interactions III. Response of suspension-cultured soybean cells to the elicitor isolated from *Phytophthora megasperma* var. *sojae*, a fungal pathogen of soybeans. *Plant Physiology* 57, 775 - 779.

28. ERB, K., GALLEGLY, M.E. & LEACH, J.G. (1973). Longevity of mycelium of *Colletotrichum lindemuthianum* in hypocotyl tissue of resistant and susceptible bean cultivars. *Phytopathology* 63, 1334 - 1335.

29. ERSEK, T., KIRÁLY, Z. & DOBROVOLSZKY,A. (1977). Lack of correlation between tissue necrosis and phytoalexin accumulation in tubers of potato cultivars. *Journal of Food Safety* 1, 77 - 85.

30. HADWIGER, L.A. & SCHWOCHAU, M.E. (1971). Ultraviolet light-induced formation of pisatin and phenylalanine ammonia lyase. *Plant Physiology* 47, 588 - 590.

31. HARDER, D.E., ROHRINGER, R., SAMBORSKI, D.J., RIMMER, S.R., KIM, W.K. & CHONG, J. (1979). Electron microscopy of susceptible and resistant near-isogenic (*Sr6/Sr6*) lines of wheat infected by *Puccinia graminis tritici*. II. Expression of incompatibility in mesophyll and epidermal cells and the effect of temperature on host-parasite interactions in these cells. *Canadian Journal of Botany* 57, 2617 - 2625.

32. HARDER, D.E., SAMBORSKI, D.J., ROHRINGER, R., RIMMER, S.R., KIM, W.K. & CHONG, J. (1979). Electron microscopy of susceptible and resistant near-isogenic (*Sr6/Sr6*) lines of wheat infected by *Puccinia graminis tritici*. III. Ultrastructure of incompatible interactions. *Canadian Journal of Botany* 57, 2626 - 2634.

33. HARGREAVES, J.A. (1979). Investigations into the mechanism of mercuric chloride stimulated phytoalexin accumulation in *Phaseolus vulgaris* and *Pisum sativum*. *Physiological Plant Pathology* 15, 279 - 287.

34. HARGREAVES, J.A. (1981). Stimulation of phytoalexin accumulation in cotyledons of French bean (*Phaseolus vulgaris* L.) by Triton (t-octyl phenol polyethoxy ethanol) surfactants. *New Phytologist* in press.

35. HARGREAVES, J.A. & BAILEY, J.A. (1978). Phytoalexin production by hypocotyls of *Phaseolus vulgaris* in response to constitutive metabolites released by damaged cells. *Physiological Plant Pathology* 13, 89 - 100.

36. HARGREAVES, J.A., MANSFIELD, J.W. & ROSSALL, S. (1977). Changes in phytoalexin concentrations in tissues of the broad bean plant (*Vicia faba* L.) following inoculation with species of *Botrytis*. *Physiological Plant Pathology* 11, 227 - 242.

37. HARGREAVES, J.A. & SELBY, C. (1978). Phytoalexin formation in cell suspensions of *Phaseolus vulgaris* in response to an extract of bean hypocotyls. *Phytochemistry* 17, 1099 - 1102.

38. HEATH, M.C. (1976). Hypersensitivity, the cause or the consequence of rust resistance ? *Phytopathology* 66, 935 - 936.

39. INGRAM, D.S. (1978). Cell death and resistance to biotrophs. *Annals of Applied Biology* 89, 291 - 295.

40. KAPLAN, D.T., KEEN, N.T. & THOMASON, I.J. (1979). Studies on the mode of action of glyceollin in soybean incompatibil-

ity to the root knot nematode, *Meloidogyne incognita*. *Physiological Plant Pathology* 16, 319 - 325.

41. KEEN, N.T. (1975). Specific elicitors of plant phytoalexin production : Determinants of race specificity in pathogens. *Science* 187, 74 - 75.

42. KLARMAN, W.L. & HAMMERSCHLAG, F. (1972). Production of the phytoalexin, hydroxyphaseollin, in soybean leaves inoculated with tobacco necrosis virus. *Phytopathology* 62, 719 - 721.

43. LANDES, M. & HOFFMAN, G.M. (1979). Ultrahistological investigations of the interaction in compatible and incompatible systems in *Phaseolus vulgaris* and *Colletotrichum lindemuthianum*. *Phytopathologische Zeitschrift* 96, 330 - 350.

44. LANGCAKE, P., CORNFORD, C.A. & PRYCE, R.J. (1979). Identification of pterostilbene as a phytoalexin from *Vitis vinifera* leaves. *Phytochemistry* 18, 1025 - 1027.

45. MACLEAN, D.J., SARGENT, J.A., TOMMERUP, I.C. & INGRAM, D.S. (1974). Hypersensitivity as the primary event in resistance to fungal parasites. *Nature* 249, 186 - 187.

46. MANSFIELD, J.W. & SEXTON, R. (1974). Changes in the localization of β-glycerophosphatase activity during the infection of *Phaseolus vulgaris* by *Colletotrichum lindemuthianum*. *Annals of Botany* 38, 711 - 717.

47. MAYAMA. S., REHFELD, D.W. & DALY, J.M. (1975). A comparison of the development of *Puccinia graminis* in resistant and susceptible wheat based on glucosamine content. *Physiological Plant Pathology* 7, 243 - 258.

48. MERCER, P.C., WOOD, R.K.S. & GREENWOOD, R.D. (1974). Resistance to anthracnose of French bean. *Physiological Plant Pathology* 4, 291 - 306.

49. MOESTA, P. & GRISEBACH, H. (1980). Effects of biotic and abiotic elicitors on phytoalexin metabolism in soybean. *Nature* 286, 710 - 711.

50. MÜLLER, K.O. (1958). Studies on phytoalexins I. The formation and the immunological significance of phytoalexin produced by *Phaseolus vulgaris* in response to infection with *Sclerotinia fructicola* and *Phytophthora infestans*. *Australian Journal of Biological Sciences* 11, 276 - 300.

51. MÜLLER, K.O. (1959). Hypersensitivity. In : *Plant Pathology*

Vol. I. Ed. by J.G. Horsall and A.E. Dimond, pp. 469 - 519. Academic Press.

52. PAXTON, J., GOODCHILD, D.J. & CRUICKSHANK, I.A.M. (1974). Phaseollin production by live bean endocarp. *Physiological Plant Pathology* 4, 167 - 171.

53. PUEPPKE, S.G. & VAN ETTEN, H.D. (1976). The relation between pisatin and the development of *Aphanomyces euteiches* in diseased *Pisum sativum*. *Phytopathology* 66, 1174 - 1185.

54. RAHE, J.E. (1973). Occurrence and levels of the phytoalexin phaseollin in relation to delimitation of sites of infection of *Phaseolus vulgaris* by *Colletotrichum lindemuthianum*. *Canadian Journal of Botany* 51, 2423 - 2430.

55. RAHE, J.E. (1973). Phytoalexin nature of heat-induced protection against bean anthracnose. *Phytopathology* 63, 572 - 577.

56. RAHE, J.E. & ARNOLD, R.M. (1975). Injury-related phaseollin accumulation in *Phaseolus vulgaris* and its implications with regard to specificity of host-parasite interaction. *Canadian Journal of Botany* 53, 921 - 928.

57. RAHE, J.E., KUĆ, J., CHUANG, C.-M. & WILLIAMS, E.B. (1969). Correlation of phenolic metabolism with histological changes in *Phaseolus vulgaris* inoculated with fungi. *Netherlands Journal of Plant Pathology* 75, 58 - 71.

58. RATHMELL, W.G. (1973). Phenolic compounds and phenylalanine ammonia lyase activity in relation to phytoalexin biosynthesis in infected hypocotyls of *Phaseolus vulgaris*. *Physiological Plant Pathology* 3, 259 - 267.

59. REILLY, J.J. & KLARMAN, W.L. (1980). Thymine dimer and glyceollin accumulation in u.v.-irradiated soybean suspension cultures. *Environmental and Experimental Botany* 20, 131 - 134.

60. SKIPP, R.A. & CARTER, G.A. (1978). Adaptation of fungi to isoflavonoid phytoalexins. *Annals of Applied Biology* 89, 366 - 369.

61. SKIPP, R.A. & DEVERALL, B.J. (1972). Relationships between fungal growth and host changes visible by light microscopy during infection of bean hypocotyls (*Phaseolus vulgaris*) susceptible and resistant to physiological races of *Colletotrichum lindemuthianum*. *Physiological Plant Pathology* 2, 357 - 374.

62. SKIPP, R.A., HARDER, D.E. & SAMBORSKI, D.J. (1974). Electron microscopy studies on infection of resistant (*Sr*6 gene) and susceptible near-isogenic wheat lines by *Puccinia graminis* f. sp. *tritici*. *Canadian Journal of Botany* 52, 2615 - 2620.

63. SKIPP, R.A. & SAMBORSKI, D.J. (1974). The effect of the *Sr*6 gene for host resistance on histological events during the development of stem rust in near-isogenic wheat lines. *Canadian Journal of Botany* 52, 1107 - 1115.

64. SMITH, I.M. (1978). The role of phytoalexins in resistance. *Annals of Applied Biology* 82, 325 - 329.

65. STHOLASUTA, P., BAILEY, J.A., SEVERIN, B. & DEVERALL, B.J. (1971). Effect of bacterial inoculation of bean and pea leaves on the accumulation of phaseollin and pisatin. *Physiological Plant Pathology* 1, 177 - 183.

66. STOESSL, A. (1978). Phytoalexins - a biogenetic perspective. Abstracts of 3rd Int. Congress of Plant Pathology, Munich p. 217.

67. TEASDALE, J., DANIELS, D., DAVIS, W.C., EDDY, R. Jr. & HADWIGER, L.A. (1974). Physiological and cytological similarities between disease resistance and cellular incompatibility responses. *Plant Physiology* 54, 690 - 695.

68. THEODOROU, M.K. & SMITH, I.M. (1979). The implication of a rapid method for the determination of differential interactions in French bean anthracnose. *Phytopathologische Zeitschrift* 96, 1 - 8.

69. TOMIYAMA, K., DOKE, N., NOZUE, M. & ISHIGURI, Y. (1979). The hypersensitive response of resistant plants. In : *Recognition and Specificity in Plant Host-Parasite Interactions*, Ed. by J.M. Daly and I. Uritani, pp. 69 - 84. Japan Scientific Societies Press, Tokyo.

70. TOMIYAMA, K. & FUKAYA, M. (1975). Accumulation of rishitin in dead potato-tuber tissue following treatment with $HgCl_2$. *Annals of Phytopathological Society of Japan* 41, 418 - 420.

71. VALENT, B.S. & ALBERSHEIM, P. (1977). Role of elicitors of phytoalexin accumulation in plant disease resistance. In : *Host Plant Resistance to Pests*, Ed. by P.A. Hedin, pp. 27 - 34. American Chemical Society Symposium No. 62.

72. VAN ETTEN, H.D. (1979). Relationship between tolerance to isoflavonoid phytoalexins and pathogenicity. In : *Recognition and Specificity in Plant Host-Parasite Interactions*, Ed. by J.M. Daly and I. Uritani, pp. 301 - 316. Japan Scientific Societies Press, Tokyo.

73. WARD, E.W.B. & STOESSL, A. (1976). On the question of 'elicitors' or 'inducers' in incompatible interactions between plants and fungal pathogens. *Phytopathology* 66, 940 - 941.

74. WOODWARD, M.D. (1979). New isoflavonoids related to kievitone from *Phaseolus vulgaris*. *Phytochemistry* 18, 2007 - 2010.

75. YOSHIKAWA, M. (1978). Diverse modes of action of biotic and abiotic phytoalexin elicitors. *Nature* 275, 246 - 247.

MECHANISMS CONFERRING SPECIFIC RECOGNITION IN GENE-FOR-GENE PLANT-PARASITE SYSTEMS

N.T. KEEN

Department of Plant Pathology
University of California
Riverside, California 92521, U.S.A.

INTRODUCTION

The high degree of specificity inherent in gene-for-gene systems *a priori* points to the existence of selective recognitional molecules in host plants that possess a high degree of fidelity for detection of certain unique features of incompatible pathogen races. Although elucidation of the biochemical machinery conferring this process constitutes one of the most crucial questions facing plant pathology, progress has been excruciatingly slow. This paper will attempt to review some of the hypothetical models proposed and assess the experimental evidence germane to their validity. Several similar papers have appeared (1,14,30,36), and in a related review (35) I will consider the topic from a broader and more theoretical point of view.

BIOCHEMICAL MODELS FOR GENE-FOR-GENE RECOGNITION-CRITERIA THAT THEY MUST MEET

Several models have been advanced to explain the recognitional specificity of gene-for-gene systems and have varying degrees of experimental verification. None, however, is sufficiently supported to be considered established. The candidate mechanisms have generally been proposed for different plant-parasite systems and it is therefore possible that two or more of them or other as yet undisclosed mechanisms may operate in various interactions. Since knowledge is scarce, however, it is as yet difficult to critically subject candidate mechanisms to rigorous theoretical and experimental tests. Nevertheless, we now know enough about the genetic behaviour of resistance genes and avirulence genes and about the biochemical expression of resistance in gene-for-gene systems to offer several constraints that must be accommodated by legitimate biochemical models for the recognitional mechanism(s). Some of these are listed in Table 1.

Table 1 Requirements for models explaining gene-for-gene complementarity.

1. The determinative (recognition) and expressive phases are distinct.
2. Recognition factors conferring general resistance are different than those operating in gene-for-gene systems.
3. Models must be consistent with genetic and biochemical evidence indicating that the expression of incompatibility is an active process, while compatibility is a passive failure to invoke defense.
4. In a few carefully studied gene-for-gene systems, accumulation of phytoalexins is the mechanism for expression of incompatibility. Accordingly, the specific recognitional mechanism in such systems must modulate ultimate phytoalexin accumulation.
5. Host and parasite recognitional molecules must be demonstrated to be present at the host-parasite interface during the crucial early interactions in which specificity is determined.
6. Sufficient potential structural variability must exist in the recognitional molecules to account for the number of resistance and virulence genes operative in the system.

Substantial evidence derived from several host-parasite systems firmly indicates that the recognitional event(s) (or determinative phase, as I prefer to call it (36)) chronologically occurs considerably before invocation of the biochemical events that actually stop colonization by the incompatible pathogen (expressive phase) (point 1, Table 1). Some of the data supporting this conclusion are reviewed in 36,49 and 50. Recognition therefore appears to ultimately confer resistance expression through a series of intermediate steps, probably involving hypersensitive host cell death, protein synthesis and the accumulation of phytoalexins and/or inducibly formed structural barriers.

The recognitional factors involved in gene-for-gene systems must clearly be different than those involved in determination of the resistance of a plant species to an entire species or to *formae speciales* of a parasite (general resistance). This is important because induced general resistance can, among other means, be overcome by the successful parasite via non-specific suppressor mechanisms (see 34,35). On the other hand, suppressor or 'induced susceptibility' models are inconsistent with a large body of genetic evidence obtained with gene-for-gene systems (see Ellingboe, 25,26). To the contrary, the genetic and biochemical evidence dictates that, in the studied systems, incompatibility is the specifically determined plant response and compatibility is a passive failure of specific recognition to occur and active defense to be invoked (point 3, Table 1).

The role of phytoalexins in conferring the expression of both general resistance and incompatibility in a few studied gene-for-gene systems is now established beyond reasonable doubt (7,34,36,50,57), but the mechanism will not be discussed in detail here. However, as indicated in point 4, (Table 1), the recognitional mechanism in these plants must of necessity influence the ultimate phytoalexin accumulation in the incompatible infection court.

Any proposed biochemical mechanism for gene-for-gene recognitional specificity must of course be demonstrated to be present during the relatively early interactions of host and parasite in which specificity is known to be determined (point 5, Table 1). This has led to the suspicion that the recognitional factors of host and parasite are constitutively produced surface molecules, since this would insure their contact during even the earliest interaction of host and parasite cells. It is also possible that extracellular molecules, at least of the pathogen, may act as specificity determining factors, but the burden of proof rests with establishing that the extracellular substance is in fact produced early enough and in sufficient amount in the infection court to account for specificity.

Finally, point 6 in Table 1 requires that molecules for specificity determination must possess enough structural variability to allow a relatively large number of distinct specificities. Some of the better studied gene-for-gene systems possess 30 or more defined plant resistance genes and almost as many demonstrated pathogen virulence genes. If modification of common, structurally related recognitional elements is controlled by all of these genes, then clearly the recognitional elements must be macromolecules of sufficient size and structural diversity to allow the observed number of independent specificities.

THE CANDIDATE MODELS FOR BIOCHEMICAL RECOGNITION

Defense conferred solely by constitutive, complementary molecules in host and parasite

In the first proposed mechanism for recognitional specificity, primary or secondary products of resistance and avirulence genes are assumed to exist as constitutive molecules on (or possibly in) cells of the host and parasite, respectively (mechanism 1, Table 2).

Table 2 Candidate mechanisms for recognition in gene-for-gene systems

1. Constitutive molecules - agglutinins or lectin-like molecules functioning both as specific recognitional factors and effectors (27).
2. Differentially inducible pathogen elicitors - the 'double induction' hypothesis (28,44).
3. Constitutive non-specific pathogen elicitors and specific suppressors (20,29).
4. Constitutive specific elicitors and complementary plant receptors modulating an inducible defense mechanism (1,14,18,30,36).

The association of these complementary molecules is hypothesized to constitute both the specific recognitional system and also the expressive phase. That is, interaction of the host and parasite factors directly inhibits the growth of the parasite in some as yet undisclosed manner. Sharon's group (10,45) showed that plant lectins bound to certain fungi and that wheat germ agglutinin inhibited the growth of *Trichoderma viride*. However, no specific effects of this nature have been demonstrated with gene-for-gene systems. In addition to the shortage of experimental evidence on which to assess the constitutive inhibitor hypothesis, it also suffers from the theoretical shortcoming that it conflicts with the evidence (point 1, Table 1) establishing that the determinative and expressive phases in the studied gene-for-gene systems are chronologically distinct. It is also at odds with point 4, Table 1, indicating that phytoalexins constitute the expressive mechanism for incompatibility in at least some systems. Accordingly, the constitutive inhibitor mechanism must not operate in these.

Differentially inducible pathogen elicitors

The second hypothesis in Table 2 proposes that specific constitutive substances in hosts of differing resistance genotypes interact with the invading parasite such that only the incompatible parasite race is induced to produce a non-specific elicitor that then causes invocation of an active defense reaction. This 'double induction' hypothesis has not to my knowledge been supported by conclusive experimental evidence and no indications are at hand as to the chemical nature of the specific host inducers. Operation of such a mechanism is, in addition, difficult to envision in the studied gene-for-gene systems in which recognition has been demonstrated to occur very rapidly after initial contact of host and parasite. Recognition may occur in one hour or less (see 36), and it is unlikely that the complicated interchange of information postulated by the double induction hypothesis could occur so rapidly.

Constitutive non-specific pathogen elicitors and specific suppressors

The third mechanism in Table 2 has recently been experimentally supported by Doke, Kuć and collaborators with the *Phytophthora infestans* potato host-parasite system (19,20,21,29). These workers isolated high molecular weight phytoalexin elicitors from disrupted mycelium and other propagules of several races of the fungus which non-specifically elicited accumulation of the potato terpenoid phytoalexins. The same groups also isolated low molecular weight glucan molecules from homogenates and germination fluids of two fungus races that appeared to function as race-specific suppressors of hypersensitivity and phytoalexin accumulation otherwise caused by the high molecular weight elicitors. It was therefore proposed that race-specific suppressors confer gene-for-gene specificity in this host-parasite system. Doke and Tomiyama (23,24) have recently published evidence using nine resistance genotypes of potato and seven *P. infestans* races that strongly supports the specific suppressor mechanism. Cell wall preparations from all the fungus races elicited rapid aggregation of potato protoplasts, a response interpreted as diagnostic of a hypersensitive reaction. No cultivar-race specificity was seen for the wall-elicited aggregation, but pre-treatment of protoplasts with the water soluble glucans followed by the wall preparations gave clear indications that protoplasts pre-treated with the glucans only from compatible races underwent much less aggregation.

Available correlative evidence notwithstanding, the specific suppressor hypothesis conflicts with several of the constraints listed in Table 1, and these must be rationalized before its general acceptance. It is consistent with requirements 1 and 4 (Table 1), but no mechanism has been proposed whereby primary or secondary products of the dominant resistance genes in potato recognize the specific suppressors. This brings us to a major theoretical problem with the specific suppressor hypothesis - its incongruity with requirement 3 (Table 1).

As in several other studied gene-for-gene systems, the available evidence indicates that the terpenoid phytoalexins constitute the expressional mechanism for R gene action and all the known resistance genes in potato to *P. infestans* are inherited as dominant characters. If the specific suppressor mechanism operates, the potato resistance genes would be required to confer a suppressor function and should be inherited as recessive traits. Similarly, virulence genes in the pathogen should be inherited as dominant characters (and avirulence genes as the recessive) if the suppressor hypothesis is correct. This has not been adequately tested with *P. infestans*, but is unlikely to occur since virulence genes in other fungus pathogens are generally inherited as the recessive; this logic is further supported by the fact that short-term pathogen evolution is from avirulent to virulent genotypes, but not in the reverse direction (18), and a wealth of empirical data (*viz.* the rapid appearance of 'new' virulent races) confirms this behavior with *P. infestans*. Thus, the direction of pathogen evolution in gene-for-gene systems is for *loss* of avirulence gene function. In the suppressor scheme, short-term parasite evolution would be required to *add* gene function, a biologically unlikely occurrence.

Yet another point of uncertainty with the suppressor model is the lack of data fulfilling requirement 5 (Table 1). Although the suppressors have been shown to accumulate in germination fluids of cystospores, it has not been established that these concentrations are sufficient to be active in the plant. Indeed, relatively high concentrations of the glucan suppressors (1-10 mg ml^{-1}) were used in most experiments (20,24,29), and for unexplained reasons the suppressors were always applied from 15 minutes to 18 hours *before* the living incompatible fungus race or the elicitor. If the suppressor model operates as proposed, the suppressor should function when applied simultaneously with elicitor or an incompatible fungus race. Another peculiarity is the fact that the *P. infestans* glucans appear similar to those from *P. megasperma* f. sp. *glycinea* described in the next section that are reported to be highly efficient elicitors of phytoalexins in soybeans and potatoes (16). Bostock *et al.* (12) indicated that one of the elicitors from cell walls of *P. infestans* was also carbohydrate in nature. How these similar molecules can function in such different ways in the same and different plants is unclear.

Finally, since specificity in the specific suppressor model is believed to be conferred by small, branched β-1,3-glucans of 17-23 glucose units, it is essential to demonstrate that these contain enough structural variability (point 6, Table 1) to explain function of the many fungus virulence genes predicted to exist by the gene-for-gene relationship. However, structural differences have not been reported for the glucans from various *Phytophthora* spp.

Wang and Bartnicki-Garcia (54,55) isolated and characterized the glucans, called mycolaminarans, and their phosphorylated derivatives from the cytoplasm of *Phytophthora palmivora*. At least three other *Phytophthora* spp. produce similar if not identical metabolites (39,58). These all appear to be β-1, 3-linked glucans of 30-36 residues with one or two β-1,6 branches per molecule. Since they accumulate to copious intracellular concentrations during growth on glucose-rich culture media (to one third of the fungus dry weight!) and decrease upon growth on poorer substrates, they appear to function as storage carbohydrates in *Phytophthora* spp. and may also be important in cell wall biosynthesis. With the exception that the phosphorylated derivatives can be isolated at certain points in the fungus life cycle, the mycolaminarans have not been shown to possess significant structural variability in *Phytophthora* spp. Further, I am not aware of precedents with other homopolymers of relatively small size and possessing only a single linkage type for sufficient structural variation to convey the relatively large number of unique structural specificities required to account for the proposed role of the mycolaminarans as specific suppressors in the various *P. infestans* races.

The specific elicitor-specific receptor model

This model, mechanism 4, Table 2, is similar to mechanism 1, Table 2, and proposes that constitutive structural elements of the incompatible but not compatible pathogen race are recognised by specific receptor molecules in the incompatible host genotype. Unlike mechanism 1, Table 2, the initial recognition leads to subsequent invocation of an inducible defense mechanism, possibly involving hypersensitive host cell death and phytoalexin accumulation. This scheme appears to be consistent with the requirements listed in Table 1, but experimental testing of the model is as yet incomplete and there is no direct evidence indicating existence of the hypothetical specific plant receptors.

Stelzig's group (42,47) observed that mycolaminaran and related non-specific glucan elicitors from cell walls of *Phytophthora infestans* and *P. megasperma* f.sp. *glycinea* agglutinated potato cells and caused rapid depolarization of potato petiole cells. Related polysaccharides did not produce similar effects. The observations were interpreted to indicate the probable existence of receptors for the non-specific glucan elicitors on the surface of potato cells. This interesting work has not yet been extended to systems involving specific phytoalexin elicitors, but Doke and Tomiyama (22) reported that *p*-chloromercuribenzoate covalently linked to dextran inhibited expression of the hypersensitive response to *P. infestans*. Since the high molecular weight inhibitor presumably did not enter the potato cells, the experiments were interpreted to indicate the probable existence of specific receptors on the potato cells which confer hypersensitivity and that these were inactivated by the sulfhydryl inhibitor. The probable occurrence of surface receptors on potato cells for invocation of the HR is also indicated by the work with cell wall fragments and potato protoplasts (23) discussed in the previous section.

Information has recently become available concerning the hypothesized specific elicitors from incompatible and compatible pathogen races. Bruegger and Keen (13) obtained soluble fractions from the cellular envelopes of several *Pseudomonas glycinea* races that, with one exception, gave race specificity for glyceollin elicitation when bioassayed on cotyledons of two soybean cultivars. Significantly, mixtures of preparations from incompatible and compatible races were observed quantitatively to elicit phytoalexins identically to the incompatible preparation only. This failure to observe any suppressive effect by the compatible preparation was thus fully consistent with the cross protection observed in this system - *viz.*

mixtures of incompatible and compatible living bacteria result in plant reactions that are fully incompatible (37). The *P. glycinea* elicitors have not yet been obtained in pure form, but extracellular polysaccharides and lipopolysaccharides from the bacterial races uniformly lacked elicitor activity. Present indications point to the possible function of one or more outer membrane proteins as the specific phytoalexin elicitors, but this has not been conclusively established.

Evidence from two groups indicates that surface and/or extracellular glycoproteins from *Phytophthora megasperma* f. sp. *glycinea* function as race specific defense elicitors. Our laboratory (33, 38) extracted glycoproteins from isolated cell walls of several fungus races and observed that these gave race-specific phytoalexin elicitation in bioassays on soybean plants of various resistance genotypes. The glycoproteins were fractionated on gel filtration and DEAE Bio-gel columns in the presence of dissociating agents to prevent their self-association. The resulting preparations had little or no polysaccharide or protein, but contained glycoproteins, as deduced by the co-stainability of bands on SDS electrophoresis gels for carbohydrate and protein. The purified glycoprotein preparations specifically elicited glyceollin accumulation in soybean hypocotyls. They were stable to mild heating and pronase, but activity was lost following treatment with metaperiodate, thus indicating that the carbohydrate portions of the glycoproteins were the elicitor-active moieties. The glycoproteins were concanavalin A (con A) reactive in a fully hapten-reversible manner and experiments with fluorescein labelled con A demonstrated that the glycoproteins were present on the cell surface of the living fungus (38).

Wade and Albersheim (53) also obtained preparations believed to contain primarily glycoproteins from culture fluids of several *P. megasperma* f.sp. *glycinea* races and showed that these specifically protected soybean plants against inoculation with living compatible fungus races. Glycoprotein preparations from incompatible races protected the plants, but similar preparations from genetically compatible fungus races did not. Thus, the work is consistent with the possible involvement of glycoproteins as the fungus factors involved in specific plant recognition. A major difference between work of the California and Colorado groups, however, is that Wade and Albersheim reported that their extracellular glycoprotein preparations did not specifically elicit production of the soybean glyceollins in bioassays with soybean cotyledons or hypocotyls. Instead, they speculated that an unknown defense mechanism not involving glyceollin might be responsible for inhibiting growth of the normally compatible fungus race. However, no direct evidence for the existance of such a defense mechanism has been forthcoming. It is furthermore difficult to see why the expression of incompatibility in the soybean - *P. megasperma* f. sp. *glycinea* system need involve a primary mechanism other than production of phytoalexins since substantial evidence indicates that they are in fact responsible for inhibition of fungus growth (see 36, 57).

A complicating factor in interpreting the role of phytoalexin elicitors in the soybean - *P. megasperma* f. sp. *glycinea* system is the presence in fungus cell walls and cytoplasm of branched β-1,3-glucans that function as efficient non-specific phytoalexin elicitors in soybeans (6). However, due to their non-specificity, the lack of information establishing their presence in the infection court, and the fact that they do not protect soybean plants against co-inoculation with compatible fungus races, it is unlikely that they have any role in specific recognition in this host-parasite system. It should also be noted that the *P. megasperma* f.sp. *glycinea* glucans are similar if not identical to those from *P. infestans* discussed in the previous section that are believed to function as specific suppressors of phytoalexin accumulation in potato. Yet another anomaly is the fact that β-glucans from *P. megasperma* f. sp. *glycinea, Colletotrichum lindemuthianum* and *Fusarium oxysporum* f. sp. *lycopersici* are efficient phytoalexin elicitors in green bean and soybean (2,6), but are less efficient in potato and tomato (4, 42). The significance of this is unclear, but it clouds interpretation of the possible role of the glucan elicitors as recognitional elements in plant disease defense. It is possible that they have importance as secondary factors in host-parasite interactions, but their presence in the infection court during initial contact of host and parasite cells has not been demonstrated.

The surface/extracellular glycoproteins from *P. megasperma* f.sp. *glycinea* represent the extant model for the fungus factors that are specifically recognized by incompatible genotypes of soybean, but there are several unresolved questions concerning their role. As noted, there is disagreement between the California and Colorado groups concerning their function as specific glyceollin elicitors. The second concerns the question of structural integrity of the extracellular glycoproteins and those extracted from cell walls. In our work, the only efficient approach to solubilizing the cell wall surface glycoproteins was with NaOH solution at 0°C. Although preparations made in this way exhibited specificity for phytoalexin elicitation, the differences seen in hypocotyl bioassays comparing incompatible and compatible plant genotypes were not large, ranging from 2 to 10x, with variability occurring between different glycoprotein preparations. Further, concentrations of the crude or fractionated glycoproteins between 0.2 and 1 mg ml^{-1} were necessary for elicitor activity in hypocotyl bioassays. These factors question whether the base extraction technique caused considerable structural alterations of the glycoprotein molecules. Similarly, one could question whether the structures of Wade and Albersheim's extracellular glycoproteins were affected by autolytic enzymes of the fungus. Ayers *et al.* (6) also extracted glucan and glycoprotein elicitors by autoclaving fungus cell walls for several hours in water and reported that these did not exhibit specificity for phytoalexin elicitation. However, the harsh extraction treatment would also seem suspect of producing serious structural alterations, especially of the glycoproteins. Recent work by Tamai *et al.* (48) with Baker's yeast surface glycoproteins is informative relative to this discussion. Antisera prepared to intact yeast cells or to isolated cell walls gave the strongest immunochemical reaction with the surface mannan glycoprotein of intact cells, slightly less specificity for the surface mannan extracted with pronase and considerably less to the mannan extracted by autoclaving cell walls. The autoclaved preparations were also observed to be very size disperse, again suggesting the occurrence of artefacts during extraction by this method.

Overcoming the above problems may be aided by the recent work of Yoshikawa *et al.* (56). They made the important discovery that incubation of *P. megasperma* f. sp. *glycinea* cell wall suspensions on cut surfaces of soybean cotyledons resulted in the rapid (within 2 minutes) release of soluble phytoalexin elicitors from the cell walls. The liberated elicitor activity mostly adsorbed to anion exchange media, was sensitive to periodate and was stable to mild heating. These properties indicated that the liberated elicitor activity was carbohydrate but not β-glucan. It will be interesting to test whether the liberated elicitors contain the carbohydrate portions of the fungus surface glycoproteins.

Anderson (3) recently prepared extracellular, high molecular weight fractions from young cultures of three races of *Colletotrichum lindemuthianum* that contained covalently-linked protein and carbohydrate when analyzed on SDS polyacrylamide gel electrophoresis. Preparations from two different isolates of the compatible β race were found to be about 100-fold less efficient elicitors of phytoalexins in *cv.* Red Kidney bean cotyledons than corresponding preparations from the avirulent α race. Similar preparations from the intermediate-reacting γ race were about 10-fold less efficient elicitors than were those from the α race. Although further fractionation of the putative glycoprotein specific elicitors is needed, the work with *C. lindemuthianum* thus far appears consistent with the specific elicitor model.

ASSESSMENT OF THE ROLE OF GLYCOPROTEINS AS RACE SPECIFIC RECOGNITIONAL FACTORS IN FUNGUS PATHOGENS

Despite the shortage of sufficient experimental data, the direct elicitor-receptor model (mechanism 4, Table 2) is supported by evidence from three apparent gene-for-gene systems, is consistent with points 1,2,3, and 4 in Table 1 and would appear to also meet requirements 5 and 6. It should also be noted that carbohydrate containing surface molecules have recently attracted considerable attention due to their role in specific cell-cell interactions in many different biologic systems (5,32,52). The putative specific phytoalexin elicitors reported from *P. megasperma* f.sp. *glycinea*, *Pseudomonas glycinea* and *Colletotrichum lindemuthianum* also appear to be associated with the cell surface and/or are extracellularly produced. Therefore, plant cells would conceivably be exposed to the surface molecules even during their earliest contact with the incompatible parasite race (point 5, Table 1).

Glycoproteins appear to occur universally on the surface of eucaryote cells. With fungi, similar if not identical glycoproteins are found both on the plasma membrane surface and, of more importance here, on the cell wall surface (43). In yeasts, Tronchin *et al.* (51) and Cassone *et al.* (15) showed that the surface mannan glycoprotein ensheaths the underlying structural polymers of the cell wall. Such glycoprotein masking of underlying macro-molecules is recognized as an important pathological mechanism involved in the lack of detection of certain mammalian parasites (52) and tumor cells (17) by the immune systems. This masking may also be an important feature in micro-organisms that are plant pathogens - *P. megasperma* f.sp. *glycinea* and *Colletotrichum lindemuthianum*, similarly to yeasts, possess structural β-1,3-glucans in cell walls that are efficient phytoalexin elicitors (6,31). However, their apparent ensheathment by the fungal surface glycoproteins diminishes the possibility that the wall glucans of such organisms would be recognized by plant cells during early pathogenic interactions.

The possible role of surface and/or extracellular glycoproteins as specific defense elicitors in the studied fungus systems would also seem consistent with point 6, Table 1. Although detailed structural studies have not been reported with plant pathogenic fungi, the surface glycoproteins of various yeasts have been studied in considerable depth. Of note, the carbohydrate portions of these molecules have been observed to exhibit a great deal of species - and strain-specific variation (8,9). Ballou's pioneering work (8) with yeasts, for instance, has established that the surface mannan glycoproteins are the major antigenic determinants of various strains and species, and serological strain specificities are related to distinct structural features of the carbohydrate portions. Further, the large 'outer-chain' carbohydrates of the yeast mannan glycoproteins have the potential for a large number of unique structures. Since similar fluidity in the carbohydrate structures and antigenic specificities of surface glycoproteins from other fungi including plant and animal pathogens have also been reported, it is clear that the potential exists for race-specific structural variation in the surface carbohydrates of different fungus races. Ziegler and Albersheim (59) examined the carbohydrate structure of the extracellular glycoprotein, invertase, from three *P. megasperma* f.sp. *glycinea* races and noted unique structural features for each of them. Although it was not established that these differences were in fact related to the race phenotype, the work supports the possibility that the carbohydrate structures of surface and extracellular glycoproteins may confer this character.

In their creative and pioneering review, Albersheim and Anderson-Prouty (1) pointed out that if surface carbohydrates function in the above manner, then the pathogen's avirulence genes most likely encode the primary structures of specific glycosyl transferase enzymes. No direct tests of this idea have appeared with plant pathogens, but recent information with yeasts and mammalian cells supports the feasibility of the proposal (11,41,46). The major characteristic of these transferases germane to our discussion is their high degree of specificity for addition to unique sugars in specific anomeric and linkage configurations to nascent carbohydrates that are destined to become the glycosyl components of surface glycoproteins. As predicted, yeast mutants lacking specific glycosyl transferases form surface and extracellular glycoproteins that are defective relative to the wild-type since they lack the specific sugar residues in question (9,41). Those of us concerned with plant pathogens should, in due course, arrive at the position to test rigorously the cardinal feature of the Albersheim-Anderson theory - that dominant avirulence alleles lead to functional glycosyl transferases, while the corresponding recessive virulence allelles of the pathogen lead to non-functional transferase enzymes and thus to glycoproteins which lack the structural features recognized by the as yet hypothetical receptors of the incompatible plant genotype.

The author thanks Anne Anderson and Masaaki Yoshikawa for making manuscripts available before publication and J. Kuć for critically reading this paper. The author's research is supported by the National Science Foundation.

REFERENCES

1. ALBERSHEIM, P & ANDERSON-PROUTY, A.J. (1975). Carbohydrates, proteins, cell surfaces, and the biochemistry of pathogenesis. *Annual Review of Plant Physiology* 26, 31-52.

2. ANDERSON, A. (1978). Initiation of resistant responses in bean by mycelial wall fractions from three races of the bean pathogen *Colletotrichum lindemuthianum*. *Canadian J. of Botany* 56, 2247-2251.

3. ANDERSON, A.J. (1980). Differences in the biochemical compositions and elicitor activity of extracellular components produced by three races of a fungal plant pathogen, *Colletotrichum lindemuthianum*. *Canadian J. Microbiology* (in press).

4. ANDERSON, A.J. (1980). Studies on the generality of the glucan-

elicitor phenomenon: non-detection of glucans by tomato tissue. *Canadian J. Botany* (Submitted).

5. ASHWELL, G. & MORELL, A.G. (1977). Membrane glycoproteins and recognition phenomena. *Trends in Biochemical Science* 2, 76-78.

6. AYERS, A.R., EBEL, J., VALENT, B. & ALBERSHEIM, P. (1976). Host-pathogen interactions. X. Fractionation and biological activity of an elicitor isolated from the mycelial walls of *Phytophthora megasperma* var. *sojae*. *Plant Physiology* 57, 760-765.

7. BAILEY, J.A. (1981). Physiological and biochemical events associated with the expression of resistance to disease. In : *Active defense mechanisms in plants*, Ed. by R.K. S. Wood (in press).

8. BALLOU, C.E. (1974). Some aspects of the structure, immunochemistry and genetic control of yeast mannans. *Advances in Enzymology* 40, 239-270.

9. BALLOU, C. (1976). Structure and biosynthesis of the mannan component of the yeast cell envelope. *Advances in Microbial Physiology* 14, 93-158.

10. BARKAI-GOLAN, R., MIRELMAN, D. & SHARON, N. (1978). Studies on growth inhibition by lectins of Penicillia and Aspergilli. *Archives of Microbiology* 116, 119-124.

11. BEYER, T.A., REARICK, J.I., PAULSON, J.C., PRIEELS, J-P, SADLER, J.E. & HILL, R.L. (1979). Biosynthesis of mammalian glycoproteins. Glycosylation pathways in the synthesis of the non-reducing terminal sequences. *Journal of Biological Chemistry* 254, 12531-12541.

12. BOSTOCK, R., HENFLING, J.W. & KUĆ, J. (1978). Release and fractionation of components from mycelial cell walls of *Phytophthora infestans* which elicit the accumulation of sesquiterpenoid stress metabolites in potato. *Phytopathology News* 12, 161.

13. BRUEGGER, B.B. & KEEN, N.T. (1979). Specific elicitors of glyceollin accumulation in the *Pseudomonas glycinea* - soybean host-parasite system. *Physiological Plant Pathology* 15, 43-51.

14. CALLOW, J.A. (1977). Recognition, resistance, and the role of plant lectins in host-parasite interactions. *Advances in Botanical Research* 4, 1-49.

15. CASSONE, A., MATTIA, E., & BOLDRINI, L. (1978). Agglutination of blastospores of *Candida albicans* by concanavalin A and its relationship with the distribution of mannan polymers and the ultrastructure of the cell wall. *Journal of General Microbiology* 105, 263-273.

16. CLINE, K., WADE, M., & ALBERSHEIM, P. (1978). Host-pathogen interactions. XV. Fungal glucans which elicit phytoalexin accumulation in soybean also elicit the accumulation of phytoalexins in other plants. *Plant Physiology* 62, 918-921.

17. CODINGTON, J.F., VAN DEN EIJNDEN, D.H. & JEANLOZ, R.W. (1978). Structural studies on the major glycoprotein of the TA3-HA ascites tumor cell. p. 49-66, In:R.E. Haromon, ed., Cell surface carbohydrate chemistry. Academic Press, New York.

18. DAY, P.R. (1974). Genetics of host-parasite interaction. W.H. Freeman, San Francisco. 238p.

19. DOKE, N. (1975). Prevention of the hypersensitive reaction of potato cells to infection with an incompatible race of *Phytophthora infestans* by constituents of the zoospores. *Physiological Plant Pathology* 7, 1-7.

20. DOKE, N., GARAS, N.A. & KUĆ, J. (1979). Partial characterization and aspects of the mode of action of a hypersensitivity-inhibiting factor (HIF) isolated from *Phytophthora infestans*. *Physiological Plant Pathology* 15, 127-140.

21. DOKE, N., GARAS, N. A. & KUĆ, J. (1980). Effect on host hypersensitivity of suppressors released during the germination of *Phytophthora infestans* cystospores. *Phytopathology* 70, 35-39.

22. DOKE, N. & TOMIYAMA, K. (1978). Effect of sulfhydryl-binding compounds on hypersensitive death of potato tuber cells following infection with an incompatible race of *Phytophthora infestans*. *Physiological Plant Pathology* 12, 133-139.

23. DOKE, N. & TOMIYAMA, K. (1980). Effect of hyphal wall components from *Phytophthora infestans* on protoplasts of potato tuber tissues. *Physiological Plant Pathology* 16, 169-176.

24. DOKE, M. & TOMIYAMA, K. (1980). Suppression of the hypersensitive response of potato tuber protoplasts to hyphal wall components by water soluble glucans isolated from *Phytophthora infestans*. *Physiological Plant Pathology* 16, 177-186.

25. ELLINGBOE. A.H. (1978). A genetic analysis of host-parasite

interactions. In:*The powdery mildews*, Ed. by D.M. Spencer, p. 159-181, Academic Press, London.

26. ELLINGBOE, A.H. (1979). Inheritance of specificity: the gene-for-gene hypothesis, p. 3-17, In:*Recognition and specificity in plant host-parasite interactions*, Ed. by J.M. Daly and I. Uritani. University Park Press, Baltimore.

27. ELLINGBOE, A.H. (1980). Genetical aspects of active defense. In:*Active defense mechanisms in plants*, Ed. by R.K.S. Wood (in press).

28. FRANK, J.A. & PAXTON, J.D. (1971). An inducer of soybean phytoalexin and its role in the resistance of soybeans to *Phytophthora* rot. *Phytopathology* 61, 954-958.

29. GARAS, N.A., DOKE, N. & KUĆ, J. (1979). Suppression of the hypersensitive reaction in potato tubers by mycelial components from *Phytophthora infestans*. *Physiological Plant Pathology* 15, 117-126.

30. HADWIGER, L.A. & SCHWOCHAU, M.E. (1969). Host resistance responses - an induction hypothesis. *Phytopathology* 59, 223-227.

31. HAHN, M.G. & ALBERSHEIM, P. (1978). Host-pathogen interactions. XIV. Isolation and partial characterization of an elicitor from yeast extract. *Plant Physiology* 62, 107-111.

32. HUGHES, R.C. (1976). Membrane glycoproteins. Butterworths. London 367 p.

33. KEEN, N.T. (1978). Surface glycoproteins of *Phytophthora megasperma* var. *sojae* function as race specific glyceollin elicitors in soybeans. *Phytopathology News* 12, 221 (abstr).

34. KEEN, N.T. (1980). Evaluation of the role of phytoalexins. In: *Resistance and suceptibility for control of plant diseases*, Ed. by R. Staples and G. Toenniessen. Wiley Interscience, New York (in press).

35. KEEN, N.T. (1981). Specific recognition in gene-for-gene host-parasite systems. In *Advances in Plant Pathology*, Ed. by D. Ingram and P.H. Williams,Academic Press,London (in press).

36. KEEN, N.T. & BRUEGGER, B. (1977). Phytoalexins and chemicals that elicit their production in plants. In:*Host plant resistance to pests*, Ed. by P. Hedin. *American Chemical Society Symposium Series* 62, 1-26.

37. KEEN, N.T. & KENNEDY, B.W. (1974). Hydroxyphaseollin and related isoflavonoids in the hypersensitive resistant response of soybeans against *Pseudomonas glycinea*. *Physiological Plant Pathology* 4, 173-185.

38. KEEN, N.T. & LEGRAND, M. (1980). Surface glycoproteins: evidence that they may function as the race specific phytoalexin elicitors of *Phytophthora megasperma* f.sp. *glycinea*. *Physiological Plant Pathology* 17, 175 - 192.

39. KEEN, N.T., WANG, M.C., BARTNICKI-GARCIA, S., & ZENTMYER, G.A. (1975). Phytotoxicity of mycolaminarans - β-1,3-glucans from *Phytophthora* spp. *Physiological Plant Pathology* 7, 91-97.

40. KOTA, D.A. & STELZIG, D.A. (1977). Electrophysiology as a means of studying the role of elicitors in plant disease resistance. *Proceedings of the American Phytopathological Society* 4, 216-217 (abstr.).

41. LEHLE, L., COHEN, R.E. & BALLOU, C.E. (1979). Carbohydrate structure of yeast invertase. Demonstration of a form with only core oligosaccharides and a form with completed polysaccharide chains. *Journal of Biological Chemistry* 254, 12209-12218.

42. LISKER, N. & KUĆ, J. (1977). Elicitors of terpenoid accumulation in potato tuber slices. *Phytopathology* 67, 1356-1359.

43. MARRIOTT, M.S. (1977). Mannan-protein location and biosynthesis in plasma membranes from the yeast form of *Candida albicans*. *Journal of General Microbiology* 103, 51-59.

44. METLITSKII, L.V., D'YAKOV, YU. T., & OZERETSKOVSKAYA, O.L. (1973). Double induction - a new hypothesis for the immunity of plants to blight and similar diseases. *Doklody Akademiia Nauk SSSR* 213, 497-500.

45. MIRELMAN, D., GALUN, E., SHARON, N. & LOTAN, R. (1975). Inhibition of fungal growth by wheat germ agglutinin. *Nature* 256, 414-416.

46. PARODI, A.J. (1979). Biosynthesis of yeast glycoproteins. Processing of the oligosaccharides transferred from dolichol derivatives. *Journal of Biological Chemistry* 254, 10051-10060.

47. PETERS, B.M., CRIBBS, D.H. & STELZIG, D.A. (1978). Agglutination of plant protoplasts by fungal cell wall glucans. *Science* 201, 364-365.

48. TAMAI, Y., NAKASHIMA, T., TAKAKUWA, M. & MISAKI, A. (1980). Anti-genicities of several cell wall mannan preparations and cell envelope preparations from Baker's yeast. *Agricultural and Biological Chemistry* 44, 49-53.

49. TANI, T. & YAMAMOTO, H. (1979). RNA and protein synthesis and enzyme changes during infection, p. 273-287. In:*Recognition and specificity in plant host-parasite interactions*, Ed. by J.M. Daly and I. Uritani, University Park Press, Baltimore, Md.

50. TOMIYAMA, K., DOKE, N., NOZUE, M. & ISHIGURI, Y. (1979). The hypersensitive response of resistant plants, p. 69-84. In:*Recognition and specificity in plant host-parasite interactions*, Ed. by J.M. Daly and I. Uritani. University Park Press, Baltimore, Md.

51. TRONCHIN, G., POULAIN, D. & BIGUET, J. (1979). Études cytochimiques et ultrastructurales de la Paroi de *Candida albicans*. *Archives of Microbiology* 123, 245-249.

52. TURNER, M. (1980). How trypanosomes change coats. *Nature 284*, 13-14.

53. WADE, M. & ALBERSHEIM, P. (1979). Race-specific molecules that protect soybeans from *Phytophthora megasperma* var. *sojae*. *Proceedings National Academy of Sciences U.S.A.* 76, 4433-4437.

54. WANG, M.C. & BARTNICKI-GARCIA, S. (1973). Novel phosphoglucans from the cytoplasm of *Phytophthora palmivora* and their selective occurrence in certain life cycle stages. *Journal of Biological Chemistry* 248, 4112-4118.

55. WANG, M.C. & BARTNICKI-GARCIA, S. (1974). Mycolaminarans: storage (1→3)-β-D-glucans from the cytoplasm of the fungus *Phytophthora palmivora*. *Carbohydrate Research* 37, 331-338.

56. YOSHIKAWA, M., MADAMA, M. & MASAGO, H. (1981). Release of a soluble phytoalexin-elicitor from mycelial walls of *Phytophthora megasperma* var. *sojae* by soybean tissues. *Plant Physiology* (in press).

57. YOSHIKAWA, M., YAMAUCHI, K. & MASAGO, H. (1978). Glyceollin: its role in restricting fungal growth in resistant soybean hypocotyls infected with *Phytophthora megasperma* var. *sojae*. *Physiological Plant Pathology* 12, 73-82.

58. ZEVENHUIZEN, L.P.T.M. & BARTNICKI-GARCIA, S. (1970). Structure and role of a soluble cytoplasmic glucan from *Phytophthora*

cinnamomi. *Journal of General Microbiology* 61, 183-188.

59. ZIEGLER, E. & ALBERSHEIM, P. (1977). Host-pathogen interactions. XIII. Extracellular invertases secreted by three races of a plant pathogen are glycoproteins which possess different carbohydrate structures. *Plant Physiology* 59, 1104-1110.

DETERMINANTS OF PLANT RESPONSE TO BACTERIAL INFECTION

LUIS SEQUEIRA

Department of Plant Pathology
University of Wisconsin
Madison, Wisconsin, U.S.A.

INTRODUCTION

Throughout their entire life cycle, plants are exposed to large numbers of saprophytic and pathogenic bacteria. These bacteria colonize the surfaces of roots, stems, and leaves at all stages of growth of the plant. The large numbers of natural openings in plant organs as well as the frequent wounds caused by mechanical factors provide avenues that, under suitable environmental conditions, allow bacteria to reach the intercellular spaces or the interior of xylem vessels. Penetration must occur frequently, but, in spite of the fact that the nutritional environment in the host tissues is suitable for bacterial multiplication, only an extremely low proportion of the invading bacteria survive. An even lower proportion is capable of causing disease.

These facts suggest that in the course of evolution, plants, like animals, have developed systems to immobilize or destroy the bacteria that gain access to internal tissues. Very little is known about these defense systems, but they appear to depend, to a large extent, on initial recognition events that occur soon after the bacterium and the plant cell come in contact. It is evident that there must be general as well as highly specific systems for recognition. On the one hand, plants must have mechanisms to rid themselves of many different types of bacteria and, by inference, these mechanisms must depend on recognition of constituents that are common to all potential pathogens. On the other hand, plants also must have specific recognition systems that allow them to react differently to physiologic strains of the same bacterial species.

By analogy with other cell-cell interactions, it is likely that recognition phenomena between bacteria and the plant host involve the complementary interaction of macromolecular constituents at the surface of the two organisms. In gram-negative bacteria, these constituents must include the extracellular slime or capsule, or the outer membrane of the cell wall. In the host, the interacting macromolecules must include the cell wall polysaccharides and/or proteins.

In their interaction with plant pathogenic and other bacteria, plants exhibit responses that have many parallels with phenomena that are well known in animal systems. It is useful, from a conceptual standpoint, to consider how the determinants of recognition have evolved in both systems. In both, there has been intense selection pressure towards *compatibility* in the bacterial population and towards *incompatibility* in the host population. In this sense, compatibility must be viewed as avoidance or inhibition of the host response.

Animal microbiologists consider that pathogenic bacteria are of two different types. One group, the *extracellular parasites*, are readily engulfed and killed by fixed or circulatory phagocytes (12). The virulence of these organisms is dependent on their ability to resist or avoid phagocytosis and this is accomplished by the presence of slime or capsules. The other group, the *intracellular parasites*, are adapted to ingestion by phagocytes and multiply within the phagocytes. These parasites benefit from recognition by the host and even use the circulatory phagocytes as a means to spread throughout the host.

From the limited information available on the interaction of plant pathogenic bacteria and their hosts, it is apparent that there are also two general groups and that the components that determine the plant response are similar to those that are important in animal systems. It is the purpose of this article to examine the nature of these components in both animal and plant systems, in the hope that such an examination will help our understanding of the evolution of host-parasite interactions.

EXTRACELLULAR POLYSACCHARIDES

All species of phytopathogenic bacteria are capable of producing extracellular polysaccharides (EPS). These polysaccharides may be present either as a discrete capsule or, more commonly, as extracellular slime. Capsules can be removed from cells only with difficulty, but slime generally is soluble and diffuses readily in liquid media. As cultures age, however, the capsules may dissolve and both capsulate and non-capsulate bacteria may be present at the same time (12).

Extracellular polysaccharides are composed of neutral hexoses, such as D-glucose, D-galactose, and D-mannose, and of deoxyhexoses, such as L-fucose and L-rhamnose. Uronic acids are of wide occurrence in many types of EPS. Amino sugars, such as N-acetyl-D-glucosamine and N-acetyl-D-galactosamine, are common constituents. An unusual substituent, typical of the EPS of *Xanthomonas campestris*, is a ketal-linked pyruvate. Most EPS molecules are heteropolysaccharides which are made up of multiple repeating units of a tetra or a pentasaccharide (41).

The Function of Extracellular Polysaccharides in Extracellular Parasites

EPS is produced by most extracellular parasites, which include many important human pathogens. These organisms are readily killed once they are ingested by phagocytes and owe their virulence to their ability to resist phagocytosis. Animal microbiologists have provided ample evidence for a close relationship between virulence and the production of EPS. For example, pneumococci isolated from diseased tissues are always encapsulated; strains that lack capsules always are avirulent (11). Capsules apparently interfere with phagocytosis because they cover specific antigenic sites on the bacterial cell wall (35).

Differences in virulence of many bacteria that are pathogenic to animals often are correlated with differences in the amount of EPS that they produce in culture. For instance, differences in virulence of strains of pneumococcus to mice are related to the amount of EPS produced by the bacteria (11). Such differences have been related to the ability of these and other bacteria to resist phagocytosis *in vitro* (12).

Although the relationship between virulence and capsulation in extracellular animal parasites has been known since the turn of the century, it is surprising that similar studies with plant pathogenic bacteria are relatively recent and are few in number. One of the earliest reports was that of Kelman (24) who established an all-or-none relationship between virulence and the production of slime by *Pseudomonas solanacearum*. More recently, a similar relationship has been established for *Erwinia amylovora* (1) and *Erwinia stewartii* (5). The slime-less or non-encapsulated forms invariably are found to be avirulent. Freshly isolated cultures commonly produce copious amounts of slime, but this property may be lost readily in culture (24).

It should be pointed out that the production of EPS is not the only factor that confers virulence to a bacterium. Avirulent strains of numerous plant pathogenic bateria, for instance, retain slime production *in vitro*. A possible exception is *P. solanacearum* in which fluidal, avirulent forms are not usually produced in culture (24).

Attachment of Bacteria to Plant Cell Walls

In contrast with the wealth of information available about animal pathogens, little is known about the function of EPS in plant pathogenic bacteria. This may be attributed to the fact that plants do not have defense mechanisms involving phagocytosis or the production of antibodies. More recent evidence, however, suggests that mechanisms comparable to phagocytosis do exist in plants. As in phagocytosis, invaded bacteria may be engulfed and immobilized at the plant cell wall (17, 37). The non-encapsulated or slime-less forms of some plant pathogenic bacteria are less virulent because they are rapidly immobilized by the host cell. Incompatible EPS-producing strains (i.e. strains pathogenic on other hosts) are immobilized in the same fashion, however.

In tobacco, bacteria are immobilized as a result of a localized response by the mesophyll cell. When avirulent, slime-less strains of *P. solanacearum* are infiltrated into tobacco leaves the bacteria come to rest on the host cell wall as soon as the water in which the bacteria are suspended either evaporates or is absorbed by the host cells. Very soon thereafter, the plasma membrane of the host cell invaginates and there is an accumulation of membrane-bound vesicles in the space between the plasma membrane and the host cell wall. Fibrillar and granular material envelop the bacteria and a pellicle, probably of cuticular origin, surrounds the bacterium. The response to the presence of the bacterium is rapid (three to six hours) and is restricted to the area where the bacteria come in contact with the host cell wall. With some but not all avirulent forms of *P. solanacearum*, attachment leads to a hypersensitive response (HR) which is characterized by collapse of the plant cell and a rapid reduction in bacterial numbers (39). In contrast, virulent forms are not attached to or enveloped by the host cell wall and they remain free to multiply in the intercellular spaces (37).

Precisely how EPS prevents attachment of virulent forms of *P. solanacearum* to tobacco cell walls is not known. Certain types of EPS, such as that produced by *P. phaseolicola*, induce water-soaking (14) and it could be argued that maintenance of free water in the intercellular spaces is sufficient to prevent close contact of the bacteria with the host cell wall. As an example, attachment of bacteria to bean cell walls was prevented by water-soaking the leaves continuously (20). Similar results were obtained by Stall & Cook (40), who found that water-soaking inhibited the HR normally induced in certain pepper cultivars by *Xanthomonas vesicatoria*. Although Stall & Cook did not demonstrate by direct examination of the tissue that attachment of the bacterium was prevented, the general conclusion was that lack of close contact of the bacterium with the cell wall was sufficient to prevent induction of the HR.

Attachment of bacteria to the host cell wall may be a prerequisite for induction of the HR, but does not lead to this reaction in all instances. Saprophytic bacteria, or heat-killed bacteria, are readily attached to host cell walls, but the HR does not ensue. Although there are several unsubstantiated claims that the HR can be induced by bacterial cell wall fractions, the general consensus is that this reaction is induced only by a live bacterium.

Since many EPS-producing but incompatible strains of bacteria also are attached to and enveloped by tobacco cell walls (17, 37), a pertinent question relates to the inability of EPS in these instances to prevent attachment. In the case of *P. solanacearum*, there appear to be no differences in composition of the EPS from compatible and incompatible strains that are associated with differences in the ability of these bacteria to attach (Whatley, unpublished data). A more likely explanation is that the incompatible strains cannot produce sufficient amounts of EPS *in vivo* to interfere with attachment to the host cell walls. This, of course, does not tell us why sufficient EPS is not produced in the early stages of infection.

Lectin - EPS interactions

In an earlier publication from this laboratory we suggested that the function of EPS produced by *P. solanacearum* might be to saturate binding sites on a cell wall lectin (38). A lectin located on tobacco and potato cell walls (25) was shown to be specific for internal N-acetyl glucosamine moieties, and these are common in bacterial lipopolysaccharides (LPS). Thus, if a lectin-LPS interaction was involved in recognition and attachment, saturation of the binding sites by EPS would provide an effective means to allow the compatible bacteria to multiply in the intercellular spaces. This hypothesis received considerable support from *in vitro* experiments in which EPS from *P. solanacearum* was shown to be an effective inhibitor of bacterial agglutination by potato lectin.

A more detailed analysis of the interactions of potato lectin with both EPS and LPS from *P. solanacearum* was made by means of a binding assay in which the lectin was immobilized on a nitrocellulose filter. Ligands, such as EPS and LPS, pass through the filter, but lectin is retained without loss of activity. The amount of radioactively labelled ligand bound to the filter can be used as a measure of the lectin-binding affinity of the ligand (13). Although EPS binds to the lectin rapidly and quantitatively, it does so only at relatively low ionic strengths (Duvick & Sequeira, unpublished data). In contrast, binding of LPS to the lectin is unaffected by ionic strength of the buffer, except at very high molarity. This suggests that at the normal ionic strength in the intercellular fluids, binding of EPS to the lectin is very weak.

These findings explain our inability to prevent the HR by adding EPS to leaves infiltrated with avirulent strains of *P. solanacearum*. In contrast, LPS readily inhibits the HR under the same conditions (18).

Our results do not entirely rule out the possibility that EPS plays an important function in preventing attachment of bacteria to cell walls. They do suggest that other properties of EPS, such as its ability to cause water-soaking, might be more important than its lectin-binding properties in preventing attachment of bacteria to cell walls.

The Function of Extracellular Polysaccharides in Intracellular Parasites

Among bacteria that cause disease in animals, many are rapidly ingested by phagocytes, but they resist digestion and multiply within the phagocytes(e.g. *Mycobacterium tuberculosis*, *Brucella abortus*, and staphylococci). Thus, these organisms depend for their virulence on their ability to gain ready access to the interior of the host cell. Rapid recognition, therefore, is a very definite advantage to an intracellular parasite. In this process, the chemotactic and antigenic properties of surface polysaccharides play a major role. Extracellular polysaccharides act as attractants for phagocytes and bacteria may be ingested before there is detectable antibody to the polysaccharide (12). More commonly, however, encapsulated bacteria are phagocytized only after previous sensitization with specific antibody (opsonization). Once ingested, resistance against lysis is definitely associated with the nature of the capsular polysaccharide; rough forms generally are susceptible to digestion, but smooth forms may be resistant.

Rhizobia-legume root interactions

In a very general way, the concepts that we have described above for animal parasites may be applied to the interaction of Rhizobia with leguminous host root tissues. In order to survive and multiply, these bateria must be in close contact with the host cell. It is to their advantage to be rapidly attached to the host cell wall where they can cause specific growth alterations that result in the formation of the infection thread. The Rhizobia are encapsulated and the nature of the capsular constituents appears to be important in the highly specific process that results in nodulation.

Attachment of Rhizobia to the appropriate host generally is followed by the production of microfibrillar material, identified as cellulose, at the site of attachment (33). Although non-specific attachment of rhizobia to non-hosts is common (38), only specific attachment to the host results in the morphological changes in the host cell that precede infection and nodulation. Bacterial polysaccharides are thought to play two primary functions: a) they may be essential in the recognition process in which specific lectins are the complementary host component, and b) they may be responsible for the release of pectolytic enzymes, which are essential for the cell wall alterations that accompany infection. There is as yet no incontrovertible evidence that capsular polysaccharides are involved in either of these functions, but the evidence is stronger for the first than for the second (12).

The hypothesis that specificity in Rhizobial infections may depend on a lectin-polysaccharide interaction stems from the initial observations by Hamblin & Kent (19). They reported that an extract from beans was absorbed by *Rhizobium phaseoli* and was capable of binding treated cells to the roots of bean seedlings and to human type A red blood cells. The active agent in the extract was thought to be a lectin. These observations have been extended to *R. japonicum* and soybean lectin (SBL) by Bohlool & Schmidt (4), Bhuvanewswari *et al.* (3), Calvert *et al.* (6) and others, and to *R. trifolii* and clover lectin (trifoliin) by Dazzo (8). There is indirect evidence that trifoliin is present in the root hair region of clover seedlings, but there is no adequate proof that SBL is similarly located on soybean seedlings.

There has been controversy as to the nature of the bacterial component that is important in recognition. For example, Wolpert and Albersheim (46) reported that the lectins of legumes interacted specifically with the LPS of their symbiont rhizobia. Later work from the same laboratory, however, indicated that there are no differences in rhizobial LPS that could account for the specificity of the interaction (7). In contrast, Maier & Brill (28) found major differences in composition between the O-antigen component of LPS of a non-nodulating mutant of *R. japonicum* and that of the wild type. Similar conflicting evidence exists in the case of *R. trifolii*. Dazzo & Hubbell (10) initially thought that the binding polysaccharide from this species was EPS, because their preparations contained no 2-keto-3-deoxyoctonic acid(KDO) or endotoxin activity. More recent evidence (9) indicates that the active polysaccharide is serologically identical to the O-antigen of *R. trifolii*. Also, the interaction of plant lectins with *R. japonicum* and *R. leguminosarum* is inhibited by LPS from the same strains (23).

Evidence that SBL binds specifically to the capsular (EPS) components of *R. japonicum* and not to the outer membrane (LPS) was provided by Calvert *et al.* (6). This is in agreement with the report that specific changes in sugar composition of the capsule are associated with changes in ability of *R. japonicum* to bind to SBL (30). Further, a mutant of *R. leguminosarum* that produces less than normal amounts of EPS was found to be incapable of nodulating pea seedling roots (36).

Although the bulk of the evidence seems to favour EPS as the determinant of specificity in rhizobial-legume root interactions, it is clear that a role for LPS cannot be ruled out. In part, the controversy stems from the fact that LPS is continuously extruded by bacteria into the surrounding medium and, thus, it is difficult to separate from the capsular materials. Perhaps, as Kamberger (22) has suggested, both EPS and LPS play a role; EPS might be important for attachment to the root, while LPS may be required for the tighter binding which is necessary for infection.

Recent evidence from Bauer's laboratory (2) suggests an intriguing and possibly significant role for rhizobial EPS. Infections by *R. japonicum* normally are initiated on regions of the soybean roots that do not have root hairs at the time of inoculation. When roots are treated with bacterial EPS, the normal three hour delay between inoculation and the initiation of infection is eliminated. Thus, EPS apparently sensitizes the potential root hair cells to a state of infectibility.

Whatever role EPS may ultimately be shown to play in these interactions, it is evident that in the Rhizobia, as in the other intracellular parasites, EPS facilitates a compatible relationship. Rapid recognition by the host is essential for survival of the bacterium. This is in strong contrast with the extracellular parasites, in which EPS prevents or inhibits recognition by the host.

LIPOPOLYSACCHARIDES

In the gram-negative, plant pathogenic bacteria that produce slime or other soluble forms of EPS, the outer membrane is the cell wall component most likely to interact with the host. This outer membrane contains phospholipids, proteins, lipoproteins, glycoproteins, and LPS. The LPS molecules are intercalated on the outer face of a lipid bilayer, so that the polysaccharide chains project out of the membrane (45). Thus, LPS molecules are highly exposed and are well suited as agents for specificity because of the great diversity of the polysaccharide structure.

The classical model of the LPS molecule was proposed originally by Lüderitz *et al.* (27) for smooth strains of *Salmonella* spp. A complete LPS molecule contains three covalently linked segments, the O-specific antigen, the core region, and the lipid A region. The O-specific antigen is highly variable and the most active serologically; it contains repeating oligosaccharide units, each usually composed of hexoses, deoxyhexoses, and hexosamines. The core region characteristically contains hexoses, heptoses, hexosamines, and KDO. The lipid A region, which is the least variable component, typically contains a β-1, 6-linked disaccharide of glucosamine which is acylated by fatty acids, phosphate, and, quite often, ethanolamine. The lipid A and core regions are linked via KDO.

Several LPS molecules may be linked via the lipid A backbone, but not all molecules are necessarily complete; some may lack part or all of the O-specific antigen. Rough mutants or many different types of bacteria may lack the O-antigen portion entirely and their LPS is called R-LPS, as opposed to S-LPS of the smooth, wild type. R-LPS is much less soluble in water than S-LPS, but can be extracted by methods that employ highly selective polar solvents (27).

The Function of Lipopolysaccharides

Apart from their function as potent antigens, the most important property of the LPS of bacteria that are pathogenic to animals is its relationship to virulence. In *Salmonella*, for instance, the arrangement of sugars in the O-antigen is associated with differences in virulence (34). The ability of certain species to cause disease is related to the presence of specific repeating units in the O-antigen. Similarly, the ability of certain strains of *Escherichia coli* to kill mice by intraperitoneal injection is correlated with the presence of specific sugars in the O-antigen. The presence of the wild-type O-antigen endows certain strains with resistance to phagocytosis and antigenicity (29).

There is very strong evidence, from many different sources that strains of *Salmonella typhimurium* that contain S-LPS are virulent, whereas those that are defective and have lost certain O-antigen constituents generally are avirulent (32). Although R-LPS strains are avirulent, they are antigenically active and exhibit a new specificity, R-specificity, which is cryptic in the wild type (45).

Virulence in gram-negative bacteria is generally independent of their endotoxin activity. This activity resides in the lipid A portion of LPS, a constituent that is structurally similar in many species of gram-negative bacteria.

Relationship of LPS to Virulence in Plant Pathogenic Bacteria

The possibility that LPS plays an important role in recognition of pathogenic bacteria by a plant host was first established by Whatley *et al.* (42) for the interaction of *Agrobacterium tumefaciens* and pinto bean leaves. The LPS extracted from virulent strains was highly inhibitory to tumor formation in a pinto bean leaf bioassay. Inhibition was obtained when LPS was added before or along with the inoculum, but not when added 15 minutes later. It is assumed that LPS interferes with tumor formation by saturating sites on the injured cell wall which are essential for attachment of the bacterium. There has been no direct demonstration that this is the role of LPS, however.

Hydrolysis of the *A. tumefaciens* LPS by mild acid treatment separates the lipid A portion from the core and O-antigen polymer. Only the latter polymer was shown to be biologically active in the bean leaf assay (42). Further evidence that LPS is a determinant of virulence for the crown gall bacterium was obtained in experiments in which transformation of avirulent non-inhibitory strains was effected by introduction of the Ti plasmid. The LPS of the transformed strains became inhibitory in the tumor inhibition assay (43). Similarly, loss of the plasmid was associated with loss of inhibitory activity of LPS.

The relationship of LPS structure to virulence has been studied most extensively with *P. solanacearum*. Sequeira & Graham (38) first determined that there was a strong correlation between the lack of EPS in certain avirulent strains and their strong agglutination by potato lectin. After virulent cells were washed to remove EPS, they were agglutinated to some degree; when EPS was added back, agglutination was inhibited. These data suggested that LPS was the lectin-binding component; indeed, LPS isolated from both virulent and avirulent strains was shown to bind to potato lectin *in vitro*. The LPS from a rough strain (B-1) always agglutinated more strongly than did that of a virulent strain (K-60), suggesting that R-LPS contained lectin binding sites that were more exposed than were those in S-LPS. Because the B-1 strain, unlike K-60, caused a rapid HR after it was introduced into tobacco leaves, differences in LPS structure were thought to determine the ability of certain strains to cause the HR.

Examination of a series of virulent and avirulent strains of *P. solanacearum* indicated a strong correlation between LPS structure and ability to induce the HR (43). In general, strains with R-LPS induced the HR, those with S-LPS did not. These differences in LPS structure were demonstrated most readily by polyacrylamide gel electrophoresis in the presence of SDS. The LPS of HR-inducing strains (R-LPS) migrated rapidly, that from the non-HR-inducing strains (S-LPS) migrated more slowly. Initial analysis was carried out with 13 strains and, of these, only one (strain Q) had R-LPS but did not induce the HR. Because the HR is not induced by LPS alone, it seems likely that strain Q has additional mutations that affect its ability to cause host cell collapse.

Analyses of the LPS of 13 additional strains have now been completed. These strains include race 2 isolates which produce copious amounts of EPS in culture, but induce the HR in tobacco (Whatley, unpublished data). Overall, the analysis of the LPS of these additional strains strongly supports the hypothesis that R-LPS is necessary for induction of the HR. Evidently, mutations that result in total or partial loss of the O-antigen uncover sites that are readily recognized by the host cell and this leads to an incompatible response, the HR. Conversely, the O-antigen chain of the S-LPS of virulent strains apparently masks these recognition sites.

The loss of the O-antigen chain in the HR-inducing strains of *P. solanacearum* is reflected in the sugar composition of their LPS as well as in their resistance to lysis by certain bacteriophages. These strains have LPS with low levels of rhamnose and xylose relative to glucose. This is in contrast with the non-HR inducing strains which have LPS with high rhamnose and xylose relative to glucose (43). A possible explanation is that rhamnose and xylose are components of the O-antigen chain, while glucose is located in the core region of LPS. Similarly, the fact that phage CH 154 does not lyse the rough strains, but is inactivated by LPS from the smooth strains (43), suggests that the O-specific antigen is necessary for attachment of the phage. Mutations that alter the length of this chain apparently result in loss of these attachment sites, but uncover other sites that are important in the induction of the HR.

A relationship between LPS structure and virulence, similar to that described for *P. solanacearum*, has been established for *Xanthomonas phaseoli* var. *sojensis* (16). Although relatively few strains have been examined, it is apparent that only the avirulent forms are attached and enveloped by host cell walls. Although there were no significant differences in the amounts or sugar composition of EPS produced by five strains, the avirulent and virulent forms had R-LPS and S-LPS, respectively. The LPS from the virulent forms contained approximately 30% fucose and less than 10% glucose, while that of the avirulent forms contained no fucose and over 30% glucose (Whatley, unpublished data). Thus, differences in virulence and in the ability of these bacteria to attach to the host cell wall appear to be dependent on LPS structure.

These few examples serve to illustrate the point that in plant pathogenic bacteria, as in those that attack animals, virulence and LPS structure are intimately associated. In both groups, as is typical of extracellular parasites, lack or avoidance of recognition by a putative receptor appears to be the key to their success in multiplying within the host.

It is still uncertain what the receptor for bacterial LPS might be in plant cells. Because certain lectins agglutinate plant pathogenic bacteria and precipitate their LPS (38), it has been suggested that these proteins are the most likely receptors. There is as yet no strong evidence that lectins play this role. In the case of *P. solanacearum* LPS and potato lectin, it is now clear that much of the interaction between these molecules may be non-specific. Recent studies (Duvick, unpublished data) indicate that charge-charge interactions may explain the very strong binding between LPS and potato lectin. Potato lectin is a highly basic protein, while LPS is negatively charged. Since most agglutination or binding assays are carried out at relatively acid pH, binding of lectin and LPS occurs readily.

The non-specificity of the interaction of LPS with basic cell wall proteins does not necessarily imply that these molecules play no role in the attachment of bacteria to cell walls. Attachment is a highly non-specific process in tobacco leaves, since most bacteria are rapidly immobilized. It is tempting to speculate that differences in tightness of binding between different strains to cell walls might be associated with differences in surface charge.

Evidence for a carbohydrate-carbohydrate interaction, rather than a lectin-carbohydrate interaction, was provided in the binding of *A. tumefaciens* to plant cell walls. Lippincott & Lippincott (26) concluded that the binding site for bacterial LPS may involve polygalacturonic acid, a major component of the primary cell wall. Of the cell wall components that were tested in the pinto bean assay, only polygalacturonic acid was highly inhibitory to tumor formation. When non-binding cell walls, such as those from monocots or embryonic tissues, were treated with pectin methylesterase, they became inhibitory. It must be pointed out that no evidence for a direct interaction between LPS and polygalacturonic acid has been presented.

The Role of Lipid A

There is a general structural similarity in the lipid A portion of the LPS from different strains of gram-negative bacteria.

Lipid A is the major component of endotoxin (27) and it appears to owe its biological activity to its ability to bind to cell membranes (15). Lipid-free bacterial polysaccharides do not bind to erythrocyte membranes, for example. It seems likely that the unstabilizing effects of lipid A on mammalian membranes trigger some of the biological effects associated with endotoxin activity, such as pyrogenicity and mitogenicity (21).

One of the most interesting effects of endotoxin on animals is an enhanced non-specific resistance to infection (21). This is partly due to its effects on the reticulo-endothelial system, which result in macrophage activation. There is considerable evidence that the effects of endotoxin are mediated through activation of the membrane-bound adenylate cyclase (21). Alteration of the levels of cyclic nucleotides, such as cAMP and cGMP, by endotoxin apparently results in stimulation of a large number of host metabolic events.

In view of the above, it is perhaps not surprising that infiltration of tobacco leaves with LPS from *P. solanacearum* causes systemic, non-specific resistance to infection (18). Purified LPS preparations are highly active as protection inducers against bacterial and viral infection at concentrations of 50 μg/ml. Lipid A preparations, made soluble by condensation with bovine serum albumin, also were active protection inducers.

More recently, we have determined (Barlow & Sequeira, unpublished data) that the LPS of a strain of *S. minnesota* (R595), which contains only lipid A and KDO, is also an active protection inducer. Since this LPS is essentially devoid of the core and O-oligosaccharides, it is evident that these components of LPS are not essential for the induction of disease resistance. Mild specific chemical hydrolysis of the *P. solanacearum* LPS indicates that fatty acid sterification of lipid A is necessary for biological activity (18).

Changes in soluble peroxidase are associated with binding of bacterial LPS to the host cell wall (31). Enhanced peroxidase activity, however, was observed in shaded leaves treated with LPS, although protection did not develop in these leaves. Peroxidase changes, therefore, cannot be involved directly in the protective response. It seems more likely that this response is the result of attachment of the lipid A portion of LPS to plant cell membranes, just as it binds to erythrocyte membranes. The same type of membrane-destabilizing effects that trigger the biological activity of endotoxin on mammalian cells may be involved in its effects on plant mesophyll cells.

REFERENCES

1. AYERS, A.R., AYERS, S.B. & GOODMAN, R.N. (1979). Extracellular polysaccharide of *Erwinia amylovora:* a correlation with virulence. *Applied Environmental Microbiology* 38, 659 - 666.

2. BAUER, W.D., BHUVANESWARI, T.V., MORT, A.J. & TURGEON, G. (1979). The initiation of infections in soybean by *Rhizobium*. 3. *R. japonicum* polysaccharide pretreatment induces root hair infectibility. *Plant Physiology* 63 (suppl.), 135.

3. BHUVANESWARI, T.V., PUEPPKE, S.G. & BAUER, W.D. (1977). Role of lectins in plant-microorganism interactions. I. Binding of soybean lectin to Rhizobia. *Plant Physiology* 60, 486 - 491.

4. BOHLOOL, B.B. & SCHMIDT, E.L. (1974). Lectins: a possible basis for specificity in the *Rhizobium*-legume nodule symbiosis. *Science* 185, 269 - 271.

5. BRADSHAW-ROUSE, J., SEQUEIRA, L., KELMAN, A. & COPLIN, D. (1980). Extracellular polysaccharide and virulence of *Erwinia stewartii* in relation to agglutination by a corn agglutin. (Abstr.) *Phytopathology* 70, in press.

6. CALVERT, H.E., LALONDE, M., BHUVANESWARI, T.V. & BAUER, W.D. (1978). Role of lectins in plant-microorganisms interactions. 4. Ultrastructural localizations of soybean lectin binding sites on *Rhizobium japonicum*. *Canadian Journal of Microbiology* 24, 785 - 793.

7. CARLSON, R.W., SANDERS, R.E., NAPOLI , C. & ALBERSHEIM, P. (1978). Host-pathogen interactions. 13. Purification and partial characterization of *Rhizobium* lipopolysaccharides. *Plant Physiology* 62, 912 - 917.

8. DAZZO, F.B. (1980). Adsorption of microorganisms to roots and other plant surfaces. In : *Adsorption of Microorganisms to Surfaces*, Ed. by G. Bitton & K.C. Marshall, pp. 253 - 316. J. Wiley & Sons, New York.

9. DAZZO, F.B. & BRILL, W.J. (1979). Bacterial polysaccharide which binds *Rhizobium trifolii* to clover root hairs. *Journal of Bacteriology* 137, 1362 - 1373.

10. DAZZO, F.B. & HUBBELL, D.H. (1975). Cross-reactive antigens and lectin as determinants of symbiotic specificity in

the *Rhizobium*-clover association. *Applied Microbiology* 30, 1017 - 1033.

11. DUBOS, R.J. (1945). *The Bacterial Cell.* Harvard University Press, Cambridge, Massachusetts.

12. DUDMAN, W.F. (1977). The role of surface polysaccharides in natural environments. In : *Surface Carbohydrates of the Prokaryotic Cell*, Ed. by I. Sutherland, pp. 357 - 414. Academic Press, New York.

13. DUVICK, J.P., SEQUEIRA, L. & GRAHAM, T.L. (1979). Binding of *Pseudomonas solanacearum* surface polysaccharides to plant lectin *in vitro*. (Abstr.) *Plant Physiology* 63, 134.

14. EL-BANOBY, F.E. & RUDOLPH, K. (1979). A polysaccharide from liquid cultures of *Pseudomonas phaseolicola* which specifically induces water-soaking in bean leaves (*Phaseolus vulgaris* L.). *Phytopathologische Zeitschrift* 95, 38 - 50.

15. ESSER, A.F. & RUSSELL, S.W. (1979). Membrane perturbation of macrophages stimulated by bacterial lipopolysaccharide. *Biochemistry Biophysics Research Communications* 87, 532.

16. FETT, W.F. & SEQUEIRA, L. (1980). A new bacterial agglutinin from soybean. II. Evidence against a role in determining pathogen specificity. *Plant Physiology*, in press.

17. GOODMAN, R.N., HUANG, P.T. & WHITE, J.A. (1976). Ultrastructural evidence for immobilization of an incompatible bacterium, *Pseudomonas pisi*, in tobacco leaf tissue. *Phytopathology* 66, 754 - 764.

18. GRAHAM, T.L., SEQUEIRA, L. & HUANG, T.R. (1977). Bacterial lipopolysaccharides as inducers of disease resistance in tobacco. *Applied and Environmental Microbiology* 34, 424 - 432.

19. HAMBLIN, J. & KENT, S.P. (1973). Possible role of phytohaemagglutinin in *Phaseolus vulgaris* L. *Nature* 245, 28 - 30.

20. HILDEBRAND, D.C., ALOSI, M.D. & SCHROTH, M.N. (1980). Physical entrapment of pseudomonads in bean leaves by films formed in air-water interfaces. *Phytopathology* 70, 98 - 109.

21. KABIR, S., ROSENSTREICH, D.L. & MERGENHAGEN, S.E. (1978). Bacterial endotoxins and cell membranes. In : *Bacterial Toxins and Cell Membranes*, Ed. by J. Jclzaszewicz and T. Wadstrom, pp. 59 - 88. Academic Press, London.

22. KAMBERGER, W. (1979). Role of cell surface polysaccharides in the *Rhizobium*-pea symbiosis. *FEMS Microbiology Letters* 6, 361 - 365.

23. KATO, G., MARUYAMA, Y. & NAKAMURA, M. (1979). Role of lectins and lipopolysaccharides in the recognition process of specific legume-*Rhizobium* symbiosis. *Agricultural and Biological Chemistry* 43, 1085 - 1092.

24. KELMAN, A. (1954). The relationship of pathogenicity in *Pseudomonas solanacearum* to colony appearance on a tetrazolium medium. *Phytopathology* 44, 693 - 695.

25. LEACH, J.E., CANTRELL, M.A. & SEQUEIRA, L. (1978). Localization of potato lectin by means of fluorescent antibody techniques. (Abstr.) *Phytopathology News* 12, 197.

26. LIPPINCOTT, J.A. & LIPPINCOTT, B.B. (1978). Cell walls of crown-gall tumors and embryonic plant tissues lack *Agrobacterium* adherence sites. *Science* 199, 1075 - 1077.

27. LÜDERITZ, O., WESTPHAL, O., STAUB, A.M. & NIKAIDO, H. (1971). Isolation and chemical and immunological characterization of bacterial lipopolysaccharides. In : *Microbial Toxins*, Ed. by G. Weinbaum, S. Kadis & S.J. Ajl, Vol. 4, pp. 145 - 233. Academic Press, New York.

28. MAIER, R. & BRILL, W.J. (1978). Involvement of *Rhizobium japonicum* O-antigen in soybean nodulation. *Journal of Bacteriology* 133, 1295 - 1299.

29. MEDEARIS, D.N., CAMITTA, B.M. & HEATH, E.C. (1968). Cell wall composition and virulence in *Escherichia coli*. *Journal of Experimental Medicine* 128, 399 - 414.

30. MORT, A.J. & BAUER, W.D. (1980). Composition of the capsular and extracellular polysaccharides of *Rhizobium japonicum*. Changes with culture age and correlations with binding of soybean seed lectin to the bacteria. *Plant Physiology* 66, 158 - 163.

31. NADOLNY, L. & SEQUEIRA, L. (1980). The role of peroxidase in induced resistance in tobacco. *Physiological Plant Pathology* 16, 1 - 8.

32. NAKANO, M. & SAITO, K. (1969). Chemical components in the cell wall of *Salmonella typhimurium* affecting its virulence and immunogenicity on mice. *Nature, London* 222, 1085 - 1086.

33. NAPOLI, C.A., DAZZO, F.B. & HUBBELL, D.H. (1975). Production of cellulose microfibrils by *Rhizobium*. *Applied Microbiology* 30, 123 - 131.

34. ROANTREE, R.J. (1971). The relationship of lipopolysaccharide structure to bacterial virulence. In : *Microbial Toxins*, Ed. by S. Kadis, G. Weinbaum & S.J. Ajl, Vol. 5, pp. 1 - 37. Academic Press, New York.

35. ROTTINI, G., DRI, P., ROMEO, D. & PATRIARCA, P. (1976). Influence of *E. coli* polysaccharide on the interaction of *E. coli* K^+ and K^- with polymorphonuclear leukocytes *in vitro*. *Zbl. Bakt. Hyg., I. Abt. Orig.* A234, 189 - 201.

36. SANDERS, R.E., CARLSON, R.W. & ALBERSHEIM, P. (1976). A *Rhizobium* mutant incapable of nodulation and normal polysaccharide secretion. *Nature* 271, 240 - 242.

37. SEQUEIRA, L., GAARD, G. & DE ZOETEN, G.A. (1977). Attachment of bacteria to host cell walls : its relation to mechanisms of induced resistance. *Physiological Plant Pathology* 10, 43 - 50.

38. SEQUEIRA, L. & GRAHAM, T.L. (1977). Agglutination of avirulent strains of *Pseudomonas solanacearum* by potato lectin. *Physiological Plant Pathology* 11, 43 - 54.

39. SEQUEIRA, L. & HILL, L.M. (1974). Induced resistance in tobacco leaves : the growth of *Pseudomonas solanacearum* in protected tissues. *Physiological Plant Pathology* 4, 447 - 455.

40. STALL, R.E. & COOK, A.A. (1979). Evidence that bacterial contact with the plant cell is necessary for the hypersensitive reaction but not the susceptible reaction. *Physiological Plant Pathology* 14, 77 - 84.

41. SUTHERLAND, I.W. (1977). Bacterial exopolysaccharides - their nature and production. In : *Surface Carbohydrates of the Prokaryotic Cell*, Ed. by I. Sutherland, pp. 27 - 96. Academic Press, New York.

42. WHATLEY, M.H., BODWIN, J.S., LIPPINCOTT, B.B. & LIPPINCOTT, J.A. (1976). Role for *Agrobacterium* cell envelope lipo-

polysaccharide in infection site attachment. *Infection & Immunity* 13, 1080 - 1083.

43. WHATLEY, M.H., HUNTER, N., CANTRELL, M.A., HENDRICK, C.A., KEEGSTRA, K. & SEQUEIRA, L. (1980). Lipopolysaccharide composition of the wilt pathogen, *Pseudomonas solanacearum:* correlation with the hypersensitive response in tobacco. *Plant Physiology* 65, 557 - 559.

44. WHATLEY, M.H., MARGOT, J.B., SCHELL, J., LIPPINCOTT, B.B. & LIPPINCOTT, J.A. (1978). Plasmid and chromosomal determination of *Agrobacterium* adherence specificity. *Journal of General Microbiology* 107, 395 - 398.

45. WILKINSON, S.G. (1977). Composition and structure of bacterial lipopolysaccharides. In : *Surface Carbohydrates of the Prokaryotic Cell*, Ed. by I. Sutherland, pp. 99 - 175. Academic Press, New York.

46. WOLPERT, J.S. & ALBERSHEIM, P. (1976). Host-symbiont interactions. I. The lectins of legumes interact with the O-antigen containing lipopolysaccharides of their symbiont Rhizobia. *Biochemistry Biophysics Research Communications* 70, 729 - 737.

DEFENCE MECHANISMS OF PLANTS AGAINST VARIETAL NON-SPECIFIC PATHOGENS

ANDRÉ TOUZÉ and MARIE-THÉRÈSE ESQUERRÉ-TUGAYÉ

Centre de Physiologie Végétale, Université Paul Sabatier
118, route de Narbonne
31062 Toulouse cédex, France

INTRODUCTION

General resistance, also called horizontal or polygenic resistance, protects plants against a large number of different species of potential pathogens and pests, but is usually less efficient than the one which is given by gene-for-gene systems. However, tissues showing race non-specific resistance are less suitable than susceptible tissues for development of pathogens; penetration of the pathogens is limited and their growth and sporulation capacities are restricted. Mechanisms responsible for these phenomena, or presumably involved in them, are located within plant cell surfaces and cytoplasm.

DEFENCE MECHANISMS AT THE CELL WALL LEVEL

Until recently, plant cell walls were considered as an efficient, yet passive barrier against pathogens and unfavourable environmental conditions. The discovery that changes in cell walls components, lignins, proteins or glycoproteins do occur upon infection give another view of the role of cell surface in pathogenesis. Lignification, accumulation of hydroxyproline rich glycoproteins and inactivation of pathogen degrading enzymes will be considered in turn.

Lignification

Lignification may be an important factor in controlling the development of varietal non-specific pathogens as well as of gene-for-gene pathogens (17) and of non-pathogenic fungi (41, 42).

Hijwegen (21) investigated the importance of lignification with respect to natural resistance which occurs in some cucumbers against cucumber scab caused by *Cladosporium cucumerinum*. Both resistant and susceptible plants were inoculated and thereafter infected plants were histologically compared with uninfected controls. Lignin formation was observed in resistant inoculated plants only. It started in the cell layers just beneath the epidermal cells; epidermis itself did not become lignified. Hijwegen suggested that lignification could be an active resistance mechanism which inhibits the parasite in its progress.

In a series of papers Asada and Matsumoto (7, 8, 9) reported on the lignification of parenchyma in Japanese radish roots, *Raphanus sativus*, infected by *Peronospora parasitica*. Lignification was studied by U.V. microspectrophotometry and it was concluded that lignin accumulated in the middle lamellae of the parenchyma cell walls. Chemical analysis showed that there were differences between the lignin formed after infection and lignin present in non-infected roots. In this system, lignification took place in response to infection but no cessation of disease development occurred. Thus, if lignification has something to do with defence it is too late in time, or insufficient in magnitude.

Better evidence for the role of lignin formation in disease resistance is indicated in more recent experiments by Vance and Sherwood (51) with reed canarygrass, *Phalaris arundinacea*. In this case, localized lignin formation has been implicated as a general mechanism of resistance for excluding fungal species which are non-pathogenic to the plant, but it also confers protection against normal pathogens. In leaf discs inoculated by *Helminthosporium avenae* (a non-pathogen) and floated on water, lignin content increased; epidermal cells formed lignified papillae around the penetration peg and the fungus did not penetrate. Conversely, treatment with cycloheximide inhibited this increased lignin deposition and tissues became susceptible to fungi that did not normally infect reed canarygrass. An approximately two-fold increase in the activities of enzymes associated with lignin biosynthesis (P.A.L., cinnamic acid-CoA ligases, peroxidase) paralleled the induced lignification. When inoculated tissues were treated with cycloheximide enzyme activities were inhibited. Some further evidence was obtained (52) which supported the conclusion that lignified papillae may play an important role in the resistance response of reed canarygrass to leaf infecting fungi. Cross protection of whole plants against the pathogen *Helminthosporium catenarium* was obtained by first inoculating intact leaves with the non-pathogen *Botrytis cinerea*. The number of lesions was significantly reduced; papillae formation induced by *B. cinerea* was effective in limiting the number of successful penetrations by *H. catenarium*. Therefore, the same resistance mechanism seems to be involved.

Involvement of lignification in defence was also reported in melons systemically protected against *Colletotrichum lagenarium* by Touzé and Rossignol (49). Phenylalanine ^{14}C was fed as a precursor of lignin to protected cuttings harvested at 0, 2, 4 and 6 days following challenge inoculation. After a metabolic period of 24 hours, cell walls were isolated and subjected to oxidation by alkaline nitrobenzene. Determination of the specific radioactivities of the aromatic aldehydes obtained by this technique indicated that syringaldehyde did not markedly differ in the two cases, but that the potentialities of coniferyl and coumaryl units synthesis were enhanced in protected plants. This was particularly striking four days after inoculation when symptoms occurred. Thus, in this system, higher potentialities of lignin synthesis, along with other mechanisms, could be involved in restricting development of the fungus after immunization.

Lignification might hinder fungal progress in a number of ways. Unfortunately there is little information about the possible mechanisms of fungal growth inhibition. Along with lignin, plant cell walls contain phenolic compounds (20) which could be associated with the defence of plants through their linkage with polysaccharides by ester bonds, thus protecting these carbohydrates polymers from degradation by pathogen hydrolases.

Accumulation of hydroxyproline rich glycoproteins

Hydroxyproline-rich proteins were discovered in cell walls twenty years ago by Lamport and Northcote (24), and later shown to be glycosilated with arabinose and galactose (25, 27). Since that time, other hydroxyproline rich glycoproteins, either intra- or extra-cellular, have been reported from different sources (5, 13, 15, 22, 26). They all have common structural features namely, a) a high level of hydroxyproline, of the β-hydroxyaminoacids serine and threonine and of either glycine and alanine b) they all contain arabinose and galactose in varying amounts c) other sugars and amino sugars may also be present. The role of these glycoproteins is not yet known; they may be structural proteins or they may fulfil more physiological functions.

Involvement of these or related molecules in plant-pathogen interactions is suggested by a few examples in which pathogens are either fungi, bacteria, nematodes or viruses. Thus, upon infection of melon seedlings by *Colletotrichum lagenarium*, plant cell walls are enriched in hydroxyproline-rich glycoproteins as compared to healthy seedlings. This increase was first demonstrated by Esquerré-Tugayé and Mazau (12) who then found that similar responses occur in different plants challenged with several pathogens. Another example is the interaction of the potato cell wall and of potato tuber lectin with *Pseudomonas solanacearum* (46). Sequeira and his co-workers have reported on the higher affinity of this lectin for avirulent or unrelated bacteria than for virulent ones (45). Whether this lectin and cell wall glycoprotein are identical is not established, but in addition to the similarities already noted, both glycoproteins exhibit the same glycosilation pattern (3). The hydroxyproline content of total proteins also increases in potato roots infected by *Heterodera rostochiensis* (18). Depending on whether varieties were resistant or susceptible, the proline to hydroxyproline ratios were differently affected after infection and this was attributed to an enhanced hydroxyproline response in resistant varieties. The last example is the appearance of protein-polysaccharides complexes, induced in *Phaseolus vulgaris* by inoculation with either TNV, TMV or wounding as reported by Brown and Kimmins (10). As compared to healthy plants, two protein components were induced, the amino acid compositions of which were related to plant cell wall proteins. Fraction II, induced around 90 to 120 hours after infection or wounding, was shown to contain small amounts of hydroxyproline and arabinose. Working with SBMV, it was later postulated, by the same authors, that these complexes might function in virus localization through sealing of plasmodesmata, thus preventing virus spread. Relationships between hydroxyproline containing glycoproteins and resistance are implied in all these examples and were demonstrated with melon seedlings having different levels of cell wall glycoproteins. These levels were modified, independently of disease, by physiologically altering the extent of biosynthesis. This was achieved by treating young seedlings with ethylene or feeding them with hydroxyproline (14). Ethylene treated plants contained two to three times more cell wall glycoproteins as shown by analysis of the hydroxyproline-arabinosides of the molecule. Plants fed with hydroxyproline as a free amino acid in the culture medium, contained lower amounts of cell wall glycoprotein. It was shown that melon plants having different levels of glycoproteins in their cell walls respond differently to infection with *Colletotrichum lagenarium*. The amount of pathogen was assessed in these plants by measuring their glucosamine content (48). It was shown that cell wall glycoprotein-enriched plants contained less pathogen and were more resistant than control plants which were killed seven to eight days after inoculation; at that time about 80% of the ethylene treated plants recovered. Inversely, plants which were fed with hydroxyproline and contained less glycoprotein than controls, were still less resistant to disease in the sense that the rate of infection was higher.

This set of experiments indicated that the enrichment of cell walls in hydroxyproline-rich glycoproteins is associated with a defence response. This response occurs in several plants upon infection by different pathogens. Early triggering of this mechanism appears essential for its full efficiency. The way in which cell wall glycoproteins interact with pathogens is not yet known. They might have a lectin like activity or they might strengthen the cell wall.

Inactivation of pectolytic enzymes

These enzymes, believed to be of key importance in numerous infection processes, may be inactivated by protein inhibitors and by differential binding to cell walls.

Protein inhibitors were characterized by Albersheim and Anderson (1) who showed that crude protein extracts of the cell walls of dicots (tomato stems, sycamore cells, and bean hypocotyls) are each capable of inhibiting the polygalacturonases (PG) secreted by *Colletotrichum lindemuthianum*, *Fusarium oxysporum lycopersici* and *Sclerotium rolfsii*. The different plant extracts had differential ability to inhibit the three PGs suggesting inhibitors with different properties or several inhibitors within a given plant. The Red Kidney bean hypocotyl crude extract did not inhibit purified preparations of exopolygalacturonase, xylanase, carboxymethylcellulase, α-arabinosidase and α-galactosidase. The inhibitor purified from Red Kidney beans was shown to be a glycoprotein which inhibited the *C. lindemuthianum* endo-PG 40 times more efficiently than it inhibited the endo-PG secreted by *Fusarium oxysporum*; it had no ability to inhibit the *S. rolfsii* polygalacturonase. These observations suggested that cell walls of a given plant may contain inhibitors more effective against the endo-PG from one of its pathogens than against those of non-pathogens. However in a later paper Fisher *et al.* (16) found that the Red Kidney bean inhibitor was also very efficient against the *Aspergillus niger* endo-PG. The endo-PG inhibitors from two bean cultivars Red Kidney and Pinto were then purified (4). These proteins which purified in an identical manner were, within experimental error, equally effective inhibitors of the endo-PG secreted by α, γ and δ races of *C. lindemuthianum*. So, it is unlikely that differential resistance of beans cultivars can be based on differential inhibition of the endo-PG secreted by the various *C. lindemuthianum* races. Protein inhibitors represent a general rather than a specific mechanism for resistance. As such, these PG-protein inhibitors seem to be pre-infectional factors. However, preliminary results obtained by Barthe and Lafitte (unpublished results of our laboratory) suggest that these compounds could undergo significant post-infectional changes. *Phaseolus vulgaris* seedlings, resistant and susceptible to *Colletotrichum lindemuthianum* were used. Following inoculation, greater amounts of inhibitor were found in infected plants as compared to uninfected ones and this response appears earlier (after 24 hours) in resistant than in susceptible plants.

Besides enzyme inhibitors, differential binding to host cell walls has also been proposed as a defence mechanism by Mussel and

Strand (32). They observed that the endo-PG from the pathogen *Verticillium albo-atrum* and from the non-pathogen *Fusarium oxysporum* f. sp.*lycopersici* bind differently to purified cotton cell walls. They also found a differential binding of the pathogen endo-PG to cell walls of susceptible and resistant cotton cultivars. From these results, it seems that the cell walls from resistant plants have a greater capacity to bind the pathogen endo-PG thus rendering it ineffective in degrading the host cell wall.

Finally, inhibition and binding have been observed by Raa and co-workers (38) for the PG of *Cladosporium cucumerinum* with the cell walls of cucumber seedlings. These cell walls contain a high molecular weight substance which inhibits the polygalacturonase produced by the pathogen, *C. cucumerinum* and has no effect against PG of a saprophyte *Aspergillus niger* or of a non-host pathogen *Botrytis cinerea*. The PG secreted by *Cladosporium cucumerinum* binds also to isolated cell walls of cucumber whereas those of *Aspergillus* and *Botrytis* do not. Tromso and Tromso (50) have obtained similar results with apple cell walls and the pectinase produced by *Botrytis cinerea*, the agent of dry eye rot on apple.

Rather than a defence mechanism these authors suggested that inhibition and binding of pectic degrading enzymes to the host cell walls may be essential for the pathogenicity of the fungus. They hypothesized that, due to binding and inhibition, the degradation of pectin may be restricted to the vicinity of the pathogen and this would allow the pathogen to grow without provoking cell death or a hypersensitive defence reaction.

DEFENCE MECHANISMS AT THE CYTOPLASM LEVEL

Three examples, other than phytoalexins, will be considered.

Accumulation of soluble phenolics

Chlorogenic acid, caffeic acid, catechol, tannins and related substances are widely distributed among higher plants; as these pre-infectional metabolites are often toxic to microorganisms at physiological concentrations, they were supposed to have a causal role in protecting plants from disease. But there is also some experimental evidence that phenolics may be synthesized in response to attempted invasion by pathogens.

Quantitative variations in soluble phenols in leaves, stem and roots of the tomato varieties Marmande-susceptible, and Marporum-resistant to infection by *Fusarium oxysporum* f. sp. *lycopersici* were reported by Matta *et al.* (28). In both varieties the phenolic content did not differ markedly and also increased in response to infection. But, in resistant plants increases occurred early enough to play a role in defence, particularly in leaves and roots. In susceptible plants significant increases were only found when the pathogen had invaded the plant.

Working with another tomato variety, Bonny Best, the same authors (29) reported a good correlation between cross protection and increase in phenolics. Inoculation with a non-pathogenic form, *F. oxysporum* f. sp. *dianthi*, protects this cultivar against the pathogenic form *lycopersici*. The protective effect rose to a maximum after two or three days, decreased and disappeared after ten days. Variations in the phenolic content with time was strikingly similar in stems and leaves. It thus seems that phenolics might be implicated in the protective response.

A third example of phenolics and defence is given by Hunter (23). In cotton hypocotyls phenolic compounds, predominantly identified as catechin, increased in response to infection by *Rhizoctonia solani*. This increase was faster and greater concentrations were reached in older seedlings which were also more resistant. The resistance seemed to be mediated through the inhibition of the fungal polygalacturonase by oxidized catechin.

Resistance against *Fusarium oxysporum* f. sp. *lycopersici* has also been induced in tomato plants by treatment with phenolic compounds. Retig and Chet (40) prevented disease symptom expression after treatment with catechol although no significant difference was observed in the number of *Fusarium* propagules in the roots and upper stems of identical segments of control and catechol treated plants. It was suggested that catechol treatment changes the plant from a susceptible to a symptomless state. In our institute Carrasco *et al.* (11) obtained protection of tomato with quinic acid. After inoculation, the increase in soluble phenolics was more pronounced and rapid in the quinate treated plants than in controls; more than 50% inhibition of fungal development was maintained in pre-treated plants until the eleventh day after inoculation.

Production of proteinase inhibitors

Proteinase inhibitors are found in many seeds or other storage organs, sometimes in high concentrations, and also in aerial tissues of plants (43). They have been mostly studied in the Leguminosae, Gramineae and Solanaceae but they are also present in other families. They are usually small proteins having molecular weights under 50,000 and more commonly under 20,000. Among different roles that have been ascribed to these inhibitors, it should be firstly mentioned, as pointed out by Ryan (43) that the known inhibitors rarely inhibit proteolytic enzymes from plants. The possibility that these molecules could serve as protective agents directed against insect and microbial proteinases having trypsin- or chymotrypsin-like activities, as do many microbial proteinases, has received some attention. It was first demonstrated (6) that the protease of *Tenebrio molitor*, a common pest of stored grains, was inhibited by inhibitors from lima bean, soybean,and purified Bowman-Birk inhibitor. Kidney bean seeds (31) contain a trypsin inhibitor which specifically inhibits the extracellular, serine-containing protease of *Colletotrichum lindemuthianum*. The proteolytic activity of another plant pathogen, *Fusarium solani*, was lowered in the presence of inhibitors extracted from *Phaseolus vulgaris*, soybeans, lima beans, wheat and potato, while other inhibitors of animal origin such as ovomucoid were ineffective (30). Plant inhibitors have also been proved effective against non-plant pathogenic microorganisms, e.g. the inhibitor subtilisin from barley (54).

Interactions between proteases and plant proteolytic inhibitors led Ryan (44) to hypothesize that these inhibitors could be a mechanism of disease resistance. This hypothesis is supported by the fact that proteolytic inhibitors accumulate in potato and tomato plants when they are damaged by beetles (19), and when they are infected by *Phytophthora infestans* (37) with higher increases in incompatible interactions. In both cases increases occur within 24 hours. Such increases also occur in tomato leaves after mechanical wounding or after detachment and are paralleled by the appearance of protein bodies within the vacuole (47). It is very remarkable that in all reported cases, infection or wounding of a single bottom leaf induces proteinase inhibitor activity in undamaged upper leaves, thus suggesting the presence of a signal which has been called Proteinase Inhibitor Inducing Factor (PIIF). From the work of Ryan (44) and of Albersheim and co-workers (2), it seems now that this factor is an oligosaccharide of plant origin, related to the cell wall, and released under damage of plant cells.

The widespread occurrence of this factor in plants and of the response to this factor, namely the accumulation of proteolytic inhibitors, together with the reported structural similarities shared by many known inhibitors are in favour of proteolytic inhibitors as a kind of primitive immune response contributing to the overall defence of a plant.

Lytic enzymes of plants as a possible defence mechanism against parasitic fungi

In this respect, two glucanohydrolases, chitinase and β-1,3 glucanase, have received special attention because chitin and β-1,3 glucans are present in the cell walls of fungi.

In vivo lysis of fungal hyphae caused by chitinase and glucanase has been reported by Pegg and Vessey (34) and Pegg (35) in tomato plants infected by *Verticillium albo-atrum*. These two enzymic activities have also been studied in relation to *Verticillium dahliae* infection of potato grown under long and short photoperiods (36). In long days (LD, 15 hour photoperiod) plants showed resistance to symptom development and this was reflected in the total leaf area; β-1,3 glucanase and chitinase activities were higher in both long day and short day infected plants as compared to controls. However, after three weeks they were higher in LD plants agreeing with the idea of their role in reducing the intensity of invasion by the pathogen.

Wargo (53) extracted β-1,3 glucanase and chitinase from the phloem of stems and roots of different perennial woody plants, *Quercus rubra*, *Quercus alba* and *Acer saccharinum*. These enzymes hydrolysed the hyphal walls of *Armillaria mellea*, a pathogen attacking forest trees. The presence of these glucanohydrolases in healthy trees suggests a protective mechanism which may account for resistance to invasion by *Armillaria mellea* and possibly by other fungal pathogens.

More recently, Netzer and co-workers (33) showed that the increase in β-1,3 glucanase activity of a host resistant cultivar of muskmelon provides a potential defence mechanism against *Fusarium oxysporum* f.sp. *melonis*. The enzyme was present in non-inoculated susceptible and in resistant seedlings but a striking difference was found in the rate of production of β-1,3 glucanase in response to inoculation. The resistant cultivar showed a higher initial response. The activity of β-1,3 glucanase could be induced by dipping the seedlings in a solution of laminarin. Higher activities were obtained if the seedlings were subsequently transferred to water for 48 hours. The increase in activity was found in the roots and hypocotyls of both cultivars but levels were higher in the resistant cultivar. A similar pattern of enzyme activity was obtained using cell walls as the substrate. In susceptible plants treated by laminarin and subsequently inoculated, the disease symptoms were markedly lower throughout a period of 21 days, whereas all controls were wilted after 14 days. A correlation was found between the limited spread of the pathogen and the induction of β-1,3 glucanase after pre-treatment with laminarin.

In contrast, in melon seedlings infected by *Colletotrichum lagenarium* we found (39) that increased endo-β-1,3 glucanase activity was involved in the infective mechanism of the pathogen whereas chitinase activity was implicated in defence. Chitinase increased on infection, and in healthy melons treated with ethylene, a treatment which decreases susceptibility to anthracnose.

There are also many reports of enhanced peroxidase activity following interaction of plants with pathogens and this has led to the hypothesis that these enzymes may be important in the defence mechanism of the host. Cell wall-bound peroxidases are most likely involved in lignification but the implication of cytoplasmic peroxidases in disease resistance is not well established.

CONCLUSION

General resistance in plants seems to result from several mechanisms which may be effective at different stages of infection. They usually occur as pre-infectional metabolites or barriers and their biosynthesis or biogenesis are activated in response to infection. This is the case in susceptible plants but it also occurs, and sometimes even more rapidly, in resistant plants.

Early triggering of these mechanisms with sufficient magnitude seems to be essential for their efficiency.

Although general resistance does not usually confer complete protection it can retard the rate of spread of a pathogen. Thus knowledge of mechanisms that support this type of resistance may help in approaching the problem of disease control in crop plants on a more rational basis.

Furthermore, general resistance has the advantage of not being rapidly overcome by new races of fungi or bacteria.

REFERENCES

1. ALBERSHEIM P. & ANDERSON, A.J. (1971). Proteins from plant cell walls inhibit polygalacturonases secreted by plant pathogens. *Proceedings of the National Academy of Sciences of the United States of America* 68, 1815 - 1819.

2. ALBERSHEIM, P. (1981). Complex carbohydrates in active defence. In : *NATO Advanced Study Institute on Active defence mechanisms in plants*, Ed. R.K.S. Wood.

3. ALLEN, A.K., DESAI, N.N., NEUBERGER, A. & CREETH, J.M. (1978). Properties of potato lectin and the nature of its glycoprotein linkages. *Biochemical Journal* 171, 665 - 674.

4. ANDERSON, A.J. & ALBERSHEIM, P. (1972). Host-pathogen interactions. V. Comparison of the abilities of proteins isolated from three varieties of *Phaseolus vulgaris* to inhibit the endopolygalacturonases secreted by three races of *Colletotrichum lindemuthianum*. *Physiological Plant Pathology* 2, 339 - 346.

5. ANDERSON, R.L., CLARKE, A.E., JERMYN, M.A., KNOX, R.B. & STONE, B.A. (1977). A carbohydrate-binding arabinogalactan-protein from liquid suspension cultures of endosperm from *Lolium multiflorum*. *Australian Journal of Plant Physiology* 4, 143 - 158.

6. APPLEBAUM, S.W., BIRK, Y., HARPAZ, I. & BONDI, A. (1964). Comparative studies on proteolytic enzymes of *Tenebrio molitor* L. *Comparative Biochemistry and Physiology II*, 85 - 103.

7. ASADA, Y. & MATSUMOTO, I. (1969). Formation of lignin-like substance in the root tissues of Japanese radish plant infected by downy mildew fungus. *Annals of the Phytopathological Society of Japan* 35, 283 - 289.

8. ASADA, Y. & MATSUMOTO, I. (1971). Microspectrophotometric observations on the cell walls of Japanese radish (*Raphanus sativus*) root infected by *Peronospora parasitica*. *Physiological Plant Pathology* 1, 377 - 383.

9. ASADA, Y. & MATSUMOTO, I. (1972). The nature of lignin obtained from downy mildew-infected Japanese radish root. *Phytopathologische Zeitschrift* 73, 208 - 214.

10. BROWN, R.G. & KIMMINS, W.C. (1973). Hypersensitive resistance. Isolation and characterization of glycoproteins from plants with localized infections. *Canadian Journal of Botany* 51, 1917 - 1922.

11. CARRASCO, A., BOUDET, A.M. & MARIGO, G. (1978). Enhanced resistance of tomato plants to *Fusarium* by controlled stimulation of their natural phenolic production. *Physiological Plant Pathology* 12, 225 - 232.

12. ESQUERRÉ-TUGAYÉ, M.T. (1973). Influence d'une maladie parasitaire sur la teneur en hydroxyproline des parois cellulaires d'épicotyles et pétioles de plantes de Melon. *Comptes Rendus de l'Académie des Sciences de Paris* 276, 525 - 528.

13. ESQUERRÉ-TUGAYÉ, M.T. & LAMPORT, D.T.A. (1979). Cell surfaces in plant-microorganism interactions. I. A structural investigation of cell wall hydroxyproline-rich glycoproteins which accumulate in fungus-infected plants. *Plant Physiology* 64, 314 - 319.

14. ESQUERRÉ-TUGAYÉ, M.T., LAFITTE, C., MAZAU, D., TOPPAN, A. & TOUZÉ, A. (1979). Cell surfaces in plant-microorganism interactions. II. Evidence for the accumulation of hydroxyproline-rich glycoproteins in the cell wall of diseased

plants as a defence mechanism. *Plant Physiology* 64, 320 - 326.

15. FINCHER, G.B., SAWYER, W.H. & STONE, B.A. (1974). Chemical and physical properties of an arabinogalactan peptide from wheat endosperm. *Biochemical Journal* 139, 535 - 545.

16. FISCHER, M.L., ANDERSON, A.J. & ALBERSHEIM, P. (1973). Host-pathogen interactions. VI. A single plant protein efficiently inhibits endopolygalacturonases secreted by *Colletotrichum lindemuthianum* and *Aspergillus niger*. *Plant Physiology* 51, 489 - 491.

17. FRIEND, J., REYNOLDS, S.B. & AVEYARD, A. (1973). Phenylalanine ammonia lyase, chlorogenic acid and lignin in potato tuber tissue inoculated with *Phytophthora infestans*. *Physiological Plant Pathology* 3, 495 - 507.

18. GIEBEL, J. & STOBIECKA, M. (1974). Role of amino acids in plant tissue response to *Heterodera rostochiensis*. I. Protein-proline and hydroxyproline content in roots of susceptible and resistant solanaceous plants. *Nematologica* 20, 407 - 414.

19. GREEN, T.R. & RYAN, C.A. (1972). Wound-induced proteinase inhibitor in plant leaves : a possible defense mechanism against insects. *Science* 175, 776 - 777.

20. HARTLEY, R.D. & JONES, E.C. (1977). Phenolic components and degradability of cell walls of grass and legume species. *Phytochemistry* 16, 1531 - 1534.

21. HIJWEGEN, T. (1963). Lignification, a possible mechanism of active resistance against pathogens. *Netherlands Journal of Plant Pathology* 69, 314 - 317.

22. HORI, H. & SATO, S. (1977). Extracellular hydroxyproline-rich glycoprotein of suspension-cultured tobacco cells. *Phytochemistry* 16, 1485 - 1487.

23. HUNTER, R.E. (1974). Inactivation of pectic enzymes by polyphenols in cotton seedlings of different ages infected with *Rhizoctonia solani*. *Physiological Plant Pathology* 4, 151 - 159.

24. LAMPORT, D.T.A. & NORTHCOTE, D.H. (1960). Hydroxyproline in primary cell walls of higher plants. *Nature* 188, 665 - 666.

25. LAMPORT, D.T.A., (1967). Hydroxyproline-O-glycosidic linkage of the plant cell wall glycoprotein extensin. *Nature* 216, 1322 - 1324.

26. LAMPORT, D.T.A. (1970). Cell wall metabolism. *Annual Review of Plant Physiology* 21, 235 - 270.

27. LAMPORT, D.T.A., KATONA, L. & ROERIG, S. (1973). Galactosylserine in extensin. *Biochemical Journal* 133, 125 - 131.

28. MATTA, A., GENTILE, I. & GIAI, I. (1967). Variazoni del contenuto in fenoli solubili indotte dal *Fusarium oxysporum* f. sp. *lycopersici* in piante di pomodoro suscettibili e resistenti. *Annali della Facoltà di Scienze Agrarie della Università degli studi di Torino* 4, 17 - 32.

29. MATTA, A., GENTILE, U. & GIAI, I. (1969). Accumulation of phenols in tomato plants infected by different forms of *Fusarium oxysporum*. *Phytopathology* 59, 512 - 513.

30. MOSOLOV, V.V., LOGINOVA, M.D., FEDURKINA, N.V. & BENKEN, I.I. (1976). The biological significance of proteinase inhibitors in plants. *Plant Science Letters* 7, 77 - 80.

31. MOSOLOV, V.V., LOGINOVA, M.D., MALOVA, E.L. & BENKEN, I.I. (1979). A specific inhibitor of *Colletotrichum lindemuthianum* protease from kidney bean *(Phaseolus vulgaris)* seeds. *Planta* 144, 265 - 269.

32. MUSSELL, H. & STRAND, L.L. (1977). Pectic enzymes : involvement in pathogenesis and possible relevance to tolerance and specificity. In : *Cell wall biochemistry related to specificity in host-plant pathogen interactions*. Ed. by B. Solheim & J. Raa, pp. 31 - 70. Universitetsforlaget, Oslo, Norway.

33. NETZER, D. & KRITZMAN, G. (1979). β-(1,3) Glucanase activity and quantity of fungus in relation to *Fusarium* wilt in resistant and susceptible near-isogenic lines of muskmelon. *Physiological Plant Pathology* 14, 47 - 55.

34. PEGG, G.F. & VESSEY, J.C. (1973). Chitinase activity in *Lycopersicum esculentum* and its relationship to the *in vivo* lysis of *Verticillium albo-atrum* mycelium. *Physiological Plant Pathology* 3, 207 - 222.

35. PEGG, G.F. (1976). The occurrence of 1,3-β-glucanase in healthy and *Verticillium*-infected, resistant and susceptible tomato plants. *Journal of Experimental Botany* 27, 1093 - 1101.

36. PEGG, G.F. (1977). Glucanohydrolases of higher plants : a possible defence mechanism against parasitic fungi. In : *Cell wall biochemistry related to specificity in host-plant pathogen interactions*, Ed. by B. Solheim & J. Raa, pp. 305 - 345. Universitetsforlaget, Oslo, Norway.

37. PENG, F.H. & BLACK, L.L. (1976). Increased proteinase inhibitor activity in response to infection of resistant tomato plants by *Phytophthora infestans*. *Phytopathology* 66, 958 - 963.

38. RAA, J., ROBERTSON, B., SOLHEIM, B. & TROMSO, A. (1977). Cell surface biochemistry related to specificity of pathogenesis and virulence of microorganisms. In : *Cell wall biochemistry related to specificity in host-plant pathogen interactions*, Ed. by B. Solheim & J. Raa, pp. 11 - 30. Universitetsforlaget, Oslo, Norway.

39. RABENANTOANDRO, Y., AURIOL, P. & TOUZÉ, A. (1976). Implication of β-1,3 glucanase in melon anthracnose. *Physiological Plant Pathology* 8, 313 - 324.

40. RETIG, N. & SHET, I. (1974). Catechol-induced resistance of tomato plants to *Fusarium* wilt. *Physiological Plant Pathology* 4, 469 - 475.

41. RIDE, J.P. (1975). Lignification in wounded leaves in response to fungi and its possible role in resistance. *Physiological Plant Pathology* 5, 125 - 134.

42. RIDE, J.P. & PEARCE, R.B. (1979). Lignification and papilla formation at sites of attempted penetration of wheat leaves by non-pathogenic fungi. *Physiological Plant Pathology* 15, 79 - 92.

43. RYAN, C.A. (1973). Proteolytic enzymes and their inhibitors in plants. *Annual Review of Plant Physiology* 24, 173 - 196.

44. RYAN, C.A. (1978). Proteinase inhibitors in plant leaves : a biochemical model for pest-induced natural plant protection. *Trends in Biochemical Sciences* July, 148 - 150.

45. SEQUEIRA, L. & GRAHAM, T.L. (1977). Agglutination of avirulent strains of *Pseudomonas solanacearum* by potato lectin. *Physiological Plant Pathology* 11, 43 - 54.

46. SEQUEIRA, L. (1978). Lectins and their role in host-pathogen specificity. *Annual Review of Phytopathology* 16, 453 - 481.

47. SHUMWAY, L.K., YANG, V.V. & RYAN, C.A. (1976). Evidence for the presence of proteinase inhibitor I in vacuolar protein bodies of plant cells. *Planta* 129, 161 - 165.

48. TOPPAN, A., ESQUERRÉ-TUGAYÉ, M.T. & TOUZÉ, A. (1976). An improved approach for the accurate determination of fungal pathogens in diseased plants. *Physiological Plant Pathology* 9, 241 - 251.

49. TOUZÉ, A. & ROSSIGNOL, M. (1977). Lignification and the onset of premunition in muskmelon plants. In : *Cell wall biochemistry related to specificity in host-plant pathogen interactions*, Ed. by B. Solheim & J. Raa, pp. 289 - 293. Universitetsforlaget, Oslo, Norway.

50. TROMSO, A. & TROMSO, A.M. (1977). Biochemistry of pathogenic growth of *Botrytis cinerea* in apple fruit. In : *Cell wall biochemistry related to specificity in host-plant pathogen interactions*, Ed. by B. Solheim & J. Raa, pp. 263 - 266. Universitetsforlaget, Oslo, Norway.

51. VANCE, C.P. & SHERWOOD, R.T. (1976). Regulation of lignin formation in reed canarygrass in relation to disease resistance. *Plant Physiology* 57, 915 - 919.

52. VANCE, C.P. & SHERWOOD, R.T. (1977). Lignified papilla formation as a mechanism for protection in reed canarygrass. *Physiological Plant Pathology* 10, 247 - 256.

53. WARGO, P.M. (1975). Lysis of the cell wall of *Armillaria mellea* by enzymes from forest trees. *Physiological Plant Pathology* 5, 99 - 105.

54. YOSHIKAWA, M., IWASAKI, T., FUJII, M. & OOGAKI, M. (1976). Isolation and some properties of a subtilisin inhibitor from barley. *Journal of Biochemistry* 79, 765 - 773.

MECHANISMS IN NON-HOST RESISTANCE

A. MATTA

Istituto di Patologia Vegetale dell' Universita degli
Studi di Torino
10126 Torino
Via Pietro Giuria 15, Italy

INTRODUCTION

Heterogeneous microbial populations surround plants and establish with them a variety of ecological and symbiotic relations. Only a few types of such relations, namely those with an immediate economic interest such as parasitism and mutualism, have been traditional fields of phytopathological and microbiological research. Less attention has been paid to forms of external or internal commensalism or to more anomalous relations such as those resulting from the invasion of a plant by a saprophyte or a pathogen of another host species.

From a strictly nutritional point of view, plants should be attractive substrates for most heterotrophic microorganisms, as most of the substances they potentially need are present in plant tissues. This has probably enabled many microorganisms to establish themselves in the environment of plants and to exert on plants a constant selection pressure. In present natural equilibria, only few microorganisms normally establish superficial contacts with plants and grow on their exudates and even fewer are endophytes. It is axiomatic that the survival of plants has depended on their ability to interpose defence barriers to surrounding microorganisms and this resulted in a biological balance subverted mostly by human intervention.

The evolution has produced in plants a general state of resistance to the non-self. A better understanding of this resistance could help in resolving theoretical as well as applied problems.

A gap exists in this field of study but there has been a recent increase in the number of investigations of this type of resistance.

A number of questions await answers. What, for instance, are the salient aspects of the relations between microorganisms and plants which are not their habitual hosts ? What are the features of general resistance that prevent the great majority of microorganisms from becoming pathogenic on plants ? Are general resistance and the more specific varietal resistance mechanisms comparable or not ? These questions can be answered only in part because we have only a small amount of relevant information and speculation not based on sound experimental evidence is not an acceptable substitute.

This new field of research has used a terminology within which the term "non-host" and "non-pathogen" have become prominent. They are both unsatisfactory but cannot easily be replaced.

Non-hosts are plant species which do not develop visible symptoms after natural or artificial inoculation with parasites of other plant species (20). The term must not be applied to varieties resistant to certain races or to avirulent isolates of a parasite. I think that it can be extended to include plants challenged with saprophytes. Non-host is a useful term with a broad meaning but it is hardly applicable to symptomless host plants in which pathogenesis but not colonization by non-pathogens is inhibited. "Non-suscept" could be an alternative, satisfactory in this respect and with wider meaning. The terms "inappropriate", "uncongenial" or "unnatural host" could also be acceptable substitutes but for the fact that they presume that the invaders always posses an appropriate, congenial or natural host and this would not be true for saprophytes. The alternative then should be between "non-host" and "non-suscept". Both will be used, the first preferably when growth of the invader is arrested before extensive colonization occurs.

The term non-pathogens has an absolute connotation in that it means organisms unable to cause disease in any host. Nevertheless it is also used for inability to cause disease in a given plant species by pathogens of other plant species. The reference to the species, if not explicitly stated (species - non-pathogen), must be understood. Saprophytes are also conceived as non-pathogens when tested on living plants.

PRESENCE OF NON-PATHOGENS IN PLANTS

The presence of non-pathogenic microorganisms on the surface of non-suscept plants in general is natural and common. Their presence inside the plant is natural in a few cases but not common and this creates great difficulties in the study of the symbiotic relations between non-parasitic microorganisms and plants in nature.

Natural presence of microorganisms in plants

Investigations on the natural presence in apparently healthy plants of soil or phylloplane bacteria are considerably complicated by the frequent occurrence of contaminants. A supposedly, but never convincingly demonstrated, constant association of bacteria and plant cells, was a false assumption on which was based the development of various theories of "universal symbiosis" (30). Nevertheless, there are so many positive reports of isolations "from deep within the interior of outwardly stripped or flamecharred plant parts" (30) that a more or less sporadic occurrence of bacteria in healthy plant tissues must be accepted.

Particularly reliable are the results obtained by isolating from the interior of fleshy organs such as fruits, roots or tubers, in which bacteria can penetrate through the stigma, the region of insertion of the peduncle, or epidermal microlesions (13) and wounds on roots with later distribution in the xylem. The presence of saprophytes or species-non-pathogens inside plants may be due to an entirely passive process of contamination from the surface, or to an active process of penetration and colonization. Little is known about the factors that influence the more or less aggressive behaviour of different bacteria : size and motility are assumed by Ercolani and Casolari (13) to be the main characters of aggressiveness of bacteria found inside tomato fruits by enabling them to take advantage of microlesions on fruit surfaces. The extent of internal spread would depend essentially on the degree of vascularization. The findings of different kinds of non-pathogenic bacteria in the inner regions of various plant organs conflicts with the current opinion that saprophytic bacteris do not multiply and die after a few days in healthy and vigorous plant tissues. It is possible, however, that the behaviour in living tissues of saprophytes such as *Erwinia* sp., *Pseudomonas fluorescens* and *Pseudomonas aeruginosa* normally used by phytobacteriologists in infiltration experiments is basically different from that of other taxa found in nature. On the other hand it is also possible that colonization with non-pathogenic forms is strictly conditioned by the environment and by conditions in plants. Thus, multiplication of saprophytic forms often occurs in senescent or in water congested tissues.

Phytopathogenic bacteria do multiply in wrong host plants. Their number, however, is generally of the order of 1/1000 of that of homologous, virulent bacteria (39).

The interior of the plants is not the best environment for bacteria but can be fairly good for a large range of saprophytic or species non-pathogenic bacteria which seemingly can establish relations of internal commensalism. The phenomenon has been considered by Starr and Chatterjee (48) for its possible consequences in relation to the differentiation of new forms of phytopathogenic *Enterobacteriaceae* and the possible survival and transport in plants of *Enterobacteria* virulent to man and animals.

Inner plant tissues can also be entered by saprophytic and species-non-pathogenic fungi. Penetration can start from the stigma to which spores can easily stick or from other floral parts. Root hairs and root cortical layers of non-susceptible plants seem to be colonized largely by lower fungi (*Plasmodiophora, Olpidium*) (35). The xylem vessels, considerably less reactive than other tissues, constitute a biological niche particularly suitable for the survival of avirulent fungi. Important agents of vascular diseases such as various *formae speciales* of *Fusarium oxysporum* can be frequently found in the xylem of wild or cultivated plants quite different from their usual hosts. The ability of tracheiphilous pathogens to maintain themselves in non-suscept plants, besides providing a means of survival and dispersal might also favour selection of new pathological forms (35).

Artificial inoculation

The traces left by microorganisms on or in non-susceptible plants are not easily found in nature. Many of the difficulties arising during this kind of investigation are partly overcome by artificial inoculation. The use of heavy artificial inocula can be fruitful and to my knowledge is the only way to rapid progress in this field of research.

Probably the first systematic inoculation experiments with pathogenic fungi in plants not natural hosts date back to de Bary (1863) who made observations on the stomatal penetration by rust and downy mildew fungi (10). Similar experiments, subsequently extended by Gibson (1904)(14), Sempio and Caporali (1957)(43) and others, revealed that the development of non-pathogenic biotrophs considerably changes from plant to plant, and established that no definite correlation existed between the extent of growth and the systematic closeness of the plants. This was particularly evident for the bean rust fungus the growth of which was strikingly more pronounced in tomato than in alfalfa or red clover, and for *Puccinia recondita*, a parasite of wheat, which developed more on bean than on oat (43).

The ability to invade wrong host plants is much more pronounced for necrotrophic or bionecrotrophic parasites. *Phytophthora infestans*, after artificial inoculation, not only invades tissues of different organs of cabbage, bean and even monocotyledonous plants but may also sporulate profusely on them (37).

GENERAL MECHANISMS OF RESISTANCE

There is evidently a high degree of heterogeneity in the type and intensity of relations between plants and the microorganisms in their environment, and this presumably implies that many different resistance mechanisms are involved in the defence of the plant. They will be considered as follows, 1. preclusion of contact 2. alteration of surface relations 3. inhibition of penetration and colonization 4. tolerance.

Preclusion of contact

We do not have a precise picture of the number and type of microorganisms which are commonly in contact with the surfaces of living plants. Because bacteria and propagules of fungi are easily, casually and passively brought by external forces on to aerial parts of plants, the possibility that deposition may be influenced by the nature of plant surfaces has been largely disregarded. Little is known about the physical factors that attract or repulse microorganisms, for example electrical charges on leaves which could directly interfere with establishment of contact. Indirect repulsion can be caused by epicuticular layers of wax which reject water films containing microbial propagules. Failure of *Plasmopara viticola* to infect *Vitis lincecumii* is attributed to the abundant, water repellent waxy cover of the lower leaf surface (6). Factors of this kind may act generally in reducing microbial deposits on plants.

Nutritional and toxic factors can be involved in the attraction and repulsion respectively at underground plant surfaces of motile cells or of hyphae. *Phytophthora* spp. zoospores appear to react positively to nutritional as well as to electrical stimuli but it is now recognized that with few exceptions they are equally attracted towards the root tips of host and non-host plants. The involvement of this kind of stimulus in non-host resistance ought to be reassessed by comparing the behaviour of pathogenic and non-pathogenic fungi.

Animal parasites and predators, being much more active than fungi and bacteria in the search of suitable food, can more commonly and strongly be conditioned by presence and lack in plants of physical or chemical attractants or repellents. Lack of attractiveness might also be a resistance factor against pathogens transported by such animals. The spread of viruses can be prevented in plants resistant to vectors (25), and vector specificity may account for the restricted host range in nature of some viruses with a larger number of experimental hosts. It can similarly be inferred that lack of attraction for insects and nematodes may be a factor of resistance against fungi and bacteria which have these agents as vectors.

Alteration of surface relations

Although the phylloplane and rhizosphere represent well suited nutritional environments for many bacteria and fungi, competition is such that most are excluded in favour of a few which are firmly associated with the surface of a given plant. Most phylloplane bacteria are gram negative, non-sporing rods of types similar to most phytopathogenic bacteria (34). Some yeasts and a few genera of hyphomycetes such as *Cladosporium*, *Sporobolomyces* and *Pullularia* are stable residents, prevailing over other fungi in the phylloplane of different plants. However, marked differences in the number and type of microbial residents may exist among different plants; the physical nature of host surfaces and the chemical nature of exudates may regulate the composition of the microflora (46).

A specialized microflora may be a first antagonistic barrier for potentially infectious agents casually in contact with plant surfaces. Well adapted resident epiphytes may act against them by producing antibiotics or by competing for nutrients. Specific differences in the composition of the stable microflora may then indirectly affect the survival and development of potential invaders.

Surface relations in plants can be influenced by the canopy, epidermal structures and nature of the exudates.

The ecological microsystems of plants depend on their canopy and habit but little attention has been paid to the role of these factors in varietal resistance. Differences among varieties are rarely so great as to become factors in high resistance but could account for economically important reduction of epidemics. Let us recall the different severity of grey mould on grape cultivars with loose and compact clusters of fruit. Differences in the canopy are certainly greater between unrelated plant taxa and could be a decisive role in the exclusion of some potential pathogens.

The physical nature of the cuticle and epidermal hairiness influence the extent and duration of leaf wettness, and thus indirectly affect spore germination. For example, the fact that germination of spores of the cowpea rust fungus exceeded 70% in all hosts and in various non-hosts but was much less on tomato, cabbage and pea was related to uneven humidity at their leaf surfaces (18).

Spore germination and superficial growth of microorganisms can also be directly affected by the composition of exudates.

Inhibition can occur both because of the absence of stimulating and the presence of inhibitory substances. Toxic factors can be counterbalanced by nutritional factors and vice versa. It has been recently pointed out (41) that leaves with a thick and/or impermeable cuticle are less subject to loss of substances with chemotactic effects on motile spores or which stimulate germination, superficial growth and formation of appressoria.

The spores of many fungi may germinate on inert surfaces in the absence of exogenous nutrients so it is not surprising that spores also commonly germinate on the cuticular surfaces of non-host plants. On the other hand lack of spore germination on non-host plants has also been repeatedly reported, and this can be reasonably attributed to inhibitors in exudates. Substances in exudates that stimulate or decrease germination and growth have been proposed but subsequently repeatedly discounted as factors of varietal resistance. A reassessment of their role in the less specific, non-host resistance is highly desirable.

The absence of attractants for and the presence of factors that repulse motile spores and germinating hyphae might explain in some cases failure to penetrate. There is, in fact, scanty evidence for repulsion factors from stomata but observations on the behaviour of *Uromyces phaseoli* on different non-host plants revealed in some cases the arrest of germ tube growth close to stomata (43).

In contrast it is well known that zoospores of *Oomycetes* such as *Plasmopara viticola* can be non-specifically attracted towards the stomata of host as well as of non-host plants. Stomatal penetration by obligate as well as by facultative parasites of a wide range of plants unrelated to their hosts, including *Bryophytes* and *Pteridophytes*, has been shown to occur. The meaning of these findings in relation to non-host resistance must not be overlooked. They have to be considered not only qualitatively but quantitatively. Stomatal penetration in non-host plants might occur at much lower rates than on hosts and this may be quite different for different plant taxa.

The most numerous and complete investigations on non-host surface relations, namely germination and stomatal penetration, have been carried out with uredospores of rust fungi. Germination, superficial growth of the germination hypha towards a stoma, formation of an appressorium on the stoma, growth of an infection peg through the stoma and of a vesicle in the substomatal chamber take place in that order. There may be considerable intercellular development especially in the spongy parenchyma before production of haustoria.

Heath (41) has calculated that the probability that a germination hypha will meet a stoma by chance would be quite low. It is believed that growth of germination hyphae perpendicular to the parallel rows of epidermal cells and to their cuticular "furrows" enhances rapid finding of stomata especially when they are arranged along such rows. It has been suggested that surface contact stimuli regulate such growth. A thigmotropically directed growth has been detected on both host and non-host plants and this also allows stomatal contact for a high proportion of hyphae on the latter. On some non-host plants, however, directional growth is clearly suppressed and this, especially with a deficiency of nutrients, could account for failure of penetration.

Close adherence is necessary for growth at right angles to cuticular furrows. The chemical nature of the cuticle surface would prevent adherence of *Uromyces phaseoli* germ tubes on wheat leaves and thigmotrophically directed growth (53).

Formation of appressoria is also controlled in the same fungi by contact stimuli: they are produced at the edge of craters and scratches of rough artificial membranes but not on smooth membranes (53). The reason for the failure of *Uromyces phaseoli* to form appressoria on wheat and oat leaves seems to depend on the small size of the stomatic ridges. Little is known about the variability of this kind of structures within varieties. Presumably it is not such as to influence directional growth or appressoria formation. Certainly such characters differ more markedly among different plant taxa and they may then become relevant defence factors in non-host plants mainly towards specialized parasites.

INHIBITION OF PENETRATION AND COLONIZATION

Constitutive resistance factors

In stomatal penetration a parasite avoids the barrier of the cuticle. The physico-chemical properties of the cuticle can have a defensive function of primary importance when direct penetration occurs.

Hashioka's (17) investigation on penetration by a powdery mildew fungus of different non-hosts is still outstanding for the number and heterogeneity of the plants tested. He divided 22 non-hosts of *Sphaerotheca fuliginea* into five groups according to degree of resistance. The most resistant group included *Ficus elastica*, *Hoya carnosa* and *Vanilla somai* on which conidial germination occurred but no penetration followed. In these plants which are not known to be attacked by any powdery mildew, thickness of the cuticle was considered as the main resistance factor. This conclusion may often be valid but there are many exceptions.

Hashioka also observed other defence mechanisms in other groups, and that physical inaccessibility of the epidermal cells cannot be considered as the only resistance factors when there is no penetration or when penetration is accompanied by a reaction of the underlying cells.

In this as in many other cases, failure to penetrate may be due to inhibitors, pre-infectional or post-infectional. Which operates is not always easily recognized, especially during the first phases of interactions.

The presence of preformed antimicrobial substances in plant tissues independently of infection is well demonstrated and common. Such substances when possibly important in pharmacology have been referred to as antibiotics or phytoncides. In plant pathology they sometimes are referred to as prohibitins.

The general importance of prohibitins in varietal resistance has not been experimentally substantiated. They could be more important as factors in non-host resistance. After having tested the fungitoxicity of 62 alkaloids of plant species from 15 families differing in resistance to *Phymatotrichum omnivorum*, Greathouse and Rigler (16) concluded that the nature of root alkaloids may be important in resistance to this pathogen.

At least two saponins (tomatine and avenacin) and one unsaturated lactone are examples of prohibitins involved in non-host resistance. Tomatine from tomato is inhibitory to many fungi but is generally more tolerated by specific pathogens of tomato plants such as *Septoria lycopersici* which is resistant towards concentrations of the alkaloid which completely inhibit *S. lactucae*, *S. linicola* and *S. glycines* avirulent on tomato (2). It is believed that avenacin confers resistance on oat to the variety of *Gallmannomyces graminis* specialized on wheat but not to the variety *avenae* active against oats. In both the cases the ability of the virulent forms to develop in their hosts seems to depend on enzymes, absent in the avirulent forms, which degrade the prohibitin (51). The importance of saponins in general resistance is suggested by their frequency in nature and by the high specificity of the enzymes that pathogens must possess to detoxify them (40).

In spite of its wide host range, *Botrytis cinerea* normally is not a parasite of tulip which is attacked by the specialized *B. tulipae*. Resistance towards *B. cinerea* seems to be strictly related to tuliposide, an unsaturated lactone. By leaching this compound tulip tissues are made susceptible to *B. cinerea*. Moreover *B. cinerea* is more sensitive to the tuliposide than is *B. tulipae*, and can induce cells to release more of this substance (41).

Most examples involve a single or few prohibitins per plant but a plant species such as potato with its phenolics, phenolic derivatives, terpenoids and steroid glycoalkaloids (27) may contain more than one prohibitin as a chemical barrier complex which precludes many potential invaders.

Among preformed resistance factors can also be considered substances of cell walls, or those localized in the cytoplasm such as concanavalin A and phytohemagglutinins, which may bind potential invaders, and so prevent their movement and growth.

Substances in plants which agglutinate or precipitate extraneous microbodies have been investigated by Carbone and Arnaudi (8). Such substances were then regarded as pseudoantibody-like factors (phytohemagglutinins, pseudobacterioagglutinins and pseudoprecipitins). Their significance was obscured by preconceptions derived from theories of humoral immunity in animals. A different interpretation of their role has been recently elaborated from evidence that they could be implicated in plant-parasite interactions. Lectins or other macromolecules of the plant wall surface which can react with receptors in the pathogen cell wall, are now considered as possible recognition factors both in compatibility and incompatibility (44). It is known that agglutinins can recognize and bind superficial carbohydrates of bacteria and fungi, but there is little evidence on their possible occurrence as inhibitors of fungi in plants.

In contrast, there is some circumstantial evidence that agglutinins inhibit saprophytic, non-pathogenic and avirulent bacteria. *Serratia marcescens* and *Pseudomonas putida* introduced into cowpea leaves were later recovered at a much lower rate than was the pathogen *P. syringae*. The reduced recovery was caused by immobilization attributed to differences in the electrical charge on the three organisms (23). Sing and Schroth (45) observed later that immobilization of *P. putida* in bean leaves was accompanied by their attachment to the cell wall and encapsulation by fibrillar materia. *P. phaseolicola* and *P. tomato*, two pathogens respectively virulent and avirulent on bean were not immobilized. However crude water extracts of bean leaf tissues agglutinated the avirulent pathogen *P. tabaci* (1). This is more in agreement with a previous finding that the pathogen *P. pisi* is immobilized in non-host plants (15). On the other hand binding to the cell wall through plant lectins has been demonstrated for avirulent strains of *P. solanacearum* (44) and *Erwinia amylovora* (15). So it is evident that agglutination *in vitro* and encapsulation *in vivo* of bacteria is possible for saprophytes, heterologous pathogens and avirulent strains of pathogens.

In general agglutination in different plants seems to occur

more commonly for saprophytes than for pathogens. Where this confirmed, it could be inferred that evolution in bacteria towards parasitism depends on acquisition of factors or mechanisms which prevent agglutination by the plant cell wall. More comparative experiments are needed, bearing in mind that results may be contradictory because recognition sites on bacterial walls are transient and depend on the age and condition of cultures (1).

It has been hypothesized that the degree of compatibility between plants and their potential invaders depends on numbers of common antigens. The greater the number, the greater the symbiotic affinity of the host and invader (11). Proteins are not themselves inhibitory to the pathogen; they condition the degree and the quality of the reactions that take place after infection. This hypothesis has been developed by comparing antigens of different varieties of plants with those of virulent or avirulent strains of pathogens. The few available examples suggest that it could also be extended to non-host resistance.

Two species non-pathogenic fungi such as *Fusarium moniliforme* and *Verticillium nigrescens* had no antigen in common with the non-host cotton whereas cotton antisera and antigens of four virulent pathogens showed a strong and distinct precipitin band (9). On the other hand, *F. oxysporum* f. sp.*vasinfectum*, specific for cotton, has a great antigenic affinity with this plant and does not share antigens with non-host plants (Venkaturam *et al.* in (11)). The fact that a high percentage of fungi from the rhizosphere of wheat possess in their cell walls saline soluble antigens that cross react with antisera of the wheat pathogen *Gaümannomyces graminis*, whereas non-rhizosphere fungi do not, has been considered as evidence that antigens are possible determinants of wheat root-fungi associations (Holland and Choo in (11)).

The significance of this hypothesis in explaining non-host resistance and for a general picture of resistance depends on further evidence on relations between biological and antigenic affinity (or disaffinity) for many more plants and microorganisms within gradients of increasing compatibility.

Adaptive resistance factors

There is plenty of evidence that non-host resistance is also often associated with post-infectional morphological reactions which are easily detected during infection.

The critical reactions, although not easily definable, can occur early in infection. In oat and wheat, for instance, *Uromyces phaseoli* does not penetrate or establish trophic relations. This suggests preformed inhibitors, but consistently higher respiratory activity of inoculated leaves indicates that the plant is reacting actively (42).

Morphological responses such as thickening of the cell wall, papillae, haloes, and darkening have been often reported for obligate and facultative parasites on non-host plants, during attempts to pierce the cuticle and/or the cell wall.

Thickening in the cellulosic portion of the mesophyll cell wall of corn is induced by *Puccinia graminis tritici*, but its effectiveness as a defence mechanism is dubious because the fungus stops growing even in narcotized tissues where thickenings are absent (32).

Papillae are a common reaction and were considered in the past as barriers to fungal penetration. In wheat infected by *Erysiphe graminis* they follow the appearance of cytoplasmic aggregates and their development generally coincides with formation of haustoria (7). Papilla formation is a non-specific response and does not relate to vertical i.e. gene-to-gene specificity, but it could have a role in non-host resistance. It has been observed that 80 - 90% of appressoria of *E. graminis tritici* induce papillae in barley but do not form haustoria (Bushnell and Berquist, 1975 in (7)). Failure to penetrate unwounded wheat leaves by *Botrytis cinerea*, *B. allii*, *B. fabae*, *Septoria apiicola* and *Alternaria tenuis* is related to papilla production beneath appressoria (38). Whether a papilla prevents fungus growth or is produced after the fungus stops growing is still undetermined.

Papillae can be accompanied by alteration of upper epidermal walls, revealed as haloes, and of adjacent lateral walls. Anilin blue staining haloes are produced in epidermal cells around penetration pegs in different non-host plants. Similar haloes can also be observed beneath ungerminated conidia of virulent and avirulent isolates of *Verticillium dahliae* and *Fusarium oxysporum* infiltrated into the spongy tissue of tomato leaves (Matta, unpublished results).

Lack of formation of haustoria in non-hosts is associated with a dark alteration of the cell wall proximal to the first haustorial mother cell of *Uromyces phaseoli* var.*vignae*. Similar darkenings are induced by leachates of germinating hyphae but are also evident next to dead host cells in susceptible cowpea-rust combinations (21).

Papillae, haloes and other wall alterations generally delimit regions of wall impregnated by different materials. Callose deposition is common but other substances are often present. Papillae and haloes in wheat are sites of lignification (38). A few hours after their appearance they may also become sites of silicon, calcium and manganese deposition. Calcium accumulates around penetration pores, manganese and silicon in the central and outer ring zones (29). Silicon on deposits seems to be particularly important in wheat leaves inoculated with *E. graminis* f. sp. *hordei*, and in barley inoculated with the same fungus or with non-pathogenic fungi of the genera *Alternaria*, *Colletotrichum* and *Cochliobolus*.

The darkening of parts of cell walls is also characterized by deposition of electron-opaque materials in small part of phenolics and in greater part of silicon. Opal deposition as well as darkening are not specific reactions. They occur not only in susceptible combinations (21, 29), but are also incited by mechanical damage of epidermal cells. Moreover they are clearly secondary reactions as they occur three to four hours later than reagent coloured haloes (29). The fact that treatments which increase the frequency of cowpea rust haustoria in bean decrease at the same time the extent of silicon deposition indicates however its possible role in non-host resistance (22).

Ride and Pearce (38) have shown that incrustation by lignin and other materials renders the papilla and halo region resistant to attack by fungal enzymes and they inferred that lignin might also provide resistance to mechanical penetration and restrict diffusion of nutrients to and of toxins from the pathogen. Accumulation of phenolic precursors might also inhibit growth of the pathogen.

Darkening and cell collapse induced by non-pathogenic microorganisms can involve a few or many cells and become evident even to the naked eye. Bean rust uredospores in tomato give rise to infecting hyphae which reach the palisade cells from the adaxial surface, and to haustoria which are soon followed by browning sufficient to obscure the hypha itself (43). Normally haustoria either do not develop remaining atrophic and darkening, or develop and stay apparently normal even after cell disorganization (19).

Browning and collapse of cells and more complex changes such as pin point browning and large discoloured areas are also induced in non-host plants by bionecrotrophic or necrotrophic parasites such as *Phytophthora infestans* on *Brassica*, *Phaseolus*, *Lactuca* or *Dahlia* leaves (37), *Helminthosporium carbonum* on apple leaves (5), *Pyricularia oryzae* on tomato (Yoshii, 1948 in (35)). Browning of the xylem accompanied by the release of gums or tyloses in the vessel lumen is induced in different plants in greater amounts by non-pathogenic than by pathogenic forms of *Fusarium oxysporum*.

No doubt the morphological reactions of non-hosts are associated with a variety of biochemical events among which phytoalexin formation could be of particular importance in relation to resistance.

The potential ability to produce phytoalexins is expressed to different extents against different invaders. There is strong evidence that fungal and bacterial pathogens on non-hosts as well as saprophytes induce phytoalexin formation in different plants at concentrations high enough to prevent their growth *in vitro*. Thus, "treatment which increase susceptibility of the non-host tissue also result in reduced level of phytoalexin accumulation" (20). Pisatin, phaseollin, and ipomeamarone share the same general tendency to be more toxic to saprophytic and avirulent than to virulent fungi (52). *Helminthosporium carbonum*, a pathogen of corn, is an efficient inducer of phytoalexins in bean (28), and in red clover. *H. turcicum* is unable to detoxify medicarpin and maakiain (12). Even the resistance of non-hosts towards nematodes can be associated with the production of inhibitory quantities of phytoalexin (26).

Similar results suggest at first that non-host resistance is related to the magnitude of accumulation, to sensitivity of the invader and to its ability to detoxify the inhibitors. A more careful examination of the literature shows that this is not always true. Many non-pathogens are poor inducers of phytoalexins or metabolise them at the same or higher rates than do pathogens (3, 12). Studies of phytoalexins in relation to non-host resistance confirm the more general rule that phytoalexins alone cannot "determine whether an infectious agent will parasitize or not" (28).

Phenolic compounds lacking specific toxicity and of generalized occurrence as a pool of antimicrobial factors, accumulate at particularly high rates after inoculation with non-pathogenic fungi. Phenol metabolism is more rapidly activated by non-pathogens than by pathogens in potato tubers and in tomato vascular tissues (28, 36).

Interactions between microorganisms and non-host plants is also indicated by different physiological reactions such as increased respiration and increased peroxidase and/or polyphenoloxidase activity (see 35). These in turn may converge on and precede necrosis of cells which appears to be the final rather than the determinative phase of defence processes in various non-host species.

The prompt death of cells of invaded tissues clearly may prevent further spread of biotrophic parasites in non-hosts but this is much less likely to be a resistance factor against necrotrophic invaders.

Hypersensitive confluent necrosis (HR) is induced by large inocula of pathogenic leaf spotting bacteria in non-host plant tissues. Varietal resistance seems in this respect in many cases to coincide with resistance of non-host species against avirulent pathogens. In contrast, saprophytic bacteria injected into leaves do not multiply or induce hypersensitive reactions. Ability to induce HR seems to depend on at least some growth of bacteria in the tissue. Lack of growth by saprophytic bacteria has been considered a consequence of their inability to induce release of nutrients by the plant cells (55). It seems reasonable to assume that saprophytic bacteria are not adapted to conditions within plant tissues (39) but the ability of saprophytic bacteria to grow in "extracellular fluids" does not support this explanation.

On the other hand the presence of saprophytic bacteria in plant tissues is not entirely passive because some of them can induce protection against subsequent or simultaneous inoculation with virulent bacteria.

TOLERANCE

Mechanisms of resistance to non-pathogenic microorganisms which prevent or actively localize infection, and the resulting damage have now been considered. Plants showing such mechanisms are non-hosts or uncongenial hosts for potential invaders. But there are plants which allow considerable infection and colonization without showing symptoms. This is a form of "internal" commensalism in which there is resistance to the injury but not to the invader. Plants are typically "non-suscept hosts" and show a high degree of tolerance to the invader.

In its strictest sense tolerance implies unrestricted and extensive colonization by a parasitic microorganism or virus without symptom expression. Unrestricted growth is rare in plants and an effective distinction between tolerance and resistance would be difficult even in extreme cases. Moreover it is unlikely that growth in a plant of an extraneous organism ever occurs with no reaction by the plant.

The absence of visible internal or external symptoms, or of dramatic biochemical changes in reaction to development of an invader is at the moment the only criterion for classifying a plant as tolerant. Fleshy plant organs are tolerant when they harbour saprophytic bacteria, or when tobacco leaves contain fungi of the genera *Alternaria*, *Penicillium*, *Aspergillus* and *Cladosporium* which are normally found in their parenchyma (47). A singular example of tolerance is offered by plants of the *Asclepiadaceae* that are commonly invaded by non-pathogenic latex inhabiting trypanosomatid protozoa (*Phytomonas elmassiani*) (50).

Saprophytes more than the species non-pathogens seem to be capable of growth without induction of defence reactions. Thus, with few exceptions, saprophytes show little or no ability to induce resistance against subsequent invaders. This implies that saprophytes are less well supplied than are pathogens with elicitors of resistant reactions. On the other hand, elicitors of phytoalexins have been detected in saprophytic as well as parasitic fungi (see 49). It has to be shown, however, that such elicitors occur *in vivo*.

CONCLUSIONS

Plurality of non-host defence

Factors and mechanisms of non-host resistance appear to be quite variable and complex. It has been stressed (20) that no single mechanism can account for such resistance in all plants. For example, penetration by *Sphaerotheca fuliginea* of a group of various plants can be inhibited by thickness of the cuticle, the presence of extracellular inhibitors or by hypersensitive collapse of the cells (17). Johnson *et al.* (24) similarly observed that growth of *Erysiphe graminis* f. sp. *hordei* and *E. cichoracearum* on plants of taxonomically distant families was suppressed at each of several successive stages. The fact that germination of spores of both fungi was almost completely inhibited on *Polypodiaceae*, *Equisetaceae* and monocotyledonous plants whereas on other non-hosts inhibition was expressed only at the level of germination pegs or appressoria, is substantial evidence that each group of plants has its own mechanisms of defence. Cessation of fungal growth in rust fungi,depending on the host,may occur during or after germination, before or after the formation of the haustorial mother cell, or at formation of the appressoria (18).

On the other hand more than one mechanism can operate in a single plant against the same pathogen. Significant here is the simultaneous appearance of morphological or chemical barriers (32, 33) or the simultaneous accumulation of different chemically related and unrelated inhibitory compounds that may occur in plants inoculated with non-pathogens.

Differences in defences against saprophytes and species non-pathogens

Is resistance of a plant to saprophytes due to factors different from those acting against parasites of other plants ? The physical and chemical nature of the plant surface, lack of attraction stimuli, plant canopy, chemical inhibitors in exudates, cuticle or wall thickness and internal chemical inhibitors have been considered as general resistance factors. They could determine whether or not a plant species is a host for another organism and are particularly effective in rejecting saprophytes. Barriers which are seemingly unimportant in resistance to pathogenic microorganisms might be more significant against saprophytes. It is possible that the overcoming of such barriers is a preliminary step in the evolution of microorganisms towards more or less ephemeral forms of parasitic life. Co-evolution of the microorganism and the plant towards better parasitic fitness and more effective resistance mechanisms starts later as a result of continuous interactions.

Looking at things from an opposite point of view, it seems that the success or failure of any defence system depends on the strength of the aggression it meets. Failure to infect not only depends on resistance factors but also on the inability of the microorganism to overcome such factors, and most microorganisms are devoid of tools for this purpose. The concept that resistance is common and susceptibility exceptional would be inverted in absence of pathogenicity is common and pathogenicity exceptional. Independently of host properties, saprophytes may be structurally unable to colonize a plant, lacking appressoria, infection pegs, toxins or enzymes. The inability to induce the release of water and nutrients from plant cells could account for the failure of saprophytic bacteria to develop in plants (55). Also significant is the assumption of parasitic habit by saprophytes exogenously supplied with a host specific toxin (see 4). Evidence on these points is meagre and not sufficient to separate clearly non-pathogens and pathogens.

Differences in defence against pathogens in non-host and in resistant cultivars

Resistance against pathogens in non-hosts can also be related to pre-infectional chemical or morphological barriers, but most commonly involves active responses of the plant (21, 22) which are in some cases profitably exploited in "cross protection" experiments. Whether or not such responses differ from those induced on resistant host varieties mainly depends on the host and non-host which are compared.

Growth of rust fungi on non-host plants generally does not go beyond the formation of the haustorial mother cells (31), but there are cases in which haustoria are formed (18). Varietal resistance to rust fungi is generally expressed much later but can be in some cases coincident with non-host resistance (18).

Generally, growth of the pathogen is arrested earlier in non-hosts than in resistant hosts, but not always. So a clear distinction between varietal and non-host reactions is not possible. Differences are sometimes more easily expressed in quantitative rather than in qualitative terms. Systems of successive adjustments probably operate in the plant which relate to different degrees of aggression by the invader.

On the other hand it must be remembered that we use arbitrary criteria for the classification of the plants as hosts and non-hosts. In a previous NATO Institute (1976) Snyder commented that the degree of recognized specificity depends on whether a "splitter's" or a "lumper's" concept of species is followed for the plant and for the pathogen. Moreover agricultural plants are very often a complex of genes from different species. With this perspective it does not seem entirely reasonable to consider a resistant variety as more a potential host of a pathogen than any one of the resistant species in its ancestry.

Transformation of non-host into host plants

It has been seen that a pathogen of a given plant under experimental conditions may still induce reactions of incompatibility in a wide array of taxonomically unrelated plants. The question is whether this is still possible in natural conditions, or do other factors of the pre-infectional type prevent interactions with non-hosts. It can be inferred that the artificial inoculations used in research do not reveal all the resistances in a plant. Thus, inoculation by large inocula creates conditions quite different from those in nature, allowing overcoming of resistance factors that would be effective against isolated spores. Again the quantitative aspects of the problem are evident.

Methods of artificial inoculation may not only obscure certain defence factors, for instance those restricting fungal or bacterial growth very early in infection, but they may make a non-host susceptible. A point in case was made when Müller (37) showed that *Phytophthora infestans*, a pathogen confined in nature to a few solanaceous plants, under experimental conditions was able to infect plants of the *Angiospermae* division "regardless of their phylogenetic position", with final results which ranged from hypersensitive reactions similar to that exhibited by *S. demissum*, to complete susceptibility ("lack of reaction and luxuriant fructification"). Many years earlier Young (54), who claimed artificial production of 198 "new diseases" by artificial inoculation of 78 species and varieties of flowering plants with various fungi, was well aware of the fact that under the conditions of the experiments, far different from those in the field, the plants were physiologically "different".

The existence in nature of innocuous contaminants which might become opportunistic pathogens in particular environmental conditions (e.g. *Clostridium* spp. on fleshy hypogeous organs) emphasizes that the identification of the host or non-host character of a given plant species is not easy and not always possible. It is a truism to say that here, as ever, *natura non facit saltus*.

REFERENCES

1. ANDERSON, A.J. & JASALAVICH, C. (1979). Agglutination of

Pseudomonad cells by plant products. *Physiol. Pl. Path.* 15, 149 - 159.

2. ARNESON, P.A. & DURBIN, R.D. (1968). The sensitivity of fungi to α-tomatine. *Phytopathology* 58, 836 - 837.

3. BAILEY, J.A., BURDEN, R.S., MYNETT, A. & BROWN, C. (1977). Metabolism of phaseollin by *Septoria nodorum* and other non-pathogens of *Phaseolus vulgaris*. *Phytochemistry* 16, 1541 - 1544.

4. BATEMAN, D.F. (1976). (General discussion on fungal elicitors). In : *Cell wall biochemistry related to specificity in host-plant pathogen interactions*. Eds. Solheim, B. & Raa, J. p. 159, Universitesforlaget, Tromsø-Oslo-Bergen.

5. BIEHN, W.L., KUĆ, J. & WILLIAMS, E.B. (1967). Accumulation of phenols in soybean after inoculation with fungi non-pathogenic to soybean. *Phytopathology* 57, 804.

6. BOUBALS, D. (1959). Contribution à l'étude des causes de la résistance des vitacées au mildiou de la vigne et de leur mode de transmission héréditaire. *Ann. Amélior.Pl.*(1959) I, 5 - 233.

7. BUSHNELL, W.R. (1976). Reactions of cytoplasm and organelles in relation to host-parasite specificity. In : *Specificity in plant diseases*. Ed. R.K.S. Wood & A. Graniti. 131 - 145. Plenum Press, New York and London.

8. CARBONE, D. & ARNAUDI, C. (1930). L'immunità nelle piante. *Monografia Istituto Sieroterapico Milanese*, Milano.

9. CHARUDATTAN, R. & DEVAY, J.E. (1972). Common antigens among varieties of *Gossypium hirsutum* and isolates of *Fusarium* and *Verticillium* species. *Phytopathology* 62, 230 - 234.

10. DE BARY, M.A. (1863). Recherches sur le dévelopment de quelques champignons parasites. *Ann. Sci. nat. bot.*, ser. IV, 20, 1 - 148.

11. DEVAY, J.E. (1976). Protein specificity in plant disease development : protein sharing between host and parasite. See 7, 199 - 212.

12. DUCZEK, L.J. & HIGGINS, V.J. (1976). The role of medicarpin and maackiain in the response of red clover leaves to *Helminthosporium turcicum*, *Stemphylium botryosum* and *S. sarcinaeforme*. *Can. J. Bot.* 54, 2609 - 2619.

13. ERCOLANI, G.L. & CASOLARI, A. (1966). Ricerche di microflora in pomodori sani. *Ind. Conserve* 1, 15 - 22.

14. GIBSON, C.M. (1904). Notes on infection experiments with various Uredineae. *New Phytol.* 3, 184 - 191.

15. GOODMAN, R.N., PI-YU HUANG, HUANG, J.S. & THAIPANICH, V. (1976). Induced resistance to bacterial infection. In : *Biochemistry and cytology of plant-parasite interaction.* Ed. K. Tomiyama, J.M. Daly, I. Uritani, H. Oku & S. Ouchi. 35 - 42. Kadanshu Ltd., Tokyo.

16. GREATHOUSE, G.A. & RIGLER, N.E. (1940). The chemistry of resistance of plants to *Phymatotrichum omnivorum.* V. Influence of alkaloids on growth of fungi. *Phytopathology* 30, 475 - 485.

17. HASHIOKA, Y. (1938). The mode of infection by *Sphaerotheca fuliginea* in susceptible, resistant and immune plants. *Trans. nat. Hist. Soc. Formosa* 28, 47 - 60. In : *Review appl. Mycol.* 17, 579.

18. HEATH, M.C. (1974). Light and electron microscope studies of the interactions of host and non-host plants with cowpea rust *Uromyces phaseoli* var. *vignae. Physiol. Pl. Path.* 4, 403 - 414.

19. HEATH, M.C. (1976). Ultrastructural responses of host and non-hosts to rust fungi. see 7, 147 - 148.

20. HEATH, M.C. (1979). Nonhost immunity. In : *Plant disease control : resistance and susceptibility.* Ed. C. Koehler, R.C. Staples & G.H. Toennissen. Wiley Interscience N.Y. in press.

21. HEATH, M.C. (1979a). Partial characterization of the electron-opaque deposits formed in the non-host plant, French bean, after cowpea rust infection. *Physiol. Pl. Path.* 15, 141 - 148.

22. HEATH, M.C. (1979b). Effects of rootstock, actinomycin D, cycloheximide and blasticidin S on non-host interactions with rust fungi. *Physiol. Pl. Path.* 15, 211 - 218.

23. HILDEBRAND, D.C. & SCHROTH, M.N. (1968). Removal of Pseudomonads from plant leaves and measurement of their *in vivo* β-glucosidase synthesis. *Phytopathology* 58, 354.

24. JOHNSON, L.E.B., ZEYEN, R.J. & BUSHNELL, W.R. (1978). Defence patterns in inappropriate higher plant species to two

powdery mildew fungi. *Phytopath. News* 12, 89.

25. JONES, A.T. (1979). Further studies on the effect of resistance to *Amphorophora idaei* in raspberry on the spread of aphid-borne viruses. *Ann. appl. Biol.* 92, 119 - 123.

26. KEEN, N.T. & RICH, J. (1976). Phytoalexins against nematode pathogens. See 7, 270.

27. KUĆ, J. (1973). Metabolites accumulating in potato tubers following infection and stress. *Teratology* 8, 333 - 338.

28. KUĆ, J. (1976). Phytoalexins and the specificity of plant-parasite interaction. See 7, 253 - 268.

29. KUNOH, H. & ISHIZAKI, H. (1976). Silicon accumulation of "halo" areas of barley leaf induced by powdery mildew infection. See 15, 56 - 65.

30. LANGE, R.T. (1966). Bacterial symbiosis with plants. In : *Symbiosis*. Ed. S.M. Henry, Vol. 1, 99 - 170. Academic Press, New York, London.

31. LEATH, K.T. & ROWELL, J.B. (1966). Histological study of the resistance of *Zea mays* to *Puccinia graminis*. *Phytopathology* 56, 1305 - 1309.

32. LEATH, K.Y. & ROWELL, J.B. (1969). Thickening of corn mesophyll cell walls in response to invasion by *Puccinia graminis*. *Phytopathology* 59, 1654 - 1656.

33. LEATH, K.T. & ROWELL, J.B. (1970). Nutritional and inhibitory factors in the resistance of *Zea mays* to *Puccinia graminis*. *Phytopathology* 60, 1097 - 1100.

34. LEBEN, C. (1965). Epiphytic microorganisms in relation to plant disease. *A. Rev. Phytopath.* 3, 209 - 230.

35. MATTA, A. (1971). Microbial penetration and immunization of uncongenical host plants. *A. Rev. Phytopath.* 9, 387 - 410.

36. MATTA, A., GENTILE, I. & GIAI, I. (1969). Accumulation of phenols in tomato plants infected by different forms of *Fusarium oxysporum*. *Phytopathology* 59, 512 - 513.

37. MÜLLER, K.O. (1950). Affinity and reactivity of angiosperms to *Phytophthora infestans*. *Nature, Lond.* 166, 392 - 395.

38. RIDE, J.P. & PEARCE, R.B. (1979). Lignification and papilla formation at sites of attempted penetration of wheat leaves by non-pathogenic fungi. *Physiol. Pl. Path.* 15, 79 - 92.

39. RUDOLPH, K. (1976). Models of interaction between higher plants and bacteria. See 7, 109 - 126.

40. SCHLÖSSER, E. (1976). Specificity and role of β-glycosidases in saponin dependent host-parasite interactions. See 7, 250.

41. SCHÖNBECK, F. (1976). Role of preformed factors in specificity. See 7, 237 - 250.

42. SEMPIO, C. & BARBIERI, G. (1966). Aumento respiratorio prodotto da *Uromyces appendiculatus* in specie refrattarie. *Phytopath. Z.* 57, 145 - 158.

43. SEMPIO, C. & CAPORALI, L. (1958). L'*Uromyces appendiculatus* su fagiolo e su altre specie : virulenza e specializzazione. *Ann. Fac. Sci. agr. Univ. Perugia* 13, 233 - 277.

44. SEQUEIRA, L. (1978). Lectins and their role in host-pathogen specificity. *A. Rev. Phytopath.* 16, 453 - 481.

45. SING, V.O. & SCHROTH, M.N. (1977). Bacteria-plant surface interactions : active immobilization of saprophytic bacteria in plant leaves. *Science* 197, 759 - 761.

46. SINHA, S. (1971). The microflora on leaves of *Capsicum annuum, Solanum melongena, Solanum tuberosum* and *Lycopersicum esculentum*. In : *Ecology of leaf surface microorganisms*. Ed. T.F. Preece & C.H. Dickinson. 175 - 189. Academic Pres, London, New York.

47. SPURR, H.W. & WELTY, R.E. (1975). Characterization of endophytic fungi in healthy leaves of *Nicotiana* spp. *Phytopathology* 65, 417 - 422.

48. STARR, M.P. & CHATTERJEE, A.K. (1972). The genus *Erwinia* : enterobacteria pathogenic to plants and animals. *A. Rev. Microbiol.* 26, 389 - 426.

49. STEKOLL, M. & WEST, C.A. (1978). Purification and properties of an elicitor of castor bean phytoalexin from culture filtrates of the fungus *Rhizopus stolonifer*. *Pl. Physiol.* 61, 38 - 45.

50. THOMAS, D.L., McCOY, R.E., NORRIS, R.C. & ESPINOZA, A.S. (1979). Electron microscopy of flagellated protozoa associated with "marchitez sorpresiva" disease of african oil palm in Ecuador. *Phytopathology* 69, 222 - 226.

51. TURNER, E.M.C. (1961). An enzymic basis for pathogenic specificity in *Ophiobolus graminis*. *J. Exp. Bot.* 34, 169 - 175.

52. URITANI, I., OLEA, K., KOJIMA, M., KIRN, W.K., OGUNI, I. & SUZUKI, H. (1976). Primary and secondary defence actions of sweet potato in response to infection by *Ceratocystis fimbriata* strains. See 15, 239 - 252.

53. WYNN, W.K. (1976). Appressorium formation over stomates by the bean rust fungus : response to a surface contact stimulus. *Phytopathology* 66, 136 - 146.

54. YOUNG, F.A. (1926). Facultative parasitism and host ranges of fungi. *Am. J. Bot.* 13, 502 - 520.

55. YOUNG, J.M. & PATON, A.M. (1971). Development of pathogenic and saprophytic bacterial populations in plant tissue. In : *Proc. 3rd Int. Conf. Pl. Pathogenic Bacteria*. Ed. H.P. Maas Geesteranus. 77 - 80·PUDOC, Wageningen.

THE ABSENCE OF ACTIVE DEFENSE MECHANISMS IN COMPATIBLE HOST-PATHOGEN INTERACTIONS

MICHÈLE C. HEATH

Botany Department, University of Toronto
Toronto, Ontario M5S 1A1
Canada

INTRODUCTION

By definition, a susceptible plant is one that cannot successfully defend itself against pathogen attack. Such a situation is relatively rare, and any one plant species is commonly susceptible to infection by only a few of the thousands of micro-organisms known to cause disease in higher plants. Nevertheless, susceptibility to only one pathogen is sufficient to cause unacceptable yield losses in a commercial crop, and it is this that plant pathologists are ultimately trying to prevent. Thus it is important to know why defense mechanisms which apparently confer resistance of a plant towards one plant pathogen seem unable to control infection by another.

One possibility, particularly for fungal pathogens (i.e. fungi known to cause disease in at least one species of higher plant), is that the fungus in question is tolerant of, or insensitive to the potential defense mechanisms of its host plant, be they pre-existing or induced by infection. Such may be the case where, for example, some pathogens show unusual tolerance towards high concentrations of the phytoalexins produced by their host species (e.g. 31, 35). However, in many, if not most, compatible host-pathogen interactions, putative defense reactions are absent, or seem to be less vigorous, or take place more slowly than in resistant plants. In these situations, it is possible that the virulent organism, to which the plant is susceptible, either does not trigger these defense reactions to an appreciable extent, or actively suppresses their expression. Few studies have tried to distinguish between these alternatives, and to my knowledge, there are currently only three cases where there is experimentally substantiated evidence that defense reactions are not initiated in susceptible plants.

NON-ACTIVATION OF DEFENSE MECHANISMS

The best documented example of non-activation of defense mechanisms in susceptible plants, described elsewhere in this volume by Sequeira, involves certain gram-negative bacteria. In a susceptible host the extracellular polysaccharides of the bacteria are not "recognized" and the bacteria do not become attached to the plant cell walls; apparently in consequence, no immediate defense reactions of the plant are triggered.

The two other examples of possible non-activation of defense reactions in susceptible plants, described in more detail by Keen in this volume, involve the production by *Colletotrichum lindemuthianum* and *Phytophthora megasperma* var. *sojae* of "specific elicitors" which are less effective in causing phytoalexin accumulation in susceptible than in resistant host cultivars. However, the role of these elicitors, or of the phytoalexins, in pathogenesis is still by no means clear and, particularly for *P. megasperma* var. *sojae* - soybean interactions, is complicated by the possibility that the different levels of phytoalexin accumulation in resistant and susceptible cultivars may be due to differential rates of degradation rather than differences in rates of synthesis (38). Moreover, *P. megasperma* var. *sojae* produces non-specific elicitors of phytoalexin accumulation, active in both resistant and susceptible host cultivars as well as in some non-host species (3), which one might expect to mask the effects of any specific elicitors which are released *in vivo* (34). Also, it is as yet unknown what role "specificity factors" (molecules from an incompatible race of the pathogen which protect soybeans against a compatible race) (32), play in determining cultivar specificity. Thus, the non-activation of phytoalexin accumulation as a determinant of susceptibility remains to be proven.

SUPPRESSION OF DEFENSE MECHANISMS

In contrast to the sparsity of well-documented examples of non-activation of defense reactions in susceptible host plants, many studies suggest that virulent fungi may actively suppress the expression of these reactions during successful infection. Daly (4) has pointed out that the action of host-selective toxins (active only in susceptible host cultivars) may be viewed as interfering with the normal resistance of the host, presumably by eliciting premature cell death. Similarly, the degradation of phytoalexins to non-toxic derivatives may also be regarded as a form of active suppression of resistance, and there are a number of diseases where the host range of a particular fungal pathogen seems related to its ability to metabolize certain phytoalexins (e.g. 20, 33). But there is at least one example where the ability to degrade a plant's known phytoalexins does not make the fungus a successful pathogen (10).

More recently, there have been indications that virulent fungi may use other means to suppress expressions of resistance in their hosts. During germination, *Mycosphaerella pinodes* releases a low molecular weight compound (possibly a peptide) that seems to prevent the accumulation of the phytoalexin, pisatin, in pea leaves, whether accumulation follows inoculation with fungi for which pea is not a host, treatment with ultraviolet light, or application of an elicitor released by *M. pinodes* during germination (23). As yet there is little direct evidence on how *M. pinodes* benefits from this "suppressor", since other studies suggest that this fungus elicits the rapid accumulation of pisatin in susceptible pea leaves (19). Nevertheless, it is possible that the action of the suppressor is crucial only during penetration of the epidermis (23) and that its effect is highly localized. Another pea pathogen, *Erysiphe pisi*, also may suppress pisatin accumulation during initial stages of infection (22, 27). However, as yet there seems to be little direct experimental evidence of such suppression.

Suppressors of putative defense reactions also have been found in exudates from infection hyphae of *Uromyces phaseoli* var. *typica* (the bean rust fungus) (Heath unpublished) and in crude extracts from susceptible, rust-infected, French bean leaves (16). Like the *M. pinodes* suppressor, the active component from infected leaves appears to be of relatively low molecular weight (<5000 daltons). When injected into uninoculated French bean leaves, it reduces the frequency of silicon-containing deposits (14) which normally develop on mesophyll cell walls after infection by the incompatible *U. phaseoli* var. *vignae* (the cowpea rust fungus). These deposits seem to prevent the cowpea rust fungus from forming the first haustorium, and a reduction in their frequency corresponds to increase in the proportion of infection sites at which a haustorium is formed (16, Heath unpublished). A plausible reason why *U. phaseoli* var. *typica* should suppress a reaction triggered by *U. phaseoli* var. *vignae* is that it too elicits but concomitantly suppresses the same response to allow the first-formed haustorium to develop (16). In support of this hypothesis, the normally compatible *U. phaseoli* var. *typica* elicits silicon deposition with increasing frequency in older French bean leaves, in association with an increase in the number of infection sites at which no haustoria form. Significantly, suppression by fungal exudates and extracts from rusted French bean leaves is less effective in older leaves (Heath, unpublished).

Perhaps the best characterized suppressors of host defense reactions are those produced by *Phytophthora infestans*. These are anionic and non-ionic glucans, of about 10 - 20 glucose units, which contain β-1→3 and β-1→6 linkages (6). All races of the fungus seem to produce these compounds which suppress the browning and terpenoid accumulation in potato tubers elicited by infection with incompatible races of the pathogen, or by application of the non-specific elicitor which can be extracted from all races of the fungus. However, each suppressor is most effective in the cultivar susceptible to the fungal race from which the suppressor was prepared (12).

All these reports suggest that defense reactions are suppressed during successful pathogenesis. Moreover, it can be argued that, because several fungi produce compounds which elicit putative defense reactions in their *susceptible* hosts (1, 2, 3, 5, 9, 12, 23, 30, 36), the virulent fungus *must* have some means of counteracting this activity, assuming, of course, that these compounds are released from the fungus during infection. Therefore, it is worth considering how these suppressors may act. One possibility is that each suppressor affects only one specific host reaction, and will therefore increase susceptibility only to those organisms which trigger this same response. Alternatively, the plant may be conditioned towards a general susceptibility which precludes the expression of *any* defense reaction and therefore renders the tissue susceptible to most fungi towards which it is normally resistant. That conditioning to general susceptibility is possible is suggested by the indiscriminate susceptibility of plant callus tissue to plant pathogens under certain environmental conditions (e.g. 13, and Ingram in this volume), regardless of the response of the whole plant. Similarly, abscisic acid treatment of potato tubers increases their susceptibility, not only to previously incompatible races of *Phytophthora infestans*, but also to *Cladosporium cucumerinum* and *Helminthosporium carbonum* which do not normally attack potatoes. This effect does not seem to be due to a specific suppression of terpenoid accumulation since this still occurs after infection by the latter two fungi (Kuć, personal communication).

Another indication that tissue can be conditioned towards general susceptibility to fungal infection is the large number of reports that successful infection by a compatible fungus renders the tissue more susceptible to a normally incompatible race or species (16, 24 and references therein). In all cases, the compatible fungus has been one which establishes some type of biotrophic relationship with its host, and it seems reasonable to infer that the changes in host metabolism known to occur soon after infection by rust and powdery mildew fungi (e.g. 26, 28) preclude affected cells from subsequently reacting in a resistant manner. However, these double inoculation experiments do not provide definitive evidence that suppression of defense reactions is involved in initially establishing the compatible relationship, as has been suggested (24). Indeed, different processes may be involved in the initial establishment of the compatible fungus, and subsequent successful pathogenesis. For example, unusually high numbers of infection hyphae of incompatible *Puccinia helianthi* and *Uromyces phaseoli* var. *vignae* develop haustoria in French bean leaves when they are close to established colonies of the bean rust fungus (16). However, injection of the suppressor from rusted bean leaves increases haustorium production only by *U. phaseoli* var. *vignae* which does trigger the silicon response, and not by *P. helianthi* which does not (16). Significantly, *P. helianthi* has to be much closer than *U. phaseoli* var. *vignae* to colonies of the bean rust fungus before any effect on haustorium production is seen (16). This supports the view that *P. helianthi* haustoria are formed only in cells already "switched" to a general susceptible condition as a consequence of the compatible relationship established by the bean rust fungus. However, it seems likely that this "switching" does not occur until the bean rust fungus forms the first haustorium, and that this haustorium cannot develop unless the deposition of silicon-containing material on the plant cell wall is *specifically* prevented by the action of the suppressor. Thus two types of "suppression" of defense reactions can be found in rusted French bean leaves; one may be the cause and the other the consequence of successful infection.

INDUCED SUSCEPTIBILITY

If fungi can actively interfere with the expression of potential defense reactions of their host plants, they have, in essence, "induced susceptibility". For biotrophic fungi, Daly (4) has applied this term to the theoretical situation in which the metabolic state induced in the susceptible plant, rather than, or as well as, preventing the affected cells from responding in a resistant manner, may be *essential* for successful fungal growth. Thus resistance to infection may be due more to the inability of the fungus to establish the "correct" metabolic state in the plant, than to the "defense reactions" which are triggered if compatibility is not established. One prediction from this hypothesis is that abolition of a resistant plant's "defense reactions" would not significantly increase the susceptibility of the tissue. This seems to be the case for the non-host resistance shown by French bean and cabbage to *Puccinia helianthi* and *Uromyces phaseoli* var. *vignae* respectively; intercellular fungal growth does not continue much beyond the development of the infection hypha even after the blocks to initial haustorium development are removed by pre-inoculation treatments with heat or metabolic inhibitors (15). Conceivably these rust fungi lack the ability to induce the compatible state in these non-hosts, or perhaps the residual effects of the treatments prevent the necessary metabolic "co-operation" by the plant. However, if such co-operation exists in compatible interactions, one might expect treatments which interfere with plant metabolism to prevent successful infection. Although examples of the reduction of resistance by heat and metabolic inhibitors abound (e.g. 15), so far there are very few examples (e.g. 24 and chapter by Bailey in this volume) where such treatments have any effect on susceptibility. Indeed at least three studies, involving *Phytophthora infestans* (21), *Puccinia coronata* var. *avenae* (29) and *Uromyces phaseoli* var. *typica* (15), suggest that compatible interactions are not significantly affected by these treatments. Obviously the idea that biotrophic fungi need to induce an essential metabolic state of susceptibility rather than merely prevent defense reactions needs to be investigated further.

SPECIFIC "EVENTS" FOR DETERMINING BOTH RESISTANCE AND SUSCEPTIBILITY

Taken as a whole, the experimental evidence to date provide some support for, and certainly no clear evidence against the hypothesis that susceptibility of the host plant is mediated by some activity of the fungus. However, the whole concept of "induced susceptibility" in its broadest form implies a *specific* interaction between the fungal pathogen and its susceptible host. At first sight, this conflicts with the current understanding of the gene-for-gene relationship for resistance or susceptibility of the host and the virulence or avirulence of the pathogen, since the genetic data suggests that resistance, rather than susceptibility, involves the specific interaction between plant and pathogen (11). What is often overlooked is that the gene-for-gene relationship has been demonstrated only in interactions involving cultivars of the host species. No such relationship has been demonstrated between pathogens and non-host plants, i.e. *species* considered not be hosts for the pathogen in question, and as several investigators have pointed out (34, 37), it seems most unlikely that every non-host plant possesses specific, different, genes to "recognize" each of the thousands of potential plant pathogens to which it is resistant. It seems more probable that each plant possesses one or more defense reactions which are almost certain to be non-specifically triggered by a pathogen during attempted invasion (18, 34, 37). Thus the basis of host *species* specificity logically lies in the ability of the pathogenic organism to overcome specifically the basic defense mechanisms of its host (18). In theory, this may be accomplished in a multitude of ways. For example, it has already been mentioned that a fungus may develop specific tolerance to the defense reaction(s) it elicits such as accumulation of phytoalexins, or that some bacteria may modify their defense elicitors to an inactive state. Alternatively the effects of the elicitors may be nullified by one of the forms of "induced susceptibility" already discussed. However, one common denominator in all these examples is that the pathogen has to accommodate itself specifically to its host species, whereas resistance of non-host species is non-specifically triggered by some inherent feature of the pathogen.

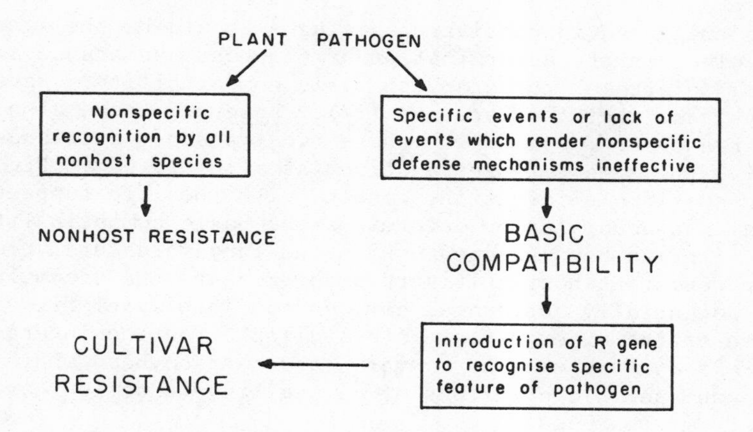

Figure 1 Diagrammatic representation of the concept that basic compatibility between a plant pathogen and its host species depends on the specific "accommodation" of the pathogen to its host so as to render potential, non-specific, defense mechanisms ineffective; cultivar resistance also involves specific interactions between host and parasite and is superimposed on basic compatibility

Once non-host-type of defense mechanisms are overcome, a "basic compatibility" (11) is established between the pathogen and its host species. There is then a strong selection pressure on the host, or the plant breeder, to devise a mechanism by which some feature of the pathogen can be recognized in order to trigger a defense reaction (Fig. 1). This sequence of events automatically results in a gene-for-gene interaction between the host and pathogen, and as Person and Mayo (25) point out, it also implies that the "avirulence gene", whose expression is recognized by a resistant host cultivar, is unlikely to govern a process initially designed to condition avirulence. Indeed, I cannot see why it has to be directly involved in pathogenesis at all- it could merely be some feature, such as a cell wall component or secretory product, that the host plant can readily devise a means of recognizing. This feature now becomes a "specific elicitor" of defense reactions active only in cultivars possessing the gene controlling its "recognition".

Within this general concept of cultivar resistance superimposed on basic compatibility, there are unnumerable ways in which either situation may be brought about. There is no reason why the defense reactions triggered in cultivar resistance should be the same as those involved in the determination of non-host resistance and, therefore, "overcome" to establish basic compatibility. Several studies suggest that they are not (17). However, there also seems to be no reason *prima facie* why these two types of resistance cannot involve the same defense reactions as illustrated in Fig. 2, where non-specific elicitation, coupled with specific suppression of phytoalexin accumulation determines the basic compatibility of a fungus with its host species, while the fungal feature "recognized" in resistant host cultivars also triggers the accumulation of these potentially antifungal compounds. This model fits at least some of the experimental data available for the interaction between *Phytophthora megasperma* var. *sojae* and soybean where specific and non-specific elicitors of glyceollin production have been isolated from the fungus.

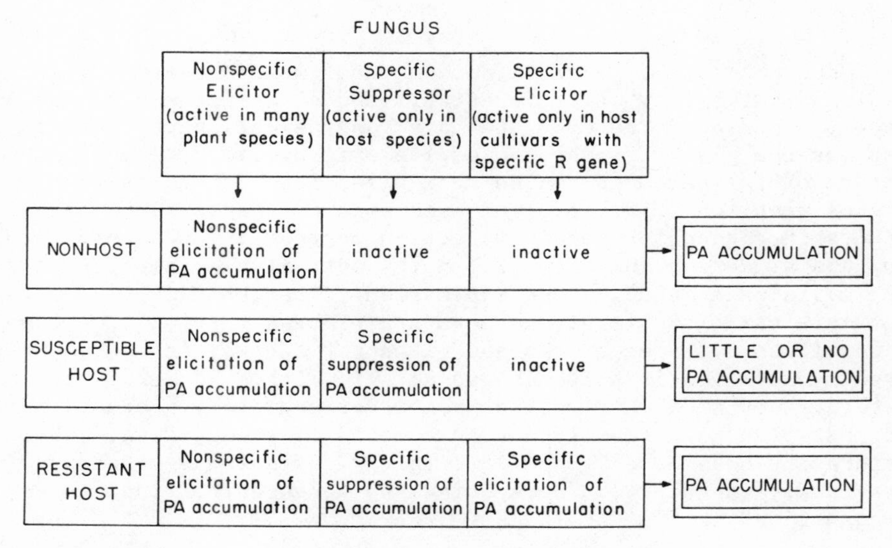

Figure 2 Hypothetical situation where basic compatibility is established by the specific suppression of phytoalexin (PA) accumulation, while cultivar resistance (controlled by gene R in the host) involves the specific elicitation of this same response

Although forms of induced susceptibility seem more likely to be involved in establishing "basic compatibility" than cultivar resistance, induced susceptibility can be reconciled with the gene-for-gene hypothesis of cultivar resistance if it is based on the induction of a metabolic state necessary for fungal growth (4). However, the involvement of suppressors of defense reactions does not readily fit into this scheme. If a suppressor is needed to nullify specifically the effects of each host gene for resistance, then a mutation of the pathogen towards virulence to a previously resistant cultivar must seemingly involve the production of a new, specific suppressor to interfere with the action of the host gene.

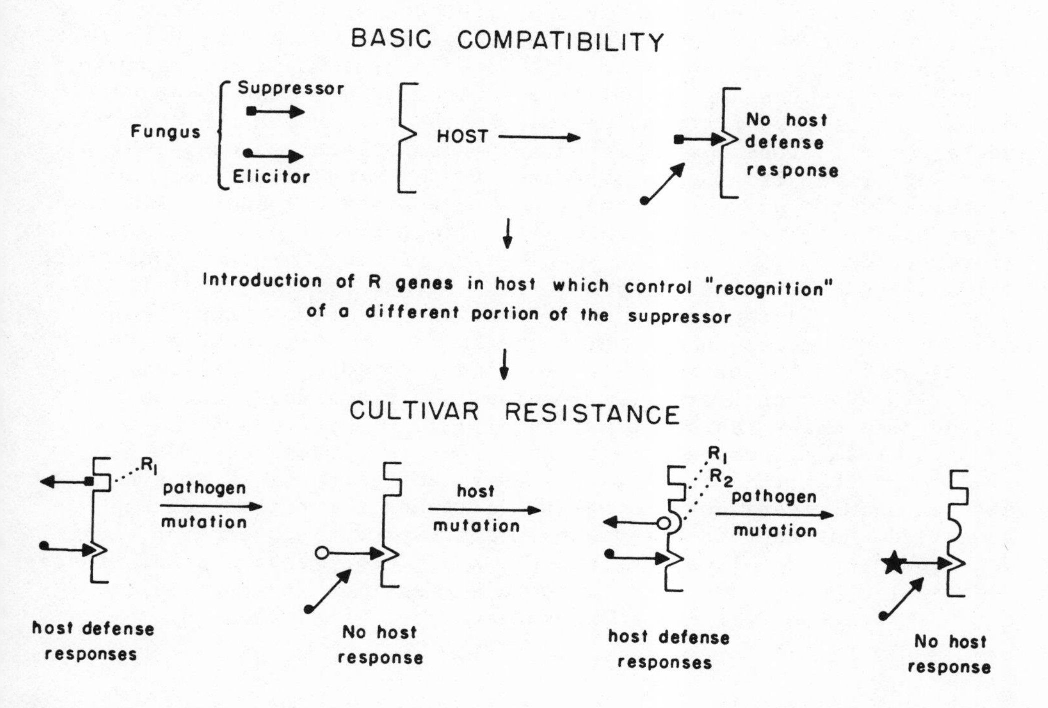

Figure 3 Hypothetical situation where cultivar resistance involves the "recognition" and "inactivation" (controlled by the host genes for resistance R) of the fungal suppressor involved in establishing basic compatibility between the fungus and its host species. Since the "recognized" portion of the suppressor is not that which is involved in its interference with elicitor action, any pathogen mutation which changes this non-specific part of the molecule may render it "unrecognizable" to R gene products without affecting its ability to suppress specifically defense responses

To produce a molecule with such specificity by random mutation must be a rare event, yet cultivar resistance is commonly overcome in the field. Nevertheless, we have seen that the suppressors produced by *Phytophthora infestans* seem to show cultivar specificity. One model which can explain this observation, while circumventing the theoretical objection to the involvement of suppressors in cultivar resistance, is illustrated in Fig. 3. The basic assumption of this model is that the fungal feature that the resistant cultivar specifically recognizes in order to trigger defense reactions is the suppressor involved in establishing basic compatibility. For simplicity, the suppressor is illustrated as competing with the non-specific elicitor for a receptor site in the plant, and with higher affinity for this site. However, this is not an essential feature of the model and other modes of interference with the action of an elicitor could be envisaged. The crucial part of the model is that the portion of the suppressor that is recognized by the products of the genes governing cultivar resistance is *not* that which determines the ability of the suppressor to interfere with the action of the elicitor. Thus *any* mutation in the suppressor molecule which does not affect its suppressing activity may render it "unrecognizable" to the product of the gene for resistance, and will allow it to continue its interference with elicitor action. This model therefore does not require a specific suppressor for each host gene for resistance, and is in accord with the observations that 1. all races of *P. infestans* contain non-specific elicitors of necrosis and terpenoid accumulation which are active in cultivars, resistant or susceptible to the fungal races used to produce the elicitor (7, 12), 2. suppressors from incompatible races have some activity in the respective resistant cultivar (12), 3. the addition of suppressor to microsomal fractions of potato tissue seems to block binding of the elicitor (6), 4. suppressor extracts which are active in cultivars containing several genes for resistance apparently do not contain a suppressor corresponding to each of these genes (Kuć, personal communication), and 5. suppressors of all races reduce elicitor - caused cytoplasmic aggregation in protoplasts from cultivars containing no known genes for resistance (8).

CONCLUDING REMARKS

These models, although they may be applied at the moment to the interactions between *Phytophthora infestans* or *P. megasperma* var. *sojae* and their respective host plants, may prove to be incorrect in detail. Nevertheless, they illustrate that the general concept of specific interactions governing host species susceptibility (basic compatibility) *and* host cultivar resistance can be used to explain data which otherwise seem difficult to integrate. If this concept is correct in essence, then situations may exist, such as in the models proposed, where the host-pathogen interactions governing both basic compatibility and host cultivar resistance are intimately related, and experiments focussed on cultivar resistance alone could provide equivocal or misleading results. Indeed, it may prove that, without a clear understanding of basic compatibility (host species specificity), the elucidation of the basis of cultivar specificity will be extremely difficult, if not impossible.

REFERENCES

1. ANDERSON, A.J. (1978). Isolation from three species of *Colletotrichum* of glucan-containing polysaccharides that elicit browning and phytoalexin production in bean. *Phytopathology* 68, 189 - 194.

2. ANDERSON, A. (1978). Initiation of resistant responses in bean by mycelial wall fractions from three races of the bean pathogen *Colletotrichum lindemuthianum*. *Canadian Journal of Botany* 56, 2247 - 2251.

3. CLINE, K., WADE, M. & ALBERSHEIM, P. (1978). Host-pathogen interactions. XV. Fungal glucans which elicit phytoalexin accumulation in soybean also elicit the accumulation of phytoalexins in other plants. *Plant Physiology* 62, 918 - 921.

4. DALY, J.M. (1972). The use of near-isogenic lines in biochemical studies of the resistance of wheat to stem rust. *Phytopathology* 62, 392 - 400.

5. DANIELS, D.L. & HADWIGER, L.A. (1976). Pisatin-inducing components in filtrates of virulent and avirulent *Fusarium solani* cultures. *Physiological Plant Pathology* 8, 9 - 19.

6. DOKE, N., GARAS, N.A. & KUĆ, J. (1979). Partial characterization and aspects of the mode of action of a hypersensitivity-inhibiting factor (HIF) isolated from *Phytophthora infestans*. *Physiological Plant Pathology* 15, 127 - 140.

7. DOKE, N. & TOMIYAMA, K. (1980). Effect of hyphal wall components from *Phytophthora infestans* on protoplasts of potato tuber tissues. *Physiological Plant Pathology* 16, 169 - 176.

8. DOKE, N. & TOMIYAMA, K. (1980). Suppression of the hypersensitive response of potato tuber protoplasts to hyphal wall components by water soluble glucans isolated from *Phytophthora infestans*. *Physiological Plant Pathology* 16, 177 - 186.

9. DOW, J.M. & CALLOW, J.A. (1979). Leakage of electrolytes from isolated leaf mesophyll cells of tomato induced by glycopeptides from culture filtrates of *Fulvia fulva* (Cooke) Ciferri (syn. *Cladosporium fulvum*). *Physiological Plant Pathology* 15, 27 - 34.

10. DUCZEK, L.J. & HIGGINS, V.J. (1976). The role of medicarpin and maackiain in the response of red clover leaves to *Helminthosporium carbonum, Stemphylium botryosum,* and *S. sarcinaeforme*. *Canadian Journal of Botany* 54, 2609 - 2619.

11. ELLINGBOE, A.H. (1976). Genetics of host-parasite interactions. In : *Encyclopedia of Plant Physiology,* Ed. by R. Heitefuss & P.H. Williams, pp. 761 - 778. Springer, Berlin, Heidelberg, New York.

12. GARAS, N.A., DOKE, N. & KUĆ, J. (1979). Suppression of the hypersensitive reaction in potato tubers by mycelial components from *Phytophthora infestans*. *Physiological Plant Pathology* 15, 117 - 126.

13. HABERLACH, G.T., BUDDE, A.D., SEQUEIRA, L. & HELGESON, J.P. (1978). Modification of disease resistance of tobacco callus tissues by cytokinins. *Plant Physiology* 62, 522 - 525.

14. HEATH, M.C. (1979). Partial characterization of the electron-opaque deposits formed in the non-host plant, French bean, after cowpea rust infection. *Physiological Plant Pathology* 15, 141 - 148.

15. HEATH, M.C. (1979). Effects of heat shock, actinomycin D, cycloheximide and blasticidin S on non-host interactions with rust fungi. *Physiological Plant Pathology* 15, 211 - 218.

16. HEATH, M.C. (1980). Effects of infection by compatible species or injection of tissue extracts on the susceptibility of non-host plants to rust fungi. *Phytopathology* 70, 356 - 360.

17. HEATH, M.C. (1980). Reactions of non-suscepts to fungal pathogens. *Annual Review of Phytopathology* 18, 211 - 236.

18. HEATH, M.C. (1981). Non-host resistance. In : *Plant Disease Control : Resistance and Susceptibility,* Ed. by R.C. Staples & G.H. Toenniessen. John Wiley & Sons, New York in press.

19. HEATH, M.C. & WOOD, R.K.S. (1971). Role of inhibitors of fungal growth in the limitation of leaf spots caused by *Ascochyta pisi* and *Mycosphaerella pinodes*. *Annals of Botany* 35, 475 - 491.

20. HIGGINS, V.J. & MILLAR, R.L. (1969). Comparative abilities of *Stemphylium botryosum* and *Helminthosporium turcicum* to induce and degrade a phytoalexin from alfalfa. *Phytopathology* 59, 1493 - 1499.

21. NOZUE, M., TOMIYAMA, K. & DOKE, N. (1977). Effect of blasticidin S on development of potential of potato tuber cells to react hypersensitively to infection by *Phytophthora infestans*. *Physiological Plant Pathology* 10, 181 - 189.

22. OKU, H., SHIRAISHI, T. & OUCHI, S. (1975). The role of phytoalexin as the inhibitor of infection establishment in plant disease. *Naturwissenschaften* 62, 486 - 487.

23. OKU, H., SHIRAISHI, T. & OUCHI, S. (1979). The role of phytoalexins in host-parasite specificity. In : *Recognition and Specificity in Plant Host-Parasite Interactions*, Ed. by J.M. Daly & I. Uritani, pp. 317 - 333. University Park Press, Baltimore.

24. OUCHI, S., HIBINO, C., OKU, H., FUJIWARA, M. & NAKABAYASHI, H. (1979). The induction of resistance or susceptibility. In : *Recognition and Specificity in Plant Host-Parasite Interactions*, Ed. by J.M. Daly & I. Uritani, pp. 49 - 65. University Park Press, Baltimore.

25. PERSON, C. & MAYO, G.M.E. (1974). Genetic limitations on models of specific interactions between a host and its parasite. *Canadian Journal of Botany* 52, 1339 - 1347.

26. PURE, G.A., CHAKRAVORTY, A.K. & SCOTT, K.J. (1979). Cell-free translation of polysomal messenger RNA isolated from healthy and rust-infected wheat leaves. *Physiological Plant Pathology* 15, 201 - 209.

27. SHIRAISHI, T., OKU, H., YAMASHITA, M. & OUCHI, S. (1978). Elicitor and suppressor of pisatin induction in spore germination fluid of pea pathogen, *Mycosphaerella pinodes*. *Annals of the Phytopathological Society of Japan* 44, 659 - 665.

28. SIMPSON, R.S., CHAKRAVORTY, A.K. & SCOTT, K.J. (1979). Selective hydrolysis of barley leaf polysomal messenger RNA during the early stages of powdery mildew infection. *Physiological Plant Pathology* 14, 245 - 258.

29. TANI, T., YAMAMOTA, H., KADOTA, G. & NAITO, N. (1976). Development of rust fungi in oat leaves treated with blasticidin S, a protein synthesis inhibitor. *Technical Bulletin of Faculty of Agriculture, Kagawa University* 27, 95 - 103.

30. THEODOROU, M.K. & SMITH, I.M. (1979). The response of French bean varieties to components isolated from races of *Colletotrichum lindemuthianum*. *Physiological Plant Pathology* 15, 297 - 309.

31. VAN ETTEN, H.D. (1979). Relationship between tolerance to isoflavonoid phytoalexins and pathogenicity. In : *Recognition and Specificity in Plant Host-Parasite Interactions*, Ed. by J.M. Daly & I. Uritani, pp. 301 - 316. University Park Press, Baltimore.

32. WADE, M. & ALBERSHEIM, P. (1979). Race-specific molecules that protect soybeans from *Phytophthora megasperma* var. *sojae*. *Proceedings of the National Academy of Science, U.S.A.* 76, 4433 - 4437.

33. WARD, E.W.B. & STOESSL, A. (1972). Postinfectional inhibitors from plants. III. Detoxification of capsidiol, an antifungal compound from peppers. *Phytopathology* 62, 1186 - 1187.

34. WARD, E.W.B. & STOESSL, A. (1976). On the question of "elicitors" or "inducers" in incompatible interactions between plants and fungal pathogens. *Phytopathology* 66, 940 - 941.

35. WARD, E.W.B., UNWIN, C.H. & STOESSL, A. (1973). Postinfectional inhibitors from plants. VII. Tolerance of capsidiol by fungal pathogens of pepper fruit. *Canadian Journal of Botany* 51, 2327 - 2332.

36. WIT, P.J.G.M. de & ROSEBOOM, P.H.M. (1980). Isolation, partial characterization and specificity of glycoprotein elicitors from culture filtrates mycelium and cell walls of *Cladosporium fulvum* (syn. *Fulvia fulva*). *Physiological Plant Pathology* 16, 391 - 408.

37. WOOD, R.K.S. (1976). Specificity - an assessment. In : *Specificity in Plant Diseases*, Ed. by R.K.S. Wood & A. Graniti, pp. 327 - 338. Plenum Press, New York, London.

38. YOSHIKAWA, M., YAMAUCHI, K. & MASAGO, H. (1979). Biosynthesis and biodegradation of glyceollin by soybean hypocotyls with *Phytophthora megasperma* var. *sojae*. *Physiological Plant Pathology* 14, 157 - 169.

PLANT IMMUNIZATION-MECHANISMS AND PRACTICAL IMPLICATIONS*

J. KUĆ

Department of Plant Pathology
University of Kentucky
Lexington
Kentucky 40546 U S A

INTRODUCTION

On a recent trip to New York City I spent an extremely pleasant morning at the Pierpont Morgan Library examining their fine collection of art, old books and manuscripts. In a gallery showing the work of Jacob Jordaens (1593-1678) I noticed a painting depicting a scene from Aesop's fables in which a satyr is passing the time of day with peasants at a village inn. The satyr is astounded that a peasant first blows on his hands to warm them and then on his spoonful of porridge to cool it. The scene is used to illustrate the moral that those who blow first hot and then cold are not to be trusted. In preparing this chapter I fear that I will play the part of the peasant blowing hot and cold. Whether what I say can be trusted is for the reader and posterity to decide.

Viruses, bacteria and fungi cause severe diseases in plants resulting in serious economic losses. In general diseases are specific for certain crops and even certain cultivars of a crop. In this chapter, however, I will describe research which indicates that the pathogens which cause disease can be used to protect plants against disease, and that systemic resistance to viral, bacterial and fungal diseases is induced in plants by either viral, bacterial

---

* Journal paper no. 80-11-188 of the Kentucky Agriculture Experiment Station, Lexington, Kentucky 40546 USA. The author's research reported in this paper has been supported in part by grants from the Ciba-Geigy Corporation, Rockefeller Foundation and Kentucky Tobacco and Health Research Institute.

or fungal pathogens - hardly a case for specificity. This chapter will also offer evidence that the damage that pathogens cause may at times be beneficial to plants, and that the mechanims associated with injury can be mechanisms which signal defense reactions in plants.

From past experience with reviewers of my manuscripts, and occasionally with senior editors and even editors-in-chief, I recognise the importance of justifying my use of the term "immunization". Regardless of the tables and figures of data, this point of terminology will probably catalyze a great deal of debate and discussion, hopefully not more than the "science" I present. My use of the term is based on the dictionary (Webster's New Collegiate, 1979) definition of immunity: a condition of being able to resist a particular disease especially through preventing development of a pathogenic microorganism or by counteracting the effects of its products. The term is not strictly restricted to animals, a molecular mechanism, or a requirement for specificity. The term "immunization" concisely defines the phenomenon I will describe. If all other arguments fail, the term "immunization" is far shorter and less awkward than terms such as "induced systemic resistance", "induced systemic protection", "acquired systemic resistance" and "activated systemic resistance".

THE PHENOMENON

The observation that plants can be immunized against a disease by infection with the pathogen causing the disease, cultivar non-pathogenic races of the pathogen, other pathogens, pathogens of other species, and, less frequently, products of infectious agents dates back at least 100 years (6,33). In the past 10-20 years, however, careful experimentation has established the validity of plant immunization, and with the establishment of its validity, progress is slowly being made on the elucidation of molecular mechanisms responsible for the phenomenon and the application of plant immunization to the practical control of disease in the field. Plant immunization, as was once the case with phytoalexins, is no longer relegated the position of a freak in the side show of a circus. It is here to stay and to be accounted for. Yarwood (48), Cruickshank and Mandryk (8) and Müller (33) established that plants can be immunized against diseases caused by fungi. Ross further established the validity of plant immunization against virus diseases on tobacco, bean and cowpea using viruses as the inducing agents (40-42). This work was subsequently expanded to include many hosts as well as viruses, bacteria, fungi, and products of infectious agents (5,8-12,16,17,19,21,23-29,31,32,43,45,46,48). Though Hecht and Bateman (17) and Mandryk (29) reported that non-specific resistance to pathogens was induced by a fungus or virus in tobacco, the broad spectrum of effectiveness of plant immunization was not generally emphasized by earlier workers. Recent reports by Kuć and his colleagues indicate that cucurbits can be immunized against viral, bacterial, and fungal diseases by infection with viruses, bacteria or fungi (5,10,16,19,21,23,24,26,27.45). Limited infection with *Colletotrichum lagenarium* or tobacco necrosis virus protects cucumber against disease caused by a broad range of pathogens including obligate and facultative fungi, local lesion and systemic viruses, fungi and bacteria that cause wilts and those that cause restricted and non-restricted lesions on foliage and fruit. It is effective against foliar as well as root pathogens (5,10,16,19,21,23-27,45, Table 1). Immunization in cucurbits is systemic and requires a lag period between induction and challenge for expression. This is also true for the phenomenon as reported by Ross with tobacco and local lesion TMV. The induction period corresponds to the length of time required for the appearance of symptoms of the inducing infection.

Table 1 The biological spectrum of effectiveness of systemic resistance induced by foliar infection of cucumber with *Colletotrichum lagenarium* or tobacco necrosis virus.

Disease	Pathogen
Anthracnose	*Colletotrichum lagenarium*
Scab	*Cladosporium cucumerinum*
Gummy stem blight	*Mycosphaerella melonis*
Fusarium wilt	*Fusarium oxysporum* f.sp. *cucumerinum*
Downy mildew	*Pseudoperonospora cubensis*
Angular leaf spot	*Pseudomonas lachrymans*
Bacterial wilt	*Erwinia tracheiphila*
Cucumber mosaic	Cucumber mosaic virus
Local necrosis	Tobacco necrosis virus
Local necrosis	*Phytophthora infestans*

In all cases reported to date with cucurbits, macroscopic or microscopic necrosis caused by an infectious agent is associated with immunization, though not all types of necrosis results in immunization. Mechanical injury, and injury by dry ice, chemicals or fungal and plant extracts do not cause immunization in cucurbits. Though infection with *C. lagenarium* protects cucumber against the hypersensitive reaction caused by *Phytophthora infestans*, a non-pathogen of cucumber, infection with *P. infestans* does not protect cucumber against disease caused by *C. lagenarium* (19).

Removal of the first true leaf (inducer leaf) 72-96 hours after infection with fungi or bacteria and 48 hours after infection with TNV does not reduce the immunization of foliage above the site of induction (5,23,26). Similarly, removal of leaves above the inducer leaf after approximately the same lag times does not reduce the immunization of the excised leaves. If the axillary bud at the junction of petiole and stem is included when these leaves are excised from immunized plants, the leaves can be rooted and resulting plantlets are immunized.

The signal for immunization in cucurbits is graft transmissable from infected rootstock to scion. Resistance is induced in the scion if grafting precedes or follows infection of the rootstock and is not cultivar, genus or species specific (22). The experiments with grafting suggest, but do not prove, that a chemical signal is produced at the site of induction and is translocated to other tissues where it conditions resistance. Recent work by Guedes *et al.* (13) indicated that the effect of immunization is stronger above the inducer leaf than below it. Girdling the petiole of the inducer leaf prevented immunization above or below the inducer leaf (13). Girdling the petioles of leaves to be challenged, while leaving the petiole of the inducer leaf intact, prevented immunization only in challenged leaves with girdled petioles. Girdling almost totally prevented the movement of ^{14}C sucrose from the inducer leaf, which supports the possibility that the effect of girdling on immunization is due to interference in phloem transport. The above data, together with those from experiments with graft transmissability and persistence of immunization after removal of the inducer leaf, are strong evidence that immunization is the result of a signal transported from the inducer leaf. Continuous production of the signal, however, is not necessary, and once tissues have been conditioned by the signal they do not require the presence of the inducer leaf for immunization.

Infection of the first true leaf of cucumber with *C. lagenarium* or TNV, followed in 2-3 weeks by a booster inoculation with these agents, immunizes cucumber against disease caused by *C. lagenarium*, *Cladosporium cucumerinum* and *Pseudomonas lachrymans* through the period of fruiting (26 and unpublished data). A single induction immunizes for 4-6 weeks, and without the booster inoculation resistance is lost systemically after this period. Cucumbers cannot be immunized, however, once they have started to flower and set fruit (13). One interpretation of this observation is that the onset of flowering and fruit set alters the hormonal balance which makes immunization impossible. Removal of flower buds before they open does not extend the period during which immunization is possible (unpublished data). This would suggest that the programming of the plant's biological clock for reproduction turns off the ability to immunize but does not prevent the expression of the phenomenon in immunized tissues. This is further evidence that the induction of immunity and the containment of an infectious agent are distinct metabolic processes.

The concentration of inoculum used for induction and the number of lesions produced on the inducer leaf are directly related to the extent of immunization until a saturation point of inoculum or lesions is attained (3,5,23,26). A single lesion caused by *C. lagen-*

arium (14,26) and as few as eight lesions caused by TNV (23) on the inducer leaf immunize the tissues above the leaf. High concentrations of challenge inoculum will overcome, to a degree, immunization produced by low levels of inoculum used for induction, e.g. there is not a reduction in the number of lesions but there is still a reduction in the size of lesions. At times, high inoculum concentrations of *C. lagenarium* cause systemic symptoms on plants. This phenomenon has not been observed on immunized plants. The reduction of spread of a systemic disease by foliar immunization is also clearly evident with plants challeneged with cucumber mosaic virus or *Erwinia tracheiphila* (Bergstrom & Johnson, Bergstrom, unpublished data).

MECHANISMS FOR CONTAINMENT

With foliar pathogens causing restricted lesions, e.g., *C. lagenarium* or TNV and those causing non-restrictive lesions, e.g. *C. cucumerinum*, or *Mycosphaerella melonis*, immunization is generally characerized by a reduction in lesion number and size. As previously stated, inoculation with high concentrations of challenge inoculum may result in a reduction in lesion size but not number. In addition, immunization prevents or retards the spread of disease caused by systemic pathogens including viruses, bacteria and fungi.

Foliar immunization is effective against *C. lagenarium* or *C. cucumerinum* applied to the surface of foliage or infiltrated into the foliage (14,23,24,45). It reduces the multiplication of *P. lachrymans* (5) and reduces the infectivity of TNV (23,24). It is effective against wilt caused by *F. oxysporum* f. sp. *cucumerinum*, a root-infecting pathogen (10) and against wilt caused by *E. tracheiphila* a foliar pathogen (Bergstrom, unpublished data). Though not impossible, it is unlikely that a single mechanism for containment would explain all the above phenomena.

The penetration from appressoria of *C. lagenarium* into immunized cucumber is markedly reduced, whereas the germination of conidia is unaffected (14,23,24,38). However, neither the germination of conidia nor the penetration by *C. cucumerinum* into immunized cucumber is reduced. In the case of *C. cucumerinum*, penetration occurs within 18 hours, whereas, with *C. lagenarium* penetration occurs from thick-walled appressoria and requires 40-60 hours. Histological and chemical studies, which will subsequently be elaborated, suggest that the mechanisms for containment of *C. cucumerinum* in immunized plants are identical to those in resistant, non-immunized plants (14). However, with *C. lagenarium*, immunization appears to activate an additional mechanism for containment in immunized as compared to non-immunized resistant plants (38). Penetration readily occurs into resistant and susceptible plants but is associated with a hypersensitive-like reaction only in the former. Immunization of a resistant plant reduces penetration and enhances resistance (26,38). Immunization does not, however, cause an increase in hypersensitive reactions around sites of penetration in susceptible plants.

Associated with immunization is an approximately three-fold increase in peroxidase activity. The increase is systemic and is observed in immunized, unchallenged tissue distant from the inducer inoculation (14,15). Since induction with *C. lagenarium*, *C. cucumerinum*, *P. lachrymans* or TNV causes the increase, it is likely due to peroxidase activity of host origin. As with immunization, a single lesion on the inducer leaf results in a statistically significant systemic increase in peroxidase activity (14, Table 2). The systemic increase in peroxidase is associated with markedly enhanced activity

Table 2 The effect of the number of lesions caused by *Colletotrichum lagenarium* on the first true leaf of SMR 58 cucumber on peroxidase (PRO) activity and protection of the second true leaf

Number of lesions on leaf one	PRO activity in leaf two[a]	Mean number of lesions on leaf two[b]	Mean diameter of lesions on leaf two (mm)[b]
0	4.95 A[c]	14.6 A[c]	4.3 A[c]
1	6.63 B	13.3 B	3.5 B
2	7.43 C	12.5 B	2.4 C
5	10.80 D	8.6 C	1.7 D
10	10.80 D	5.4 D	1.3 DE
20	10.53 D	4.4 DE	1.3 DE
30	12.15 E	4.0 E	1.0 E

[a]Peroxidase (PRO) activity expressed as change in absorbance per minute per mg protein using guaiacol as the hydrogen donor. PRO was extracted from half leaves one week after inoculation of leaf one and just prior to challenge inoculation of the other leaf half.

[b]Number and diameter of lesions determined one week after challenge of leaf two with 15 5-μl drops of a *C. lagenarium* spore suspension (10^5 spores/ml). Challenge inoculations performed one week after inducer inoculation.

[c]Means from a total of 34 plants per treatment in four experiments. Means followed by different letters are significantly different at p=0.01. LSD PRO=0.71; LSD lesion numbers=1.2; LSD lesion sized-0.5.

of several isozymes (14,15). A booster inoculation with *C. lagenarium* or TNV markedly enhances immunization and the systemic increase in peroxidase (Table 3). Peroxidase activity increases sooner in immunized than in non-immunized plants after challenge with *C. cucumerinum* or *C. lagenarium*, and the enhanced activity is also due to the increased activity of several isozymes (14). In time, however, total peroxidase activity in challenged tissue of non-immunized plants may surpass that in immunized plants. Injury of a leaf with dry ice or

Table 3 Effect of a "booster" inoculation on induced systemic resistance and peroxidase activity in SMR 58 cucumber leaves

Treatment[a]	Peroxidase activity[b]	Mean lesion number[c]
Untreated control	3.83 A[d]	14.5 A[e]
Leaf 1 inoculated	16.20 B	4.5 B
Leaf 2 inoculated	15.86 B	5.2 B
Leaves 1 & 2 inoculated	24.68 C	0.2 C

[a]Fifteen drops of a spore suspension of *C. lagenarium* (5μl, 5×10^5 spores/ml) were placed on leaf one of SMR 58 cucumber. One week later, one-half of the controls and one-half of the plants inoculated on leaf one were inoculated on leaf two.

[b]Peroxidase activity was determined in leaf three just prior to challenge. Activity expressed as change in absorbance per minute per mg protein using guaiacol as the hydrogen donor.

[c]Mean number of lesions on leaf three one week after challenge with fifteen 5 μl drops of a 10^5 spores/ml suspension of *C. lagenarium*. Challenge inoculation was performed just after time of tissue sampling for peroxidase activity.

[d]Means of a total of six extracts per treatment of two separate experiments. Means followed by different letters are significantly different, p=0.01.

[e]Means of a total of twenty plants in two separate experiments. Means followed by different letters are significantly different, p=0.01.

rubbing with carborundum enhances peroxidase activity in the injured leaf, but the increase is not systemic. Enhanced peroxidase activity, however, may be only one manifestation of the mechanisms for containment. As stated earlier, peroxidase activity in non-immunized tissue eventually often surpasses that in immunized tissue after challenge. There is also a marked increase in peroxidase activity in cucumber foliage as it senesces and this tissue in non-immunized plants is not more resistant to disease caused by *C. lagenarium* than younger tissue. Similarly, treatment of cucumber plants with ethylene enhances peroxidase activity but does not initiate immunization. At least two interpretations of these phenomena are possible. One is that enhanced peroxidase activity is not required for immunization. This interpretation would be consistent with the recent report by Nadolny and Sequeira describing studies of immunization in tobacco (34). However, when considering an increase in peroxidase activity *in vivo*, it is important to consider the concentration and availability of the two substrates for peroxidase activity, hydrogen peroxide and an appropriate hydrogen donor, unually a phenol or an amine. The first interpretation and the report by Nadolny and Sequeira (34) would be valid if peroxidase was toxic *per se* to the infectious agent or was the sole limiting factor. A second interpretation is that peroxidase is one component of the mechanism for containment of pathogens in immunized plants. For immunization, enhanced peroxidase activity would have to be accompanied by an increase or non-limiting supply of both hydrogen peroxide and hydrogen donor. Neither senescence nor treatment with ethylene, neither of which induce immunity, may provide adequate quantities of the two latter components. In addition, the localization and time of appearance of the increased peroxidase activity, relative to the development of the pathogen, may be important considerations. The activity of phenylalanine ammonia-lyase (PAL), *p*-coumaryl CoA ligase or *p*-coumaryl reductase is not higher in immunized as compared to non-immunized tissue prior to challenge. PAL activity, however, increases sooner in immunized tissue challenged with *C. cucumerinum* but not *C. lagenarium* (14). Of the enzymes studied, only the increase in peroxidase activity corresponds closely to the onset of immunization and increase in lignification in immunized plants after challenge (14,15).

Lignin is a polymer derived from the polymerization of hydroxycinnamyl alcohols. The phenylpropanoid monomers are cross-linked to each other by carbon-carbon bonds or ether linkages between the aliphatic and aromatic components of the polymer. The hydroxycinnamyl alcohols are derived from phenylalanine. Phenylalanine is converted to *trans*-cinnamic acid via the action of phenylalanine ammonia lyase (PAL). Cinnamic acid is then converted to 4-hydroxycinnamic acid (*p*-coumaric acid) by cinnmate-4-hydroxylase in the presence of molecular oxygen and reduced pyridine nucleotides. Some grasses and fungi have the ability to convert tyrosine directly to *p*-coumaric acid by action of tyrosine ammonia lyase (TAL). *p*-Coumaric acid is hydroxylated in the 3-position to yield caffeic acid by action of a phenolase. Methylation of the 3-hydroxy group of caffeic acid by S-adenosylmethionine in the presence of an *ortho*-methyl-transferase yields ferulic acid. Ferulic acid can be further hydroxylated and methylated in the 5-position to yield sinapic acid. *p*-Coumaric, ferulic and sinapic acids are converted to lignin precursors by a three step process. The cinnamic acids are first converted to the corresponding coenzyme A (CoA) esters by hydroxycinnamyl CoA ligase (CoA ligase). The CoA esters are then reduced to the aldehyde by a reductase which requires NADP (H)+H^+. The aldehydes are further reduced to the alcohols by an aromatic alcohol dehydrogenase in the presence of NADP(H) H^+. This series of reactions is similar to the reduction of β-hydroxy-β-methylglutaryl CoA to mevalonic acid in terms of irreversibility of the first reduction step. The alcohols, and possibly the aldehyde and acid precursors, are then polymerized into lignin by peroxidase in the presence of hydrogen peroxide.

In resistant cucumber, lignification rapidly occurs after penetration by *C. cucumerinum* and it occurs in numerous cells around the site of penetration. Lignification is observed in susceptible cucumbers infected with the pathogen, but the reaction is delayed until after the pathogen has ramified through the tissue, and it is weak and diffuse (14,15,20). Lignification is also evident in susceptible foliage infected with *C. lagenarium*, but the reaction occurs after the fungus has developed in the tissue, and initially is weak and diffuse. In plants immunized by infection with *C, lagenarium*, *C. cucumerinum* or TNV and challenged with *C. cucumerinum* or *C. lagenarium*, lignification is more rapid, intense, and non-diffuse than in non-immunized plants (14,15, Table 4, Fig. 1, 2). In immunized plants challenged with *C. cucumerinum*, it includes numerous cells around the

Table 4 The effect of inoculating the first true leaf of Marketer cucumber with *Colletotrichum lagenarium* on epidermal lignification in petioles of the first leaf below the apical bud challenged with *Cladosporium cucumerinum*

Trial [a]	Induced resistance [b]		Control	
	Penetration sites [c]	Lignified sites [c]	Penetration sites [c]	Lignified sites [c]
	Total	Total (%)	Total	Total (%)
1	965	835 (87)	897	85 (9.5)
2	186	184 (99)	280	10 (3.5)
3	435	435 (100)	400	3 (1)
Totals	1586	1454 (91.7)	1577	98 (6.2)

[a]Total of fifteen plants per treatment in three separate trials.

continued....

[b]Induced resistance: leaf one inoculated with *C. lagenarium* (40 5 μl drops of a 5x10^5 spores/ml suspension) one week prior to challenge of the petiole of the first leaf below the apical bud of the plant by spraying with a 10^6 spores/ml suspension of *C. cucumerinum*. Control: leaf one untreated prior to challenge.

[c]Penetration sites: all sites of penetration into the host by *C. cucumerinum*. Lignified sites: number of sites of penetration which were associated with lignification of the host. (%): per cent of penetration sites which had lignified. Observations were made using epidermal strips prepared from the petiole of the first leaf below the apical bud 48 hours after challenge.

site of infection. In plants infected with *C. lagenarium* lignification occurs at sites of penetration but is most marked beneath appressoria where penetration is not evident. Lignification is more localized in immunized plants challenged with *C. lagenarium*. In immunized plants, both fungi generally make little further growth once lignification is apparent.

Several recent papers present strong evidence that lignification is a mechanism for disease resistance (2,18,30,35,36,39,47). Lignification can restrict development of pathogens by several possible mechanisms by (a) increasing the mechanical resistance of host cell walls (b) reducing the susceptibility of host cell walls to degradation by extracellular enzymes (c) restricting the diffusion of pathotoxins and nutrients (d) inhibiting growth of pathogens by the action of toxic lignin precursors or lignification of the pathogen. The reports that chitin, a component of the walls of many fungi, can serve as an elicitor of lignification (35) and as a matrix for lignification (44) are consistent with the suggested role of lignification in disease resistance and immunization. Lignification may also function to restrict the spread viruses in plants (30). Hammerschmidt (14) was the first to report the lignification of fungal mycelia. *C. lagenarium* and *C. cucumerinum* were lignified in the presence of coniferyl alcohol or eugenol, hydrogen peroxide and peroxidase preparations from immunized plants. Lignification was verified by staining of water and solvent-washed mycelial fragments with phloroglucinol-HCl and by measuring ionization difference spectra of extracted lignin. Mycelia from both fungi did not stain with phloroglucinol-HCl when incubated with coniferyl alcohol, and gave a barely discernible reaction when incubated with coniferyl alcohol and hydrogen peroxide. Free radicals generated by the action of peroxidase on coniferyl alcohol are highly reactive and these could be toxic *per se*. By forming covalent bonds with protein and carbohydrate components of fungal walls, the radicals and resulting polymers could inhibit growth by lignifying the fungus. In addition, the fungus may be bound to wall components of the host preventing its development throughout the tissues. Lignification has been

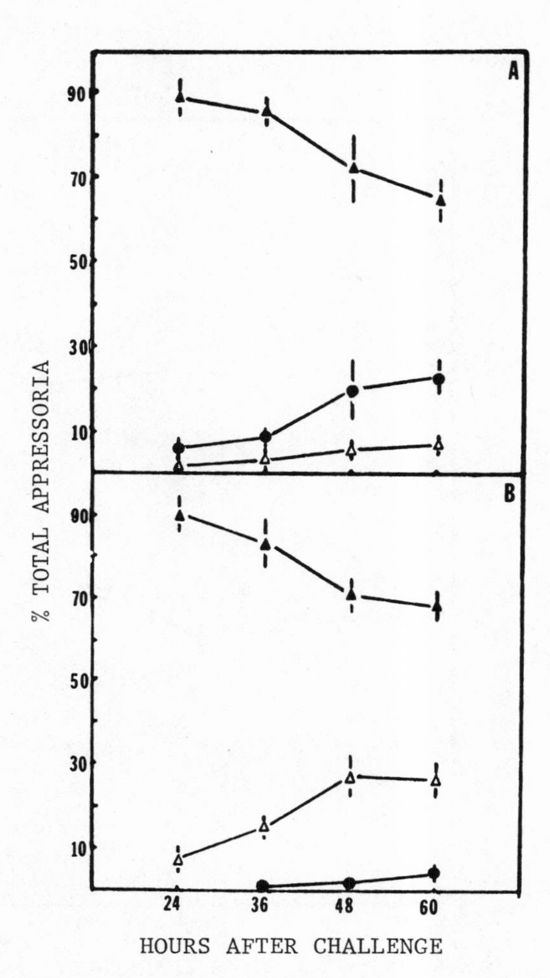

Figure 1 Penetration by *Colletotrichum lagenarium* and host cell wall lignification at intervals after inoculation of the petiole of leaf two of SMR 58 cucumber. (A) leaf one infected with *C. lagenarium*. (B) leaf one untreated. Closed triangles (▲-▲) represent appressoria without penetration hyphae visible in the host. Open triangles (Δ-Δ) represent appressoria with successful penetrations. Closed circles (●-●) represent appressoria associated with host cell lignification. Lignin was detected with phloroglucinol-HCl or toluidine blue O. Each point and vertical bar is the mean and standard error from three plants per treatment (five inoculation sites per plant) in two separate experiments.

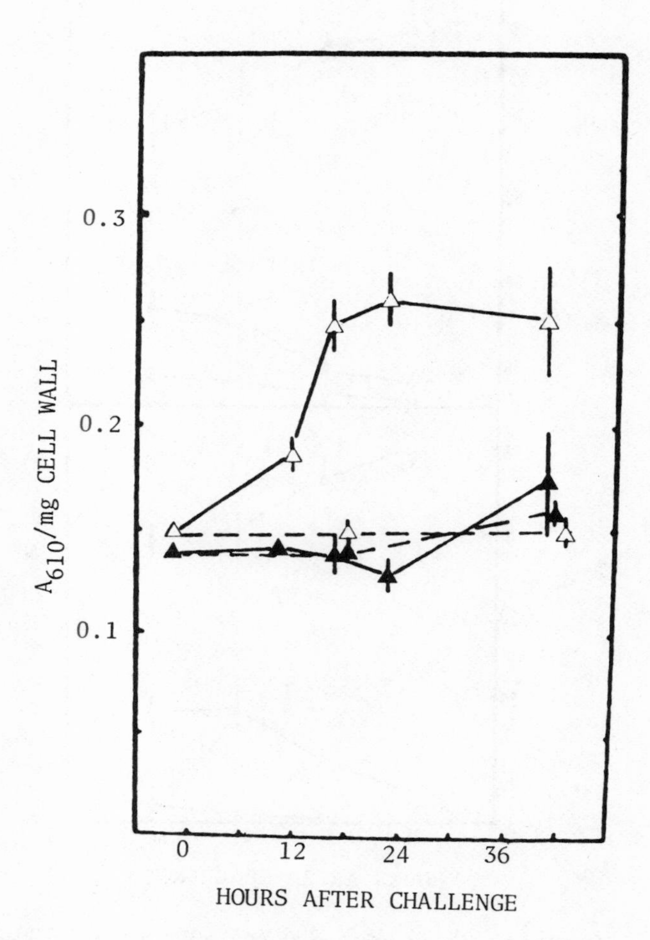

Figure 2 Lignin content of Marketer cucumber apical tissue after challenge with *Cladosporium cucumerinum*. Lignin, extracted from cucumber cell walls with hot aqueous alkaki, was determined colorometrically with Gibb's reagent (2,6-dichloroquinoneclorimide). Each point and vertical bar represent the mean and standard error for two determinations on each of three samples in three experiments. (△-△) Systemic resistance induced with *Colletotrichum lagenarium* and challenged with *C. cucumerinum*; (▲-▲) Resistance not induced, tissue challenged; (△---△) Resistance induced, tissue not challenged; (▲---▲) Resistance not induced, tissue not challenged.

reported to reduce markedly the susceptibility of plant cell walls to hydrolysis by fungal hydrolases (39).

In our studies of immunized cucumber plants, coniferyl alcohol was not consistently detected in tissues after challenge with *C. cucumerinum* or *C. lagenarium*. This was puzzling to Dr. Hammerschmidt and myself and suggested that the alcohol was oxidized by peroxidase as fast as it was synthesized *de novo* or formed by release from soluble or cell wall-bound precursors. Immunized plants did not consistently have higher concentrations of soluble or esterified *p*-coumaric or ferulic acid before or after challenge. It was interesting to observe, therefore, that confiferyl alcohol accumulated in immunized plants inoculated with *C. cucumerinum* or *C. lagenarium* in the presence of α-tocopherol acetate. Coniferyl alcohol was highly toxic to both fungi. Inhibition was evident in bioassays on silica gel plates with as little as 0.1-0.3 μg coniferyl alcohol for both fungi (14).

Thus, lignification can generate an extremely hostile environment for fungi, e.g. high toxicity of phenolic lignin precursors, toxicity of free radicals, lignification of fungus, binding of fungus to host walls, rendering host wall polysaccharides and proteins resistant to degradation by fungal hydrolases. The evidence of the role of lignification as a mechanism for the containment of fungi in immunized plants is strong but circumstantial: the fungus is in the hostile environment where lignification occurs. Investigations are in progress to ascertain whether the lignification process inhibits fungi *in vivo* and whether lignification is associated with immunization against bacterial and viral diseases. The restriction of spread of systemic diseases in immunized plants, whether viral, bacterial or fungal, could be explained by the effect of lignification on the pathogens or on their extracellular toxins or enzymes. Clearly, lignification occurs sooner and with greater intensity in immunized plants after challenge. Why this happens is the next question to be addressed.

MECHANISMS FOR INDUCTION

This portion of the chapter will be short because, until recently, all our efforts to immunize plants by abiotic agents were unsuccessful. Agents tested include: wall preparations from fungi and bacteria, extracts of or extracellular products from fungi and bacteria, phloem exudates and extracts of infected and immunized non-infected tissue. The appropriate preparations or extracts were sprayed on foliage, infiltrated into leaf panels or administered via the roots. Excised leaves and seedlings were also placed in tests solutions. Recently Dr. Garas, a postdoctoral research worker in my laboratory, has been successful in obtaining an active extract from the inducer leaf infected with *C. lagenarium* or *P. lachrymans*. The most active extracts were obtained by extraction with an acidic buffer within 48 hours after symptom appearance. Extracts of infected leaves obtained prior to symptom appearance or 72 hours or longer after symptom appearance were inactive. Infiltration of active extracts into cotyledons and the first true leaf of cucumber induced immunity in foliage above to disease caused by *C. lagenarium*. The active factor in extracts is non-dialysable, stable to boiling and stable in buffers as acidic as pH 2.5. Dr. C. Gessler, a visiting postdoctoral research worker, isolated a protein from leaves of cucumber infected with *C. lagenarium*, *C. cucumerinum*, *P. lachrymans* and TNV which has a molecular weight of *c*.16,000 d, is stable in acidic buffers, binds to DEAE cellulose, and is not detected in extracts of leaves damaged by dry ice or by rubbing with carborundum. The appearance of the protein coincides with the appearance of symptoms e.g., three to four days after inoculation with *C. lagenarium*. This protein may be identical to that reported by Andebrhan *et al.* (1). The protein was not detected in tissue systemic to the infected tissue, but the quantity of protein increased markedly in immunized tissue after challenge even though few if any visible symptoms were apparent. The amount of protein in immunized leaves after challenge was approximately equal to that in non-immunized infected leaves which had severe symptoms. Many reports of the increase of one or more proteins in plants infected with viruses, or treated with polyacrylic or salicylic acid have been presented at this conference. Experiments are in progress to test whether the protein in infected cucumber leaves is a factor in the induction of immunity.

It is possible that the inducer of immunization (signal) is produced during a rather short period after induction and it or a second factor (message) affects unopened leaves at the growing point. The unopened leaves, at an appropriate stage of development, would be immunized for 4-6 weeks. If after three weeks a booster inoculation is administered, sufficient new unopened leaves might be affected to permit immunity to persist through the period of fruiting. The possibility that a low-molecular weight, translocatable message is a component of induction is highly speculative but would not be inconsistent with the apparent immobility of the protein detected in infected leaves and the recent observation that DL-β-phenylserine and phenylthiourea fed to cucumber plants via the roots markedly reduced the severity of symptoms following inoculation with *C. lagenarium* (Preisig and Kuć, unpublished data). The concentrations which were fed to the plants were not inhibitory to the fungi, This report supports an earlier report by Hijwegen (20) who reported DL-β-phenylserine reduced the severity of disease in cucumber following inoculation with *C. cucumerinum*.

PRACTICAL IMPLICATIONS

Since immunization is the basis for preventive medicine in animals, it is surprising that the phenomenon has not been applied more vigorously to the control of plant diseases. Reports or the practical application of the phenomenon, however, are available. For many years hypovirulent strains of TMV have been used in the greenhouse (37) and field (USSR-personal observation) to protect tomato against virulent strains of the virus. Cruickshank and Mandryk (8) reported protection of tobacco against blue mold in the field following infection with *Peronospora tabacina*. Infection with the fungus, however, severely stunted the plants. Caruso and Kuć (4) demonstrated that restricted infection of cucumber, watermelon and muskmelon by *C. lagenarium* protected plants in the field from disease caused by artificial challenge with high inoculum levels of the pathogen. In some tests, lesion area on immunized plants was reduced to less than 2% of that on plants that were not immunized. Control plants surviving the challenge were protected against anthracnose. Growth of the immunized plants and yield were not affected by the immunization process (unpublished data).

Recently J.L. McIntrye (personal communication) has found that tobacco can be protected against black shank in the field using TMV as the inducer. Tobacco mosaic virus (TMV)-inoculation of a tobacco cultivar hypersensitive to the virus induced resistance against *Phytophthora parasitica* var. *nicotianae* in the field. A single TMV inoculation one week prior to transplanting reduced the incidence of black shank by 67% on 20 August 1979, but afforded no protection by 5 September. A pre-transplant inoculation and re-inoculation on 12 July provided protection through 5 September, but not by 18 September. A pre-transplant inoculation and subsequent inoculations on 12 July and 2 August significantly reduced disease incidence by about 50% throughout the growing season.

This year we expanded our investigations of immunization for the control of plant disease in the field to include blue mold of tobacco, and bacterial wilt and gummy stem blight of cucurbits. The use of TNV for the induction of immunity has increased the applicability of the technique to the field. A solution containing TNV can be applied as a high pressure spray to plants in the field, therefore, moist chambers and carefully controlled temperatures are not necessary for the expression of restricted symptoms. Thus, it no longer is necessary to rely upon induction of seedlings in the seed bed or greenhouse followed by transplanting.

Plant immunization does not control all diseases. We have not succeeded in immunizing cucurbits against powdery mildew or in using the pathogen for induction. Most of our tests in the field have been with high inoculum levels of challenge applied to the plants and the tests have been restricted to limited climatic and soil conditions. The effectiveness of immunization against diseases spread by insect vectors is being tested this year, e.g., bacterial wilt and gummy stem blight spread by the spotted or striped cucumber beetle. Immunization does not reduce yield of cucumbers; however, tobacco immunized against blue mold is stunted unless grown under conditions of high sunlight and soil nutrients (7,8). The amount of stunting of immunized, as compared to non-immunized tobacco, under low light intensity and relatively low soil nutrient levels, appears to vary with different cultivars (7).

Of course, the fact that a chemically-based defense occurs in plants is not evidence that the plants or plant products are safe or suitable for consumption by animals. As our research progresses, we hope to isolate and chemically characterize the inducer of immunization. Perhaps this substance will be effective when applied to foliage, fed via the roots, or used to treat seeds. Here again, however, the safety of the inducer to animals would remain to be proven.

I view plant immunization as an addition to the arsenal of disease control practises. Our number one defense against plant disease is still the development of resistant plants through breeding. At this time, breeding provides the only mechanism for resistance to viral diseases and the only economically sound mechanism for resistance to bacterial diseases. Plant immunization is successful against diseases caused by viruses, bacteria and fungi. It is systemic, persistent and activates resistance mechanisms in what have often been considered completely susceptible plants. It will not make chemical control obsolete, just as immunization in animals has not made the use of antibiotics, the pesticides of the animal kingdom, obsolete. It may prove to be a valuable tool for plant breeders and the chemical industry. We have initiated studies to breed plants for an enhanced facility to be immunized. Similarly, we and others are testing various synthetic chemicals for their activity as inducers. Plant immunization is strong evidence that highly effective resistance mechanisms are found in susceptible plants and that resistance often is not limited by the absence of genetic information for a mechanism but rather by whether the mechanism is expressed soon enough and with sufficient magnitude. How different is this concept for disease resistance in plants from that for disease resistance in animals?

SUMMARY

Three questions inevitably are asked when I discuss plant immunization before an audience:

(1) How is it possible that plant immunization has such a broad spectrum of effectiveness?
(2) What is the genetic basis for plant immunization?
(3) If plant immunization is so effective, why do we have serious diseases of plants?

I believe it is unlikely, but not impossible, that a single general mechanism explains immunization against diseases caused by viruses, bacteria and fungi. Furthermore, recent research indicates that restricted aphid infestation systemically protects cucumber against anthracnose (Bergstrom, unpublished data), and infection of tobacco with local lesion TMV systemically protects against aphid infestation (J.L. McIntyre, personal communication). The evidence is strong that reduced penetration through the epidermis by some fungi into immunized plants is a mechanism, but not the only mechanism, for resistance: (1) *C. cucumerinum* readily penetrates immunized plants (2) the development of fungi is restricted when infiltrated into tissues of immunized plants (3) immunization is effective against bacteria and viruses introduced through wounds (4) the multiplication of bacteria and infectivity or multiplication of viruses introduced into immunized plants are reduced.

To argue the less likely, but possible, various aspects of the lignification process can explain all aspects of immunization. Inhibition of penetration and fungal growth may be a function of the speed of lignification as related to the speed of penetration. Thus, conidia of *C. cucumerinum* readily germinate and penetrate the host before lignification has had an effect on penetration. Conidia of *C. lagenarium* germinate, form thick-walled appressoria, and penetration often requires 48 hours. In this case, the effect of lignification can be measured on penetration and growth. The lignification process could be highly toxic to fungal growth, bacterial multiplication and viral replication due to the toxicity of phenolic precursors, reactivity of free radicals, covalent binding of radicals to proteins, polysaccharides, and perhaps nucleic acids, rendering substrates unavailable, and binding the infectious agents to host walls. It appears that immunization is associated with the inhibition of pectinolytic enzymes produced by *M. melonis* (G. Bergstrom, unpublished data) and this also could be explained by the ability of free radicals and oxidation products to inactivate these enzymes.

The genetic basis for plant immunization may be a fundamental, primitive response to stress that has persisted from the earliest evolutionary development. Through the evolutionary process, infectious agents have adjusted to this basic stress response and plants have readjusted their defenses to the inroads of the infectious agents. The fundamental, broad-spectrum mechanism has persisted, is effective against many non-pathogens of a plant, and, when expressed rapidly enough, against some pathogens. The key to plant immunization may not be the development of a unique compound or process but rather the preparation or conditioning of cells to respond rapidly and with sufficient magnitude to contain infectious agents. Even in animals, the rather specific antibody-antigen response is only a small part of the overall defense against disease. Certainly it is important, but many consider it a mechanism of last resort when all non-specific mechanisms have failed. History tells us that the mechanisms for resistance in plants are equally effective for survival as are resistance mechanisms in animals.

Immunization following recovery from some diseases in animals is well documented and has probably been with us since the origin of higher animals. This aspect of immunization did not prevent smallpox and other diseases from repeatedly devastating populations of the world for centuries. The term *effective* depends upon definition. A mechanism for resistance which allows for the survival of man or a plant is effective from the long term view of perpetuating the species. An apple that is covered with scab lesions is unacceptable in today's consumer market, though the seeds of that apple may be capable of germinating and producing trees which bear more apples. A defense mechanism in man which allows for the death of multitudes periodically is no long *effective* by our standards, but the manipulation of this same *no longer effective* mechanism has produced the *highly effective* modern immunization which has *effectively* controlled some of the plagues of mankind. Plant immunization has the potential for being *highly effective* in modern agriculture.

REFERENCES

1. ANDEBRHAN, T., COUTTS, R.H.A., WAGIH, E.E. & WOOD, R.K.S. (1980). Induced resistance and changes in the soluble protein fraction of cucumber leaves locally infected with *Colletotrichum lagenarium* or tobacco necrosis virus. *Phytopathologische Zeitschrift 98*, 47-52.

2. ASADA, Y., OHGUCHI, T. & MATSUMOTO, I. (1979). Induction of lignification in response to fungal infection. In : *Recognition and Specificity in Plant-Host-Parasite Interactions*. Ed. by J.M. Daly & I. Uritani, University Park Press, pp 99-112, Baltimore.

3. CARUSO, F. & KUĆ, J. (1977). Protection of watermelon and muskmelon against *Colletotrichum lagenarium* by *Colletotrichum lagenarium*. *Phytopathology 67*, 1285-1289.

4. CARUSO, F. & KUĆ, J. (1977). Field protection of cucumber, watermelon and muskmelon against *Colletotrichum lagenarium* by *Colletotrichum lagenarium*. *Phytopathology 67*, 1290-1292.

5. CARUSO, F. & KUĆ, J. (1979). Induced resistance of cucumber to anthracnose and angular leaf spot by *Pseudomonas lachrymans* and *Colletotrichum lagenarium*. *Physiological Plant Pathology 14*, 191-201.

6. CHESTER, K.S. (1933). The problem of acquired physiological immunity in plants. *Quarterly Review of Biology 8*, 129-154, 275-324.

7. COHEN, Y. & KUĆ, J. (1980). Systemic resistance in tobacco against blue mold by soil inoculation with *Peronospora tabacina* (Accepted by *Phytopathology*).

8. CRUICKSHANK, I.A.M. & MANDRYK, M. (1960). The effect of stem infestation of tobacco with *Peronospora tabacina* Adam on foliage reaction to blue mold. *Journal of the Australian Institute of Agriculture Science 26*, 369-372.

9. DEHNE, H.W. & SCHÖNBECK, F. (1979). Untersuchungen zum einfluss der endotrophen mycorrhiza auf pflanzenkrankheiten 1. Ausbreitung von *Fusarium oxysporum* f. sp. *lycopersici* in tomaten. *Phytopathologische Zeitschrift 95*, 110.

10. GESSLER, C. & KUĆ, J. (1980). Induced resistance to *Fusarium* wilt in cucumber by soil and foliar pathogens. (submitted for publication to *Plant Disease*).

11. GOODMAN, R.N. (1980). Defenses triggered by previous invaders: bacteria. In *Plant Diseases Vol V* p 305-317. Academic Press, N.Y.

12. GRAHAM, T.L., SEQUEIRA, L. & HUANG, T.S.R. (1977). Bacterial lipopolysaccharides as inducers of disease resistance in tobacco. *Applied Environmental Microbiology 34*, 424-432.

13. GUEDES, M.E., RICHMOND, S. & KUĆ, J. (1980). Induced systemic resistance to anthracnose in cucumber as influenced by the location of the inducer inoculation with *Colletotrichum lagenarium* and onset of flowering and fruiting. *Physiological Plant Pathology 17*, 229 - 233.

14. HAMMERSCHMIDT, R. (1980). Lignification and related phenolic metabolism in the induced systemic resistance of cucumber to *Colletotrichum lagenarium* and *Cladosporium cucumerinum*. Ph.D. dissertation. University of Kentucky. 150pp.

15. HAMMERSCHMIDT, R. & KUĆ, J. (1980). Enhanced peroxidase activity and lignification in the induced systemic protection of cucumber. *Phytopathology 70*, 689.

16. HAMMERSCHMIDT, R., ACRES, S. & KUĆ, J. (1976). Protection of cucumber against *Colletotrichum lagenarium* and *Cladosporium cucumerinum*. *Phytopathology 66*, 790-793.

17. HECHT, E.I. & BATEMAN, D.F. (1964). Non-specific acquired resistance to pathogens resulting from localised infection by *Thielaviopsis basicola* or viruses in tobacco leaves. *Phytopathology 54*, 523-530.

18. HENDERSON, S.J. & FRIEND, J. (1979). Increase in PAL and lignin-like compounds as race-specific responses of potato tubers to *Phytophthora infestans*. *Phytopathologische Zeitschrift 94*, 323-334.

19. HENFLING, J.W.D.M. (1979). Aspects of the elicitation and accumulation of terpene phytoalexins in the potato tuber-*Phytophthora infestans* interaction. Ph.D. dissertation University of Kentucky, 233 pp.

20. HIJWEGEN, T. (1963). Lignification, a possible mechanism of active disease resistance against pathogens. *Netherlands Journal of Plant Pathology 69*, 314-317.

21. JENNS, A & KUĆ, J. (1977). Localized infection with tobacco necrosis virus protects cucumber against *Colletotrichum lagenarium*. *Physiological Plant Pathology 11*, 207-212.

22. JENNS, A. & KUĆ, J. (1979). Graft transmission of systemic resistance of cucumber to anthracnose induced by *Colletotrichum lagenarium* and tobacco necrosis virus. *Phytopathology 69*, 753-756.

23. JENNS, A. & KUĆ, J. (1980). Characteristics of anthracnose resistance induced by localized infection with tobacco necrosis virus. *Physiological Plant Pathology 17*, 81-91.

24. JENNS, A., CARUSO, F., & KUĆ, J. (1980). Non-specific resistance to pathogens induced systemically by local infection of cucumber with tobacco necrosis virus, *Colletotrichum lagenarium* Pass (Ell.et Halst) or *Pseudomonas lachrymans* (Sm. et Bryan) Carsner. *Phytopathologia Mediterranea* (in press).

25. KUĆ, J. & CARUSO, F. (1977). Activated co-ordinated chemical defense against disease in plants. In : *Host Plant Resistand to Pests* Ed. by P. Hedin pp. 78-89. American Chemical Soc. Symposium Series 62, American Chemical Society Press, Washington, D.C.

26. KUĆ, J. & RICHMOND, S. (1977). Aspects of the protection of cucumber against *Colletotrichum lagenarium* by *Colletotrichum lagenarium*. *Phytopathology 67*, 533-536.

27. KUĆ, J., SCHOCKLEY, G. & KEARNEY, K. (1975). Protection of cucumber against *Colletotrichum lagenarium* by *Colletotrichum lagenarium*. *Physiological Plant Pathology 7*, 195-199.

28. LOEBENSTEIN, G. (1972). Localization and induced resistance in

virus infected plants. *Annual Review of Phytopathology 10*, 177-206.

29. MANDRYK, M. (1963). Acquired systemic resistance to tobacco mosaic virus in *Nicotiana tabacum* evoked by stem infection with *Peronospora tabacina* Adam. *Australian Journal of Agricultural Research 14*, 315-318.

30. MASSALA, R., LEGRAND, M. & FRITIG, B. (1980). Effect of α-aminooxyacetate, a competitive inhibitor of phenylalanine ammonia-lyase, on the hypersensitive resistance of tobacco to tobacco mosaic virus. *Physiological Plant Pathology 16*, 213-226.

31. MATTA, A. & GARABALDI, A. (1977). Control of *Verticillium* wilt in tomato by pre-inoculation with avirulent fungi. *Netherlands Journal of Plant Pathology 83*, (Suppl. 1) 457-462.

32. McINTYRE, J.L. & DODDS, J.A. (1979). Induction of localized and systemic protection against *Phytophthora parasitica* var. *nicotianae* by tobacco mosaic virus infection of tobacco hypersensitive to the virus. *Physiological Plant Pathology 15*, 321-330.

33. MÜLLER, K.O. (1956). Einige einfache versuche zum nachweis von phytoalexinen. *Phytopathologishe Zeitschrift 27*, 237-254.

34. NADOLNY, L. & SEQUEIRA, L. (1980). Increases in peroxidases are not directly involved in induced resistance in tobacco. *Physiological Plant Pathology 16*, 1-8.

35. PEARCE, R.B. & RIDE, J.P. (1978). Elicitors of the lignification process in wheat. *Annals of Applied Biology 89*, 306.

36. PEARCE, R.B. & RIDE, J.P. (1980). Specificity of the induction of the lignification response in wounded wheat leaves. *Physiological Plant Pathology 16*, 197-204.

37. RAST, A. TH. B. (1975). Variability of tobacco mosaic virus in relation to control of tomato mosaic in glasshouse tomato crops by resistance breeding and cross protection. *Agricultural Research Report 834 of the Institute of Phytopathological Research, Wageningen, The Netherlands* 76 pp.

38. RICHMOND, S., KUĆ, J. & ELLISTON, J.E. (1979). Penetration of cucumber leaves by *Colletotrichum lagenarium* is reduced in plants systemically protected by previous infection with the pathogen. *Physiological Plant Pathology 14*, 329-338.

39. RIDE, J.P. (1980). The effect of induced lignification on the resistance of wheat cell walls to fungal degradation.

Physiological Plant Pathology 16, 187-196.

40. ROSS, A.F. & BOZARTH, R.F. (1960). Resistance induced in one plant part as a result of virus infection in another part. *Phytopathology 50*, 652.

41. ROSS, A.F. (1964). Systemic resistance induced by localized virus infections in beans and cowpea. *Phytopathology 54*, 436.

42. ROSS, A.F. (1966). Systemic effects of local lesion formation. In: *Viruses of Plants*, Ed. by A.B.R. Beemster & J. Dijkstra pp. 127-150. North Holland Publishing Company, Amsterdam.

43. SEQUEIRA, L. (1979). The acquisition of systemic resistance by prior inoculation. In: *Recognition and Specificity in Plant Host-Parasite Interactions*. Ed. by J.M. Daly & I. Uritani, pp. 231-251. University Park Press, Baltimore.

44. SIEGEL, S.M. (1957). Non-enzymic macromolecules as matrices in biological synthesis: the role of polysaccharides in peroxidase - catalyzed lignin polymer formation from eugenol. *Journal of the American Chemical Society 79*, 1628-1632.

45. STAUB, T. & KUĆ, J. (1980). Systemic protection of cucumber plants against disease caused by *Cladosporium cucumerinum* and *Colletotrichum lagenarium* by prior localized infection with either fungus. *Physiological Plant Pathology* (in press).

46. TJAMOS, E.C. (1979). Induction of resistance to *Verticillium* wilt in cucumber (*Cucumis sativus*). *Physiological Plant Pathology 15*, 223-227.

47. VANCE, C.P., SHERWOOD, R.T. & KIRK, T.K. (1980). Lignification as a mechanism of disease resistance. *Annual Review Phytopathology 18* (in press).

48. YARWOOD, C.E. (1956). Cross protection with two rust fungi. *Phytopathology 46*, 540-544.

GENETICAL ASPECTS OF ACTIVE DEFENCE

ALBERT H. ELLINGBOE

Genetics Program and Department of Botany and
Plant Pathology
Michigan State University, East Lansing
Michigan 48824, U.S.A.

INTRODUCTION

I find myself in a rather uncomfortable position in presenting this paper. The reason for this is that what I have to say does not support much of the speculation and dogma that has been presented in preceding papers. I feel like a non-believer in a congregation of believers. The belief seems to be that specificity of the interactions between host and parasite, and the restriction of development of the pathogen, are controlled by compounds not produced by transcription and translation. This I find difficult to reconcile with the known genetics of interactions. What I will try to do, therefore, is to present the kinds of results obtained from studies of the inheritance of genetic variability in host-parasite interactions, then proceed to describe some models and give examples of what the genetics tells me about those models of interactions, and conclude with some examples of uses of genetic arguments in studies of host-parasite interactions.

Inheritance of interactions

It is possible to show variability in both host and parasite for most diseases. This is commonly done by inoculating host lines

This research was supported in part by Grants from the National Science Foundation (PCM78-22898) and the U.S. Department of Agriculture (7900299)

Michigan Agricultural Experiment Station Journal Article

collected from many geographically separated regions (or from world collections of the host species) with collections of the pathogen species, also usually from many geographically separated areas. The pattern commonly obtained is illustrated in Table 1. It can be seen from these data that no two host lines react identically with the five parasite strains and that no two parasite strains give the same phenotype with the five host lines. This is the type

Table 1 An example of the type of variability observed when several host lines are inoculated with several strains of a pathogen

Parasite strain	Host line				
	A	B	C	D	E
a	1[a]	9	0	5	9
b	9	1	0	5	9
c	3	9	5	9	9
d	3	5	9	0	9
e	4	2	7	9	9

[a] Each number represents a unique phenotype. The 9 rating is considered the most compatible relationship (host susceptible, parasite virulent) and the lower numbers the different degrees of incompatibility

of variability that exists in nature and is considered naturally-occurring genetic variability in host-parasite interactions. Each host line may contain one, two, three or more *R* genes. Crosses can be made to isolate the individual *R* genes and establish isogenic host lines homozygous for *R*x*R*x and homozygous *r*x*r*x. Comparable analyses can be made to isolate single *P* genes in the parasite. When detailed analyses are made of the naturally-occurring variability, the basic pattern given in Table 2 emerges. One unique aspect of this pattern is the relationship of one gene in the host for one gene in the pathogen. A host line with an *R* gene cannot be resistant to a pathogen unless the pathogen has the corresponding *P* gene for avirulence. Conversely, a pathogen strain with a

P gene cannot be avirulent to a given host line unless the host has the corresponding *R* gene for resistance. This one-for-one, gene-for-gene relationship was first demonstrated for the flax rust disease (3), but subsequent research has indicated that the basic pattern emerges in essentially all interactions between plants and pathogens of plants. There are two apparent exceptions. The plant pathogens which produce host specific toxins appear to follow a pattern which is the reciprocal of the one presented in Table 2, i.e. one + and three - (1). The other exception is the example in *Melampsora lini* in which there is a suppressor of several *P* genes (5). The latter gives the initial appearance of two genes in the pathogen for one gene in the host, plus the appearance of a dominant gene for virulence. Detailed analyses show that only one pathogen gene confers the specificity, the other affects the expression of the gene conferring specificity. Whether the apparent reversal of the basic pattern in Table 2 is real with toxin-producing pathogens or the result of incomplete studies is unknown.

Table 2 The basic pattern of genetic interactions that emerges when there is a one allele difference in both host and parasite

Parasite genotype	Host line	
	$R1R1$	$r1r1$
$P1$	-[a]	+
$p1$	+	+

[a] "-" incompatible "+" compatible

The "-" can differ from "+" infection type, infection efficiency, rate of growth of parasite, or any other measurable difference

How should the basic pattern in Table 2 be interpreted ? The simplest interpretation is that specific recognition is for incompatibility. The product of the $R1$ gene interacts with the product of the $P1$ gene to give an incompatible relationship. The presumed failure of interaction in the other three parasite/host combinations (i.e. $P1/r1$, $p1/R1$ and $p1/r1$) leads to a compatible relationship.

A host plant like wheat has at least 30 loci involved in reaction with *Puccinia graminis* f. sp. *tritici*. What, then, is the pattern for the two or more genes in host and in parasite ? The basic pattern for two parasite/host gene pairs is presented in Table 3. There are two points to get from this pattern. One is the specificity. *P*1 recognizes only *R*1 to give an incompatible relationship. *P*2 recognizes only *R*2 to give an incompatible relationship. *P*1 does not recognize *R*2 and *P*2 does not recognize *R*1. There is a strict one-for-one relationship. The second point to get from this pattern is that if one parasite/host gene pair specifies incompatibility, it is epistatic to all gene pairs that specify compatibility.

Table 3 The basic genetic pattern for two parasite/host gene pairs

Parasite genotype	*R*1*R*1 *R*2*R*2	*R*1*R*1 *r*2*r*2	*r*1*r*1 *R*2*R*2	*r*1*r*1 *r*2*r*2
*P*1*P*2	–[a]	–	–	+
*P*1*p*2	–	–	+	+
*p*1*P*2	–	+	–	+
*p*1*p*2	+	+	+	+

[a] "–" incompatible "+" compatible

There are three other arguments that support the concept that specific recognition is for incompatibility. Only a cursory statement will be made on each. One is the temperature-sensitivity of naturally-occurring variability. Temperature-sensitive interactions usually give incompatibility at normal temperatures (permissive) and compatibility at the higher (non-permissive) temperatures. The second is the behaviour of allelic genes. Allelic (pseudoallelic ?) genes in the host are common. A *P* gene in a pathogen will recognize only one allele at a locus in the host, and treat the other alleles as a single recessive allele. The third argument pertains to induced mutations. Mutations to virulence and susceptibility are easy to obtain compared to mutations to avirulence on plants with single *R* genes and mutations to resistance. This is consistent with specific recognition for incompatibility, and compatibility as the result of no mutual recognition of the gene products.

The principal point that I want to make from the above

discussion is that in at least 95% of the genetic variability in host-parasite interactions studied in considerable detail, there is a strict one gene in the host to one gene in the parasite type of relationship. Cursory investigations have frequently suggested deviations from this one-to-one relationship, but subsequent, more detailed analyses, have shown that the one-to-one relationship is the basic pattern. This point needs to be stressed because it bears directly on a number of hypotheses presented at this meeting and elsewhere in the literature.

WHAT DOES THE GENETICS TELL ME ABOUT THE MOLECULAR INTERACTIONS ?

In this section I want to give a brief discussion of four models to explain the variation in host-parasite interactions that has been observed within one host species and within one pathogen species. The models do not have their origin with me. I just intend to show what the genetics would look like if each of these models was correct. These models are but the four most commonly expoused in the plant pathology literature.

The first model is that the primary product of a *P* gene in the pathogen interacts with the primary product of the corresponding *R* gene in the host to form a dimer(1) (Figure 1). The monomers are not active in bringing about an incompatible relationship but the dimer is. (Genetically, the possibility that the active molecule is a dimer cannot be separated from the possibility that the active molecule is a monomer that is affected by having interacted with the product of the corresponding gene (see Figure 2). There is specific recognition in the formation of a biologically active dimer, and the dimer is the cause for incompatibility. This model would predict a strict genetic one-for-one relationship when inheritance of the interactions is examined in both host and parasite. A strict one-for-one relationship has been observed when detailed inheritance studies have been performed.

A second frequently espoused concept is that the specificity of host-parasite interactions is brought about by the interaction either between glycoproteins or between a glycoprotein and a complex carbohydrate (4, 8); the latter is usually considered analogous to the ABO blood group in man. Let us consider a glycoprotein-protein interaction. How are glycoproteins assembled ? There would probably be one structural gene which coded for the protein (see Figure 3). There would probably be one enzyme if only one sugar moiety were attached to the protein. There would probably be an additional enzyme if a long chain of one sugar were attached to the protein. Additional enzymes would be needed if different sugars were to be added, if the type of linkage were important, or if branching at specific sites were important. The assembly of complex carbohydrates has been shown to involve many

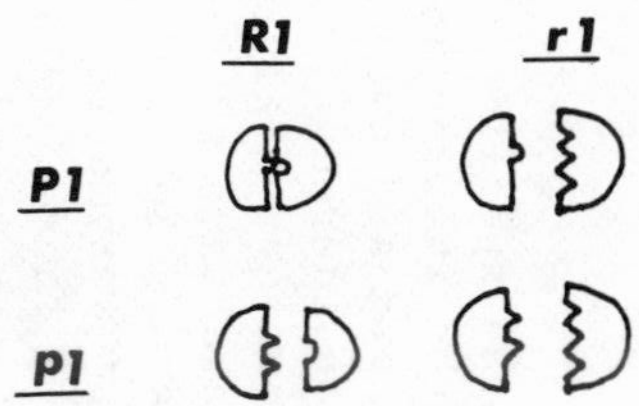

Fig. 1 Diagram showing that the primary product of *P1* interacts with the product of *R1* but that with the other three parasite/host genotypes the products do not interact

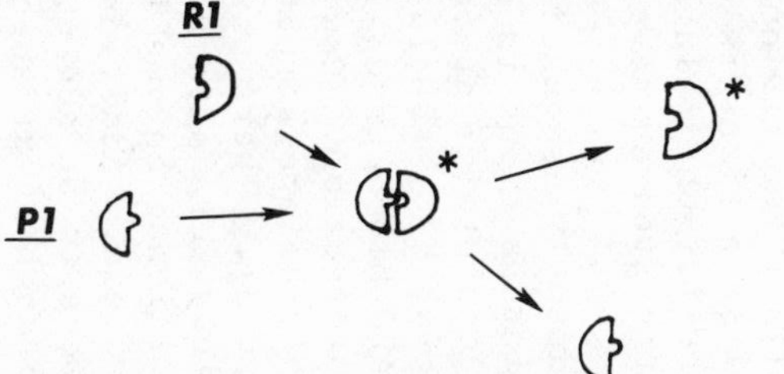

Fig. 2 Diagram showing that the active molecule is either the dimer formed from parasite and host primary products or a monomer activated by having interacted

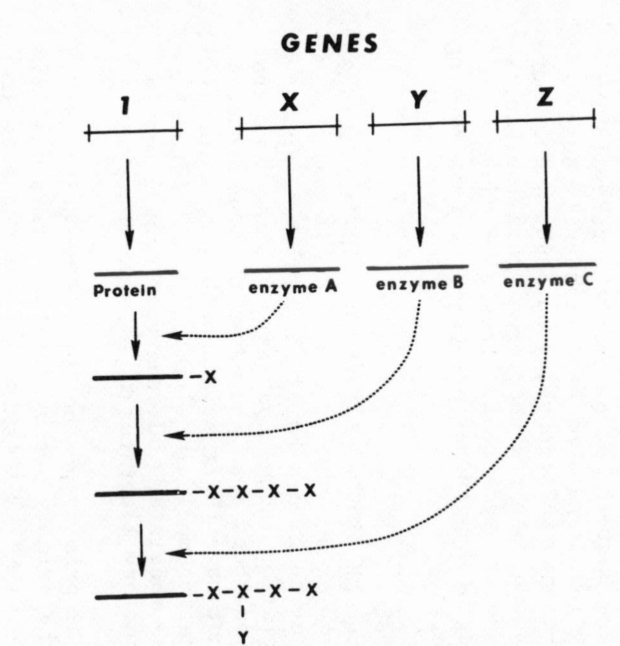

Fig. 3. A model for the genetic control of the synthesis of a glycoprotein whose specificity depended on the carbohydrate attached to the protein. Gene 1 is the structural gene for the protein. Gene 2 codes for enzyme A that recognizes the protein to attach sugar X to the protein. Gene 3 codes for enzyme B that attaches many units of sugar X with a specific linkage to make a long chain carbohydrate. Gene 4 codes for enzyme C that attaches sugar Y to a particular unit of X

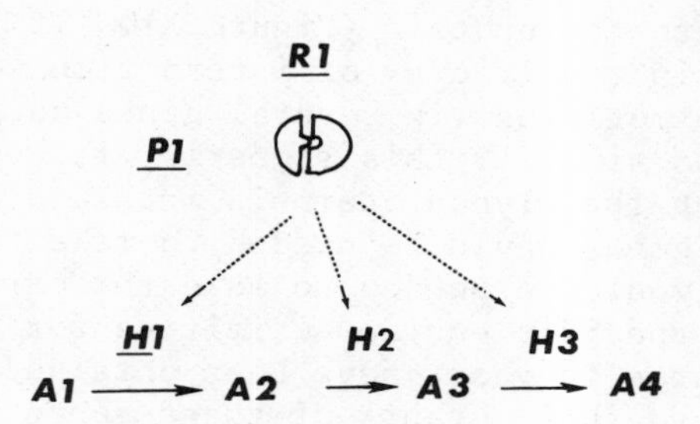

Figure 4 A model for host-parasite interactions whereby the products of parasite gene *P*1 interacts with the product of host gene *R*1 to activate host genes *H*1, *H*2 and *H*3. The products of *H*1, *H*2 and *H*3 are enzymes involved in concatenate reactions to synthesize phytoalexin A4 which restricts the growth of the pathogen

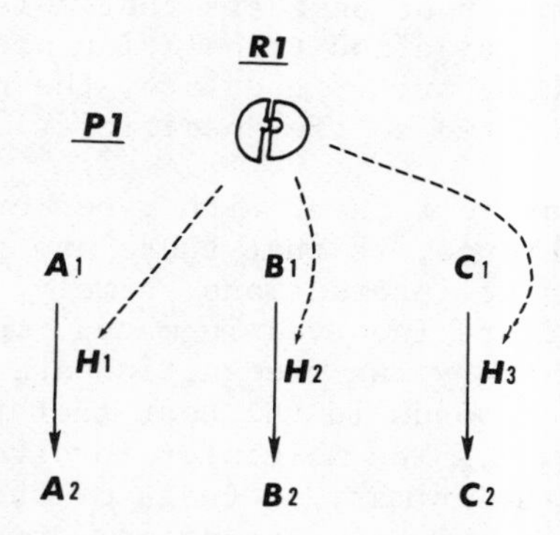

Figure 5 A model for host-parasite interactions whereby the products of parasite gene *P*1 interacts with the product of host gene *R*1 to activate host genes *H*1, *H*2 and *H*3. The products of *H*1, *H*2 and *H*3 are enzymes, each of which is involved in the final step in synthesizing phytoalexins A2, B2 and C2 each of which is capable of restricting the growth of the pathogen

enzymes (7). The number of genes to assemble a glycoprotein could be very large. There would be at least one gene for the protein and one gene for each of the enzymes involved in the attachment of the carbohydrates to the protein (Figure 3). If the intact molecule is important in specificity of interaction with its corresponding molecule, then mutations in several genes could lead to a loss of specificity. To simplify this scenario, let us assume that the molecule with which the glycoprotein is interacting is a simple protein. Several genes would be needed to make the glycoprotein but only one gene would be needed to make the corresponding protein. Mutants that lack specific enzyme activities for synthesizing complex carbohydrates in yeast have been obtained (7), so there is no reason to believe that mutants of these genes will be lethal. Therefore, if the basis for specificity were interactions involving glycoproteins, the carbohydrate portions of which was the basis for specificity, there would be predictions of deviations from a one gene-for-one gene relationship. In the example given in Figure 3, if there were four genes involved in synthesizing the glycoprotein in the parasite and one for the corresponding protein in the host, the prediction is there would be a four gene to one gene relationship, a situation not encountered in the genetic analysis of naturally-occurring host-parasite genetic variability. The failure to find 2:1, 3:1, 1:2, 1:3, etc. ratios of genes in the pathogen to genes in the host suggests that molecules not the primary product of transcription and translation are not involved in host-parasite specificity or, if they are, the post translational modification is not related to the specificity of the molecule.

A concept that has lost favor with geneticists, but is very dominant at this conference, is that there are two types of genes involved in host-parasite interactions, namely, genes that code for the molecules that are involved in mutual recognition, and genes that are turned on by the recognition and that make the enzymes that make the compounds in the host that restrict the development of the pathogen. The molecules involved in specificity are considered to be the primary products of the genes involved in the gene-for-gene interactions. The primary products interact, and the result of that interaction is not the restriction of growth of the pathogen, but the setting in motion of the synthesis of compounds that are toxic to the pathogen. Let us assume that the toxic compounds are phytoalexins.

In the first model the assumption will be that one or a few closely related phytoalexins are produced by a series of concatenate reactions. This is illustrated in Figure 4. For incompatibility, there is a need for an *R* gene whose product can interact with the product of the *P* gene in the pathogen and for an allele of each of the *H*1, *H*2, and *H*3 loci that produces a product that functions in the biosynthesis of the phytoalexin A4. A mutation in *R*1 that led to a loss in recognition of *P*1 would not be able to promote the synthesis of enzymes coded by *H*1, *H*2, and *H*3, and, therefore, would lead to susceptibility to a pathogen of genotype *P*1. Mutations in *H*1, *H*2, and *H*3 would also lead to the loss in ability to synthesize A4 and, therefore, would lead to susceptibility. A mutation in any one of four genes in the host could lead to susceptibility. A mutation in one gene in the pathogen could lead to virulence. This is a one gene to four gene relationship, a pattern that has never been observed in the genetic analysis of naturally-occurring variability.

One other variation of this model for the phytoalexins is presented in Figure 5. This model assumes that the product of *P*1 interacts with the product of *R*1 to activate three parallel pathways, each giving rise to a molecule (e.g., A2, B2 and C2) that is capable of restricting the growth of the pathogen. A mutation in *R*1 that led to a loss of recognition of *P*1 would not lead to the accumulation of any of the three toxic compounds. A mutation of the gene that codes for enzyme *H*1 would not lead to susceptibility because compounds B2 and C2 would restrict the development if genes *R*1, *H*2, and *H*3 were still producing functional gene products. In the presence of *R*2, *h*2, and *h*3, a mutation of *H*1 to *h*1 would change the plant from resistant to susceptible to a parasite with genotype *P*1. The expected genetic pattern of a host with the three induced parallel systems would behave as one unique single gene (*R*1) and triplicate factors (*H*1, *H*2, and *H*3). Basically it is a one gene to four gene pattern but the epistatic relationships have changed from the preceding model. The phenotype of the last two models is given in Table 4. Each of these individuals could be identified by test crosses.

The results from genetic analyses of naturally-occurring variability do not support the expected results with the last three models. The results are consistent with the first model presented. Why do we not see genetic variability of the type predicted for models 2, 3 and 4 ? Could it be that mutations in genes like *H*1, *H*2, and *H*3 are lethal ? On what basis would we predict that genes which modify phenolics would be lethal ? Why do we not see at least variation in the ability to be affected by the *P*2/*R*2 genotype ?

Mutations that affect the structure of polysaccharides in yeast have been obtained (7). The mutants are not lethal. The mutations map at several loci. If model 2 were correct (i.e., specificity resides in glycoproteins) why do we not see many genes involved in conferring specificity ? How can we rationalize the existence of multiple alleles if the specificity is controlled by glycoproteins ?

These are only but a few of the questions that can be asked of the four models presented. The genetics are consistent with the first model. The postulate of the many necessary genes to explain the last three models is not substantiated by genetic analysis.

Table 4 The expected phenotype of seventeen of thirty-two possible parasite/host genotypes with the models presented in Figures 4 and 5

Parasite genotype	Host genotype	Phenotype of reaction	
		concatenate	parallel
*P*1	*R*1*H*1*H*2*H*3	– [a]	–
*P*1	*R*1*H*1*H*2*h*3	+	–
*P*1	*R*1*H*1*h*2*H*3	+	–
*P*1	*R*1*H*1*h*2*h*3	+	–
*P*1	*R*1*h*1*H*2*H*3	+	–
*P*1	*R*1*h*1*H*2*h*3	+	–
*P*1	*R*1*h*1*h*2*H*3	+	–
*P*1	*R*1*h*1*h*2*h*3	+	+
*P*1	*r*1*H*1*H*2*H*3	+	+
*P*1	*r*1*H*1*H*2*h*3	+	+
*P*1	*r*1*H*1*h*2*H*3	+	+
*P*1	*r*1*H*1*h*2*h*3	+	+
*P*1	*r*1*h*1*H*2*H*3	+	+
*P*1	*r*1*h*1*H*2*h*3	+	+
*P*1	*r*1*h*1*h*2*H*3	+	+
*P*1	*r*1*h*1*h*2*h*3	+	+
*p*1	all 16 host genotypes	+	+

[a] "–" incompatible "+" compatible relationship

The genes in the host and parasite determine the phenotype of the interaction. The apparent ubiquity of the gene-for-gene relationship in host-parasite systems is a fact that must be considered in building models of the molecular basis of host-parasite interactions. I, for one, find it very difficult to accept hypotheses based only on comparative physiology and comparative biochemistry when it is clear that the hypotheses are inconsistent with the genetic variability they are intended to explain.

USE OF GENETIC ARGUMENTS IN STUDIES OF HOST-PARASITE INTERACTIONS

Genetics theory has been very useful in the elucidation of basic life processes. Genetic arguments have also been very useful in viewing a problem so that it can be analyzed experimentally. In this section, four examples of possible uses of genetic arguments for studies of host-parasite interactions will be given.

Use of temperature-sensitive mutants

I have long been concerned that almost all genetic variability follows the gene-for-gene relationship. One can rationalize the evolution of this relationship if the beginning point is a compatible relationship followed by the change to incompatibility followed by the change back to compatibility (1). The gene-for-gene relationship, therefore, appears to be superimposed on a basic compatibility. Mutations of genes whose function is crucial to successful parasitism (but not avirulence) might be lethal, particularly for an obligate parasite. How then does one get variability of the genes involved in basic compatibility ? One way is to induce conditional mutations. Geneticists have usually preferred temperature-sensitive mutants, primarily high temperature-sensitive mutants. Most high temperature-sensitive mutants are missense mutations that yield a protein product that has biological activity at a normal temperature but loses its activity at a higher temperature, whereas the wild type has activity at both temperatures. The basis for the sensitivity is usually the stability of the tertiary structure of the molecule. A conditional mutant is also convenient because its effect is usually reversible when the temperature is changed from high to low.

Three types of temperature-sensitive mutants have been obtained in *Colletotrichum lindemuthianum* (2). Wild type cultures will grow on agar media or produce disease in bean plants at both 22° and 28°C. Class I mutants can grow on media supplemented with yeast extract at 22° but not at 28°C, or they can produce disease at 22° but not 28°C. These mutants should be useful to study the ontogeny of interactions. For example, by temperature shift experiments it should be possible to determine the relationship of growth of the parasite to the physiological changes associated with the interaction. Class II mutants are temperature-sensitive on agar media but not in the host. Obviously the host can provide something that we cannot provide in agar media. These mutants should be useful to determine what is transferred from host to parasite. Class III mutants, the most interesting to me, are not temperature-sensitive in agar media but they are in the host. The mutations, therefore, are not of genes whose function is crucial to basic life processes, or they would have been temperature-sensitive on agar media. They appear to be mutations of genes whose function is crucial to successful parasitism and pathogenesis. How many loci are involved we do not know.

Identification of the number of functions

A common practice in biology has been to get large numbers of mutants to get an estimate of the number of genes that control a particular process. For example, if one were to produce 100 adenine-requiring mutants in an organism, it could be predicted with reasonable certainty that the mutations would map in a maximum of eight loci. That is because of the number of enzymatically controlled steps that can be affected and still recovered with the addition of adenine. Unfortunately, no such effort has been made with a plant pathogen, though production of a host specific toxin would be an ideal candidate for such an analysis.

Identify primary products of genes

The plant pathology literature contains many examples of detailed bioassays to be used to demonstrate the role of a particular molecule in host-parasite interactions. A very direct analysis to determine if a protein is the product of an *R* gene is to determine the relationship of a mutant of that gene to the primary structure of the protein suspected as being the product of that *R* gene. A mutation of a structural gene should lead to the production of an altered primary structure of the gene product. The proof of the relationship of a mutation (identified as a biological effect) to the primary structure of a protein can be used to show the role of the protein in host-parasite interactions even though there is no way to establish a bioassay for the effect of the isolated protein on host or parasite.

Use of mutants to test the phytoalexin hypothesis

Phytoalexins have been studied for more than twenty years. It has been assumed that these low molecular weight compounds that accumulate in infections are the basis of restricting the development of pathogens. There seem to be two basic hypotheses. One is that the successful pathogen does not stimulate the formation of sufficient phytoalexin, fast enough, to restrict its growth whereas the unsuccessful pathogen stimulates the host to make sufficient phytoalexin, fast enough, to restrict its growth. The second hypothesis is that both successful and non-successful pathogens stimulate the formation of phytoalexins but the successful pathogen can degrade the phytoalexins. If the phytoalexins are, in fact, the basis of restriction of growth of the pathogen, there are two obvious experiments. One is to produce a mutation to resistance to the phytoalexin. If the phytoalexin is the basis of restriction of growth of the pathogen, a mutation to resistance should allow the pathogen to grow more extensively. If the phytoalexin is the result, not the cause of the incompatibility that led to the restriction of growth of the pathogen, the mutation to resistance should have little or no effect on growth of the pathogen in host tissue. A second experiment would be to produce mutations in the pathogen that eliminate or reduce the ability of the organism to degrade the phytoalexin. These two experiments are intended to give different results. The first is intended to convert a non-pathogen to a pathogen (mutation to increased virulence). The second is intended to convert a pathogen to a non-pathogen (mutation to decreased virulence). The fundamental question is whether or not phytoalexins do play a role in determining resistance or susceptibility.

The above are but four simplified types of uses of genetics in studies of host-parasite interactions. There are many more approaches of these types that depend almost entirely on genetic theory.

What I have tried to do in this paper is to review some of the basic genetics of host-parasite interactions, what the basic genetics tells me about possible models to explain the genetic variability in these interactions, and some obvious uses of genetics to test models put forth by plant pathologists.

The basic genetics of interactions between a host and a parasite were first worked out for plant pathogens. The basic pattern seems to hold for all plant pathogens, be they fungi, viruses, nematodes, bacteria. The basic pattern appears to hold for pathogens of animals that are not antigenic. The pattern seems sufficiently universal to have been called interorganismal genetics (6).

The genetic variability has been exploited for practical gain. A major effort around the world is to exploit genetic variability in hosts and parasites to control diseases. Physiologists, biochemists, ultrastructuralists have used the naturally-occurring genetic variability for their comparative studies. But for all the studies with genetic variability, we still do not know the primary product of a single *R* gene in a host or a single *P* gene in a pathogen. Furthermore, the effort to find the primary product of an *R* or *P* gene is minimal.

The data from studies on inheritance of genetic variability do not support many of the theories of host plant resistance. I suspect that it will be necessary to get closer to the gene primary products before questions of whether resistance is active or passive can be answered.

REFERENCES

1. ELLINGBOE, A.H. (1976). Genetics of host-parasite interactions. In : *Encyclodpedia of Plant Physiology*, New Series, Vol. 4, Physiological Plant Pathology, Eds. R. Heitefuss and P.H. Williams. Springer-Verlag, Berlin, Heidelberg, New York.

2. ELLINGBOE, A.H. & GABRIEL, D.W. (1977). Induced conditional mutants for studying host/pathogen interactions. In : *Induced mutations against plant diseases*. International Atomic Energy Agency, Vienna.

3. FLOR, H.H. (1955). Host-parasite interaction in flax rust - its genetics and other implications. *Phytopathology* 45, 680 - 685.

4. KEEN, N.T. (1980). Mechanisms conferring specific recognition in gene-for-gene plant-parasite systems (this Symposium).

5. LAWRENCE, G.J., MAYO, G.M.E. & SHEPHERD, K.W. Interactions between genes controlling pathogenicity in flax rust. *Phytopathology* in press.

6. LOEGERING, W.Q. (1978). Current concepts in Interorganismla Genetics. *Annual Review of Phytopathology* 16, 309 - 320.

7. RASCHKE, W.C., KERN, K.A., ANTOLIS, C. & BALLOU, C.E. (1973). Genetic control of yeast mannan structure-isolation and characterization of mannan mutants. *Journal of Biological Chemistry* 248, 4660 - 4666.

8. WADE, M. & ALBERSHEIM, P. (1979). Race-specific molecules that protect soybeans from *Phytophthora megasperma* var. *sojae*. *Proceedings National Academy of Science, U.S.A.* 76, 4433 - 4437.

ACTIVE RESISTANCE OF PLANTS TO VIRUSES

B.D. HARRISON

Scottish Crop Research Institute
Invergowrie
Dundee, Scotland, U.K.

INTRODUCTION

The aim of this contribution is to survey the range of phenomena that seem relevant to active resistance of plants to viruses and to provide some information that may serve as a background against which some of these phenomena can be examined in more detail by other contributors. However, a little thought soon leads to the realisation that viruses pose some conceptual problems that do not exist for fungal and bacterial pathogens, and these must first be discussed.

CONCEPTUAL AND PRACTICAL DIFFICULTIES

The main problem in applying to plant viruses the concepts of active resistance established for plant pathogenic fungi and bacteria is that, unlike the fungi and bacteria, plant viruses have no metabolic systems of their own and rely for their replication on synthetic systems of host cells. The host is inextricably involved in virus multiplication. Moreover viruses are not separated by a semi-permeable membrane from the contents of host cells. Also, they invade plants systemically by way of pre-existing pathways in plants, plasmodesmata and phloem sieve tubes, whereas the larger size of plant pathogenic fungi and bacteria means that they must use other methods for invading tissue, and they do not produce fully systemic infections. Furthermore the ways in which the plant is first penetrated by fungi and bacteria differ radically from those in which viruses enter plants because penetration of host cells by viruses depends on a third agent, in most instances a specific vector. The vector may be an insect, mite, nematode or fungus which makes a break in the cuticle and outer wall of a plant cell and delivers virus particles into the cell. In other examples it is man who makes wounds suitable for infection, or it may be pollen which delivers virus particles *via* a pollen tube into a cell of the floral tissue. There is also some difference between viruses on the one hand, and fungi and bacteria on the other, in the concept of what is a host. A plant in whose cells the virus genome replicates is considered a host whereas a plant that allows a fungus to achieve substantial growth may be considered a non-host if the development of the fungus then stops.

The existence of these obvious differences gives cause to consider how one should apply, or perhaps whether one should apply to plant viruses, the concept of active resistance. Indeed, with many possible examples of active resistance to plant viruses it is hard to say exactly what role the host plays, although host metabolic processes are so intimately involved in virus replication that almost any facet of the behaviour of viruses may have a virus-induced, host-dependent component. For example, if active resistance to viruses is taken to include any virus-induced change in the host that results either in a decreased effect of infection or in decreased susceptibility to infection, in what category should one place the many species that fail to become infected with a given virus or the large majority of plant virus diseases that do not kill the host plant ? Has the plant resisted infection or been spared from serious ill-effects by its active defence mechanisms, or has the resistance been passive or has the virus regulated its own replication in such a way that the host survives without great damage ?

The reliance of viruses for survival on methods of transmission which may be possible only during limited periods of the year means that a virus that kills its hosts quickly, or has a strong adverse effect on the hosts' competitive ability, will survive much less readily than one without these attributes. In contrast, many fungi produce spores, in some instances long-lived, only a few days after infection and are thereby disseminated to fresh hosts. A special example of the establishment of a relatively benign relationship between host and virus is seen in tobacco plants infected with nepoviruses such as tobacco ringspot and raspberry ringspot viruses. The inoculated leaves and the first leaves to be invaded systemically develop necrotic and chlorotic rings and lines, but leaves produced subsequently are almost symptomless, although infected. It seems that cells invaded after they have begun to differentiate may become severely affected whereas those invaded at an earlier stage develop into almost symptomless tissue. Some recent work with raspberry ringspot virus shows that when leaves are small their virus content is similar irrespective of whether they will later develop symptoms. However, whereas virus continues to accumulate in leaves that develop symptoms, the virus concentration remains almost static in the leaves that remain symptomless (3). Clearly some regulatory process is operating but whether or not this is the consequence of a virus-induced reaction of the host is not known.

VIRUS REPLICATION

Before proceeding further, it is probably worth examining the series of events involved in virus multiplication, with a view to identifying processes in which the host may play an important role. The genomes of plant viruses are of several distinct types (Table 1). Many, like that of tobacco mosaic virus (TMV), consist of linear, single-stranded RNA, but the genomes of fijiviruses and phytoreoviruses are double-stranded RNA, those of caulimoviruses are double-stranded DNA, and the genomes of geminiviruses are single-stranded DNA. Moreover, the DNA genome molecules typically are circular.

Table 1 Types of genomes of plant viruses

		Virus group	Example
Single-stranded RNA			
Positive (plus) strand	1 piece	Tobamovirus	Tobacco mosaic virus
		Tymovirus	Turnip yellow mosaic virus
	2 pieces	Tobravirus	Tobacco rattle virus
		Comovirus	Cowpea mosaic virus
	3 pieces	Bromovirus	Brome mosaic virus
		Ilarvirus	Tobacco streak virus
Negative (minus) strand	1 piece	Plant rhabdo-virus	lettuce necrotic yellows virus
Double-stranded RNA	10 pieces	Fijivirus	Sugarcane Fiji disease virus
	12 pieces	Phytoreovirus	Wound tumor virus
Single-stranded DNA		Geminivirus	Maize streak virus
Double-stranded DNA		Caulimovirus	Cauliflower mosaic virus

Of the RNA genomes, the double-stranded ones of fijiviruses and phytoreoviruses are in ten or twelve segments, and single-stranded RNA genomes may be in one piece as in tobamoviruses and tymoviruses, two pieces as in tobraviruses, comoviruses and nepoviruses, or three pieces as in bromoviruses, cucumoviruses and ilarviruses. Furthermore, whereas these single-stranded RNA genomes all seem to be of the plus-strand type - they have the same nucleotide sequence as virus messenger RNA - the RNA genomes of plant rhabdoviruses are the minus strand. However, the RNA transcriptase which produces rhabdovirus messenger RNA is not host- but virus-coded, and is contained in the virus particles (11), so there is no reason to treat the rhabdoviruses differently from plus-strand viruses in this discussion.

Virus genomes therefore illustrate a series of variations on a theme, and so do their modes of replication. However, because it seems that a large majority of plant viruses have single-stranded RNA genomes of the plus-strand type, attention will be centred on these. Their replication involves the synthesis of copies of the virus genome, its translation into polypeptides and the assembly of genome nucleic acid and coat protein(s) to form new nucleoprotein particles. Synthesis of copies of the genome RNA requires the action of a replicase or replicases, first to produce molecules of complementary nucleotide sequence, the minus strands, and then to produce plus strands. There is controversy about the nature of plant virus replicases and none of them has been identified unequivocally and analysed. One view is that they are host enzymes (21), a second view is that they are virus-coded enzymes, but perhaps the most popular hypothesis is that they are complex structures with some virus-coded components and some host-coded components. If this last view is correct, one can easily imagine ways in which the plant might react to infection so as to regulate it, and this could be considered a type of active defence.

The processes involved in the path that leads from virus genome to virus-coded protein differ somewhat from one group of viruses to another, and the following examples represent variations on this theme. The unipartite genome of TMV is translated into two proteins that probably have most of their amino acid sequence in common (31). The genome RNA is also cleaved by unknown means to produce two sub-genomic fragments, each of which is translated into a single polypeptide (5). The smallest polypeptide is the virus coat protein but it is not clear which of the other polypeptides is the product of the local-lesion gene.

In brome mosaic virus, two of the three genome parts are translated directly into polypeptides and the third is translated into a polypeptide and also cleaved to yield a sub-genomic fragment which is translated to produce the coat protein (37).

In cowpea mosaic virus there is apparently no cleavage of either of the two parts of the genome, but instead there seems to be post-translational cleavage of the large primary translation products (9, 40), and there is some evidence that this cleavage is effected by a virus-coded protease, supposedly produced by cleavage of the translation product of RNA-1 (32).

In these systems, it is not proved whether cleavage of the virus RNA molecules, and post-translational cleavage of the virus-coded polypeptides is controlled by host- or virus-coded enzymes. If it is effected by host-coded or partially host-coded enzymes, here is another way in which the host can regulate virus replication. However, some of these viruses have wide host ranges, and it might be a little surprising if such taxonomically diverse plant species all produced nucleases or proteases with such similar but subtle specificities. Very little is yet known about the way virus synthesis is regulated. Somehow, viruses with multipartite genomes produce the right proportions of particles containing the different genome parts, and somehow most viruses are prevented from replicating to extents that would cause great damage to the host. That this is not simply because supplies of precursors are exhausted can be seen when a virus is inoculated to a plant already chronically infected with another, unrelated, virus. In most instances, the second virus can multiply extensively, although usually not quite as well as in previously healthy tissue. This sort of observation implies that the synthesis of each virus is regulated independently, something that seems more likely to be virus-controlled than host-controlled. Hence, an important reason for studying the possibly multiple functions of virus-coded proteins, both those incorporated and those not incorporated in virus particles, is that such studies may reveal ways in which virus replication is regulated.

Information of a different sort is provided by ultrastructural studies of virus-infected cells. These have shown that special structures are induced by infection with many viruses and are typical of the group to which the virus belongs (28). For example, cells infected with nepoviruses and comoviruses contain inclusions composed of vesiculated membranous material, those infected with caulimoviruses contain proteinaceous bodies in which virus particles are embedded, and cells infected with potyviruses contain pinwheels. Whether the formation of any of these structures involves a reaction of the cell that minimizes the effects of infection is unknown. However, it has been noted that pinwheels (structures that typically end at the plasma membrane and probably are composed largely of a virus-coded protein (33)) are commoner in cells of plants the leaves of which develop weak macroscopic symptoms than when severe symptoms are produced. It has been suggested that the pinwheels are a response to a defence mechanism of the host (6).

In cells infected with tymoviruses the chloroplasts develop peripheral vesicles that are associated with the replication of virus RNA (25, 42) and in those infected with some tobraviruses, vesicles appear in the mitochondrial outer membrane (17). Here too are changes in the host that may in some way result in the regulation of virus multiplication, or they may simply be changes induced by the virus that are an essential part of its replication cycle.

SEVERITY OF DISEASE

Three kinds of factors are involved in determining the severity of disease produced by infection with viruses. One kind consists of the well-known environmental factors. For example, at 20°C TMV causes necrotic local lesions in *Nicotiana glutinosa* plants and does not become systemic, whereas at 35°C it does not induce necrotic local lesions and causes a systemic mottle. Moreover, when plants with such a systemic mottle are transferred to 20°C they quickly collapse with a systemic necrosis. Clearly the regulatory systems that operated at the two temperatures were not flexible enough to cope with rapid changes of temperature.

The second major influence is that of the genes of the virus. There are many examples of viruses that occur in a variety of strains, which *inter alia* cause green mosaic, yellow mosaic, systemic necrosis or effects of other types in hosts of the same genotype. Similarly, some strains may induce necrotic local lesions whereas others do not. There are also many examples of effects of host genes : for instance a virus isolate can cause only mild symptoms in hosts of one genotype, and severe symptoms in hosts of another genotype. Two genes in potato that control such differences in reaction to potato virus X are described by Cockerham (7, Table 2). Using host genes in all possible combinations one can indeed

Table 2 Effect of host genes on the reaction of potato to potato virus X (Reference 7)

Genotype	Strain of potato virus X			
	Group 1	Group 2	Group 3	Group 4
NxNb	N	N	N	s
nxNb	N	N	s	s
Nxnb	N	s	N	s
nxnb	s	s	s	s

N = Hypersensitive response s = Infection without necrosis

classify the strains of a virus into a series of groups on the basis of their ability to induce necrosis in each type of host. However, I cannot help feeling that the gene-for-gene hypothesis, which these pigeonholes may be taken to illustrate, is often applied in a scientifically naive way. Why should one suppose that the effects of a particular gene in the virus can be counteracted by mutations in only one specific gene in the host, or *vice versa*. It seems more reasonable to suppose that there is more than one gene that can be effective in this way, with the proviso that plant viruses do not have so many genes to choose from, probably three to six in many instances, and rarely, if ever, more than twelve.

It is worth noting that in several instances alleles of host genes for hypersensitive reactions and for virus localization in inoculated leaves are dominant to those conferring non-necrotic susceptibility and the ability to be invaded systemically (8, 20, 36). This might be taken to suggest that hypersensitivity and virus localization are dependent on the production of an active host-coded protein.

The influence of the host genotype on virus infection need not, of course, be determined by only one or two genes. For example, tolerance of virus infection can be determined jointly by many host genes (20). Tolerance is also a reaction typically found among plants in natural communities, and it is one that presumably can evolve when a particular host and virus are associated for a long period. Whether it involves virus-induced defence mechanisms is not clear.

LOCALIZATION OF INFECTION

There are various types of localization of a virus within the host. One type is seen in plants infected with luteoviruses, which seem essentially to be restricted to phloem tissue. The reason for this restriction is not established, and the ease with which luteoviruses infect mesophyll protoplasts (24) suggests that there is nothing inadequate biochemically about many of the cells that fail to become infected in the intact plant. More likely, the restriction of these viruses to phloem results from their inability to move from phloem cells to other cells, and we have still much to learn about the mechanism of cell-to-cell movement.

A second type of localization is illustrated by the freedom of the seed embryo from infection with most viruses. However, there are many exceptions to this rule, and groups of viruses such as nepoviruses (26) and ilarviruses typically infect embryos and are seed-borne to progeny plants. In some nepoviruses, the ability to be transmitted through seed depends mainly on a genetic determinant in the larger of the two parts of the virus genome (16), but the mechanism by which this gene is expressed is not known. However, the genetic constitution of the host can also have an important influence on the frequency of seed transmission (39) and it is not impossible that virus-induced host reactions are involved.

A different kind of localization is that occurring in the mesophyll in some virus-host combinations. Here it is not a question of inability to invade a tissue of a different type but of inability to invade and multiply in adjacent cells of the same tissue. Such localization is often accompanied by the formation of necrotic lesions and, in general, hypersensitive reactions to viruses seem to be closely linked to virus replication and not to be triggered by an early event in virus infection. For example, temperature-shift experiments with temperature-sensitive mutants of tomato black ring virus showed that no necrosis developed in leaves transferred from a permissive to a non-permissive temperature for virus replication if the transfer was more than one day before lesions would have appeared at the permissive temperature (10). If the transfer was at one day before lesion appearance, lesions developed but were much smaller than normal and did not expand subsequently.

There are also many examples of virus localization without necrosis in inoculated leaves; inoculated leaves may develop chlorotic lesions or remain symptomless and the virus still fail to infect the plant systemically. Factors other than necrosis therefore can cause virus localization in inoculated leaves. In this connection, Nishiguchi, Motoyoshi and Oshima (29) have described a mutant of TMV that multiplies normally in tomato mesophyll protoplasts at 28°C but in intact leaves at 28°C is confined to the initially infected epidermal cells, although at 22°C it readily spreads to other cells. In contrast, other strains of the virus quickly spread from cell to cell in tomato leaves at 28°C. Quite how the mutant is localized at 28°C is not clear, but further study of this phenomenon may perhaps reveal a virus-induced reaction of the host that results in minimizing the spread of infection.

EVIDENCE FOR ACTIVE HOST INVOLVEMENT IN RESTRICTING THE EFFECTS OF VIRUSES

There are several phenomena in which active resistance of the host seems more directly implicated than in many of the phenomena I have referred to already. One of these, acquired systemic resistance, is found in the upper leaves of plants the lower leaves of which have developed necrotic local lesions after inoculation with virus. The main effect is a decrease in size of lesions produced by mechanical inoculation of the upper leaves with the same or another virus: lesion number also may be decreased in some instances (15, 34, Table 3). These effects are found, too,

Table 3 Effect of previous infection of half-leaves with tobacco mosaic virus on number and size of lesions produced in opposite half-leaves inoculated at intervals afterwards (Reference 34)

Days between inoculations	Lesions produced by second inoculation (%)*	
	Number	Diameter
2	88	90
4	91	73
6	37	56
8	17	49

* Half-leaves of Samsun NN tobacco were inoculated with tobacco mosaic virus. Figures are the number and diameter of lesions produced by a second inoculation of half-leaves opposite those inoculated previously, expressed as a percentage of the values for lesions produced in half-leaves opposite previously uninoculated half-leaves.

in French bean primary leaves opposite those inoculated. Ross (35) has argued that these effects result from the systemic movement of a substance or substances from inoculated to the non-inoculated leaves. He found that the effects were not produced when a section of the petioles of inoculated leaves was killed within a day after inoculation. Also, when the main vein of non-inoculated leaves was cut, resistance developed only in the part of the leaf basal to the cut. These experiments do not distinguish between the possible effects of a substance moving from inoculated to non-inoculated leaves and one moving from non-inoculated to inoculated leaves. However, when the inoculated leaves were removed seven days after inoculation, resistance was demonstrated 18 - 25 days later in leaves that probably were no more than leaf initials when inoculated leaves were removed, and this was considered to be strong circumstantial evidence that the substance moved from inoculated to non-inoculated leaves (4, 34).

The consequence of acquired systemic resistance is that the hypersensitive reaction in leaves is enhanced: lesions stop expanding and virus stops accumulating sooner than would happen in control leaves. The mechanism of this induction of resistance is, however, far from clear. The phenomenon is apparently also induced by some non-virus pathogens that produce necrotic local lesions in leaves, such as the fungus *Thielaviopsis basicola* (18). Attempts have been made to extract agents that interfere with virus infection from the leaves that have acquired systemic resistance, and some success has been claimed (27), but the case for the causal involvement of these materials does not seem to have been fully made out.

Other workers have found that the activity of various enzymes increases in leaves with necrotic local lesions or with acquired systemic resistance. For example, the activity of peroxidase approximately doubled in leaves of Samsun NN tobacco inoculated with TMV as compared with mock-inoculated leaves, and the increase occurred at the time the necrotic lesions appeared (38). Moreover in non-inoculated leaves the activity of peroxidase also increased, but somewhat later, and this increase coincided with the development of acquired systemic resistance. These and other results led to the suggestion that acquired systemic resistance was directly and causally related to an unusually rapid accumulation of quinones at or near the site of virus synthesis.

Other enzymes that do not seem involved in acquired systemic resistance, such as phenylalanine ammonia lyase, nevertheless are greatly activated in leaves with necrotic lesions, and the increase in their activity corresponds with the appearance of lesions (12). However phenylalanine ammonia lyase activity is localized in narrow rings around the lesions. Another change is the accumulation of phytoalexins in leaves with necrotic lesions induced by virus infection (2). The old question remains. Are the changes in these substances or activities causes or effects of the reactions of the plants to viruses, or do they simply reflect other changes in the cell that occur when the viruses multiply ?

Other virus-induced changes in the protein content of leaves have been detected by polyacrylamide gel electrophoresis of leaf extracts. The best studied examples are the $b_1 - b_4$ proteins that occur in substantial amounts in leaf extracts of *Nicotiana tabacum* Xanthi-nc infected with TMV but are not detectable in extracts of virus-free leaves (14), and the similar proteins, called IV-I, found in extracts of TMV-infected leaves of *N. tabacum* Samsun NN (44). Accumulation of these proteins is induced by infection with any one of several local lesion-forming viruses and also by injection of leaves with polyacrylic acid (13). Hence they must be host-coded proteins. Recently they have been named pathogenesis-related proteins (1), although this term does not seem to be very well chosen in view of the evidence that the same proteins can be induced to accumulate by treating leaves with polyacrylic acid. Some controversy has centred on whether or not these proteins exist in small amounts in healthy plants (43).

For our present purpose it does not matter too much. Clearly their accumulation is a host response to infection.

A more important question is whether accumulation of these proteins confers resistance to infection by a subsequently applied inoculum. Certainly the development of resistance, as measured by a decrease in number and size of the lesions that form, is correlated with the accumulation of the proteins (13). This intriguing question is discussed in other contributions to this book.

Reference has already been made to the localization of infection, often but not necessarily associated with the hypersensitive reaction. The production of necrotic local lesions clearly is a virus-induced reaction of the host and, if responsible for the localization of infection, must qualify as an example of active resistance. As already mentioned, lesion development is correlated with increased peroxidase activity and with the accumulation of specific proteins identifiable by polyacrylamide gel electrophoresis. It also is accompanied by the accumulation of calcium pectate between cells in and close to TMV lesions in *Nicotiana glutinosa* leaves (45), and by the accumulation of other substances including lignin in bean leaves inoculated with tobacco necrosis virus (23). Ultrastructural studies too have drawn attention to changes in cell walls in the vicinity of lesions. For example, Hiruki & Tu (19) studied three zones of the lesions caused by potato virus M in *Phaseolus vulgaris* leaves, necrotic at the lesion centre, semi-necrotic outside this, and non-necrotic at the lesion periphery. In the semi-necrotic zone they found the cells had a roughened plasma membrane, that membrane-bound vesicles and tubules occurred between the protoplast and cell wall and that secondary thickening of the cell wall had sealed off the plasmodesmata. In the non-necrotic cells at the lesion periphery a thickening of callose was found on the inner wall of cells adjacent to the semi-necrotic zone. Hiruki & Tu (19) therefore proposed that localization of the virus in lesions is an essentially physical process, depending on the blocking of plasmodesmata with callose, and the subsequent secondary thickening of cell walls. There are therefore both physical and physiological hypotheses for virus localization in mesophyll cells, and in this connection it is interesting to note that resistance expressed in intact plants as the hypersensitive response was not evident when isolated protoplasts from comparable leaves were inoculated with the same virus *in vitro* (30). To be expressed the hypersensitive response seems to need cells with walls, perhaps in critical kinds of juxtaposition.

VIRUS SPECIFIC AND NON-SPECIFIC EFFECTS

An important distinction between the phenomena discussed in this paper is that some are characterised by activity against many viruses whereas others are directed against a specific virus. For example, acquired systemic resistance is not virus-specific; nor is the production of abnormal proteins identifiable by polyacrylamide gel electrophoresis, and the activation of host enzymes. Many viruses can induce these changes and they are associated with effects on many viruses. Such non-virus-specific effects suggest effects with physical causes, or effects that result from changes in the metabolism of plant cells.

Certain kinds of genetically determined resistance also operate against many viruses. For example, the tobacco variety TI245 has a measure of resistance to infection with several mechanically inoculated viruses, and this resistance is apparently determined by multiple genes (41). However, it is not known whether active defence is involved.

Many other kinds of genetically determined resistance, particularly those involving only one or two host genes, are directed against specific viruses. They may result from interference with the action of particular virus genes, and their further analysis should prove interesting. In addition there is the phenomenon of cross-protection which is the inability of a virus strain to induce additional symptoms when it is inoculated to plants already infected with the same or another strain of the same virus. This is dealt with fully elsewhere in this volume. Too little seems to be known about the processes involved for an active defence mechanism of the host to be identified with the phenomenon.

CONCLUDING COMMENTS

In this broad survey an attempt has been made to identify stages in the multiplication cycle at which the host could have a virus-induced regulatory role. There are several such stages, but in most instances it seems at least equally possible that the regulatory influence is supplied by the virus itself. Thus there is no necessity to invoke active defence mechanisms to explain the failure of viruses to infect all plant species, or to reach great concentrations in cells and kill their hosts.

The main evidence for active defence mechanisms against viruses comes from studies of two phenomena, the localization of virus infection in leaf tissue and acquired systemic resistance. To what extent these are natural phenomena is not clear, and it would be helpful to know, for example, whether acquired systemic resistance is effective against vector-inoculated virus. It seems that polyacrylic acid-induced resistance has rather slight effects against infection by aphid-transmitted potato virus Y (22).

In regard to the localization of infection, one suspects that the tendency in wild plant communities is to select not for this ability but for tolerance of infection. In crop plants, the story is of course quite different, several genes for localization of virus within the mesophyll having been used, mostly with considerable and lasting success. Whether it will also be possible to exploit the potential value of elicitors of acquired systemic resistance must depend on the outcome of further research.

REFERENCES

1. ANTONIW, J.F., RITTER, C.E., PIERPOINT, W.S. & VAN LOON, L.C. (1980). Comparison of three pathogenesis-related proteins from plants of two cultivars of tobacco infected with TMV. *Journal of General Virology* 47, 79 - 87.

2. BAILEY, J.A. & INGHAM, J.L. (1971). Phaseollin accumulation in bean *(Phaseolus vulgaris)* in response to infection by tobacco necrosis virus and the rust *Uromyces appendiculatus*. *Physiological Plant Pathology* 1, 451 - 456.

3. BARKER, H. (1980). Studies on the behaviour of raspberry ringspot virus in plant leaves and protoplasts. Ph.D. Thesis, University of Dundee.

4. BOZARTH, R.F. & ROSS, A.F. (1964). Systemic resistance induced by localized virus infections: extent of changes in uninfected plant parts. *Virology* 24, 446 - 455.

5. BRUENING, G., BEACHY, R.N., SCALLA, R. & ZAITLIN, M. (1976). *In vitro* and *in vivo* translation of the ribonucleic acids of a cowpea strain of tobacco mosaic virus. *Virology* 71, 498 - 517.

6. CHAMBERLAIN, J.A. (1974). The relation between tolerance and the production of pinwheel inclusions in plants infected with ryegrass mosaic virus. *Journal of General Virology* 23, 201 - 204.

7. COCKERHAM, G. (1955). Strains of potato virus X. *Proceedings of the Second Conference on Potato Virus Diseases, Lisse-Wageningen, 1954*, 89 - 90.

8. COCKERHAM, G. (1970). Genetical studies on resistance to potato viruses X and Y. *Heredity* 25, 309 - 348.

9. DAVIES, J.W., AALBERS, A.M.J., STUIK, E.H. & VAN KAMMEN, A. (1977). Translation of cowpea mosaic virus RNA in a

cell-free extract from wheat germ. *FEBS Letters* 77, 265 - 269.

10. FORSTER, R.L.S. (1980). Production and properties of mutants of tomato black ring virus. Ph.D. Thesis, University of Dundee.

11. FRANCKI, R.I.B. & RANDLES, J.W. (1972). RNA-dependent RNA polymerase associated with particles of lettuce necrotic yellows virus. *Virology* 47, 270 - 275.

12. FRITIG, B., GOSSE, J., LEGRAND, M. & HIRTH, L. (1973). Changes in phenylalanine ammonia-lyase during the hypersensitive reaction of tobacco to TMV. *Virology* 55, 371 - 379.

13. GIANINAZZI, S. & KASSANIS, B. (1974). Virus resistance induced in plants by polyacrylic acid. *Journal of General Virology* 23, 1 - 9.

14. GIANINAZZI, S., MARTIN, C. & VALLEE, J.C. (1970). Hypersensibilité aux virus, temperature et protéines solubles chez le Nicotiana Xanthi n.c. Apparition de nouvelles macromolécules lors de la répression de la synthese virale. *Compte Rendu de l'Académie des Sciences de Paris* 270, 2383 - 2386.

15. GILPATRICK, J.D. & WEINTRAUB, M. (1952). An unusual type of protection with the carnation mosaic virus. *Science* 115, 701 - 702.

16. HANADA, L. & HARRISON, B.D. (1977). Effects of virus genotype and temperature on seed transmission of nepoviruses. *Annals of Applied Biology* 85, 79 - 92.

17. HARRISON, B.D., STEFANAC, Z. & ROBERTS, I.M. (1970). Role of mitochondria in the formation of X-bodies in cells of *Nicotiana clevelandii* infected by tobacco rattle viruses. *Journal of General Virology* 6, 127 - 140.

18. HECHT, E.I. & BATEMAN, D.F. (1964). Non-specific acquired resistance to pathogens resulting from localized infections by *Thielaviopsis basicola* or viruses in tobacco leaves. *Phytopathology* 54, 523 - 530.

19. HIRUKI, C. & TU, J.C. (1972). Light and electron microscopy of potato virus M lesions and marginal tissue in Red Kidney bean. *Phytopathology* 62, 77 - 85.

20. HOLMES, F.O. (1965). Genetics of pathogenicity in viruses and of resistance in host plants. *Advances in Virus*

Research 11, 139 - 161.

21. IKEGAMI, M. & FRAENKEL-CONRAT, H. (1978). RNA-dependent RNA polymerase of tobacco plants. *Proceedings of the National Academy of Sciences of the United States of America* 75, 2122 - 2124.

22. KASSANIS, B. & WHITE, R.F. (1975). Polyacrylic acid-induced resistance to tobacco mosaic virus in tobacco cv. Xanthi. *Annals of Applied Biology* 79, 215 - 220.

23. KIMMINS, W.C. & WUDDAH, D. (1977). Hypersensitive resistance: determination of lignin in leaves with a localized virus infection. *Phytopathology* 67, 1012 - 1016.

24. KUBO, S. & TAKANAMI, Y. (1979). Infection of tobacco mesophyll protoplasts with tobacco necrotic dwarf, a phloem-limited virus. *Journal of General Virology* 42, 387 - 398.

25. LAFLÈCHE, D. & BOVÉ, J.M. (1971). Virus de la mosaique jaune du navet: site cellulaire de la réplication du RNA viral. *Physiologie Végétale* 9, 487 - 503.

26. LISTER, R.M. & MURANT, A.F. (1967). Seed-transmission of nematode-borne viruses. *Annals of Applied Biology* 59, 49 - 62.

27. LOEBENSTEIN, G., RABINA, S. & VAN PRAAGH, T. (1966). Induced interference phenomena in virus infections. In : *Viruses of Plants*, Ed. by A.B.R. Beemster & J. Dijkstra, pp. 151 - 157. North-Holland, Amsterdam.

28. MARTELLI, G.P. & RUSSO, M. (1977). Plant virus inclusion bodies. *Advances in Virus Research* 21, 175 - 266.

29. NISHIGUCHI, M., MOTOYOSHI, F. & OSHIMA, N. (1980). Further investigation of a temperature-sensitive strain of tobacco mosaic virus: its behaviour in tomato leaf epidermis. *Journal of General Virology* 46, 497 - 500.

30. OTSUKI, Y., SHIMOMURA, T. & TAKEBE, I. (1972). Tobacco mosaic virus multiplication and expression of the N gene in necrotic responding tobacco varieties. *Virology* 50, 45 - 50.

31. PELHAM, H.R.B. (1978). Leaky UAG termination codon in tobacco mosaic virus RNA. *Nature* 272, 469 - 471.

32. PELHAM, H.R.B. (1979). Synthesis and proteolytic processing of cowpea mosaic virus proteins in reticulocyte lysates.

Virology 96, 463 - 477.

33. PURCIFULL, D.E., HIEBERT, E. & McDONALD, J.G. (1973). Immunochemical specificity of cytoplasmic inclusions induced by viruses in the potato Y group. *Virology* 55, 275 - 279.

34. ROSS, A.F. (1961). Systemic acquired resistance induced by localized virus infections in plants. *Virology* 14, 340 - 358.

35. ROSS, A.F. (1966). Systemic effects of local lesion formation. In : *Viruses of Plants*, Ed. by A.B.R. Beemster & J. Dijkstra, pp. 127 - 150. North-Holland, Amsterdam.

36. ROSS, H. (1958). Inheritance of extreme resistance to virus Y in *Solanum stoloniferum* and its hybrids with *Solanum tuberosum*. *Proceedings of the Third Conference on Potato Virus Diseases, Lisse-Wageningen, 1957*, 204 - 211.

37. SHIH, D.S. & KAESBERG, P. (1976). Translation of the RNAs of brome mosaic virus: the monocistronic nature of RNA 1 and RNA 2. *Journal of Molecular Biology* 103, 77 - 88.

38. SIMONS, T.J. & ROSS, A.F. (1970). Enhanced peroxidase activity associated with induction of resistance to tobacco mosaic virus in hypersensitive tobacco. *Phytopathology* 60, 383 - 384.

39. SMITH, F.L. & HEWITT, W.B. (1938). Varietal susceptibility to common bean mosaic and transmission through seed. *Bulletin of California Agricultural Experiment Station* No. 621, 18 pp.

40. STUIK, E. (1979). Protein synthesis directed by cowpea mosaic virus RNAs. Doctoral Thesis, Agricultural University, Wageningen.

41. TROUTMAN, J.L. & FULTON, R.W. (1958). Resistance in tobacco to cucumber mosaic virus. *Virology* 6, 303 - 316.

42. USHIYAMA, R. & MATTHEWS, R.E.F. (1970). The significance of chloroplast abnormalities associated with infection by turnip yellow mosaic virus. *Virology* 42, 293 - 303.

43. VAN LOON, L.C. (1976). Specific soluble leaf proteins in virus-infected tobacco plants are not normal constituents. *Journal of General Virology* 30, 375 - 379.

44. VAN LOON, L.C. & VAN KAMMEN, A. (1970). Polyacrylamide disc electrophoresis of the soluble leaf proteins from *Nicotiana tabacum* var. 'Samsun' and 'Samsun NN'. II. Changes in protein constitution after infection with tobacco mosaic virus. *Virology* 40, 199 - 211.

45. WEINTRAUB, M. & RAGETLI, H.W.J. (1961). Cell wall composition of leaves with a localized virus infection. *Phytopathology* 51, 215 - 219.

LOCALIZED RESISTANCE AND BARRIER SUBSTANCES*

G. LOEBENSTEIN, SARA SPIEGEL and A. GERA

Virus Laboratory, Agricultural Research Organization
The Volcani Center
Bet Dagan, Israel

INTRODUCTION

Infection by viruses causing diseases of economic importance is in general of the systemic type, whereby the virus invades a substantial amount of host tissue from the primary point of entry. However, in several virus-host interactions the infection remains localized and the virus moves into and multiplies in only a small group of cells near the point of entry. This local-lesion response has been well known for many years; Holmes (22) was the first to recognize the possibilities for quantitative work. Although the use of the local-lesion assay is probably one of the most common techniques employed in plant virus laboratories, the physiology of lesion formation and the mechanism of restriction of virus movement to adjacent tissues are still poorly understood.

The following types of localized infections are known : (a) necrotic lesions of the non-limited type, such as in beans where lesions caused by most strains of TMV continue to grow indefinitely; (b) self-limiting necrotic local lesions such as TMV in *Datura stramonium*, where lesions reach their maximum size three days after inoculation; (c) chlorotic local lesions, such as PVY in *Chenopodium amaranticolor*, where infected cells lose chlorophyll; (d) ring-like patterns or ringspots that remain localized, such as in *Tetragonia*

---

* Publication of the Agricultural Research Organization, Bet Dagan. No. E-225. The following abbreviations are used : CMV (Cucumber mosaic virus), PVX (Potato X virus), PVY (Potato Y virus), TMV (Tobacco mosaic virus), TNV (Tobacco necrosis virus), TSW (Tomato spotted wilt virus).

expansa infected with TSW; (e) starch lesions, such as TMV in cucumber cotyledons, where no symptoms are observed on the intact leaf, but when it is decolorized with ethanol and stained with iodine, lesions become apparent; (f) microlesions (with a mean size of 1.1 x $10^{-2}mm^2$), such as the U_2 strain of TMV on Pinto bean leaves; (g) infections that remain localized without showing any symptoms, such as tobacco etch virus in *Primula malacoides* or TMV in *Tagetes erecta* (23); and (h) subliminal symptomless infections not detectable as starch lesions, such as TMV-infected cotton cotyledons, and where virus content is 1/200,000th of that produced in a systemic host, may also be due to localization of virus (7).

In the local lesion a discontinuous distribution of virus between cells is observed, while in many systemic infections cells are usually infected uniformly. Thus, in local necrotic lesions incited by TMV in *C. amaranticolor*, virus particles in crystalline arrays were found in about one third of the cells. These were interspersed with other cells in which little or no virus was detected (39). A similar pattern was found in *Nicotiana glutinosa*, although far fewer cells (about 2%) harbor visible virus and much less is found in each cell (40). This uneven distribution and the apparent resistance of many cells to virus multiplication may be connected with the confinement of the infection to the lesion. In starch lesions a gradient of particle density from the center to the periphery of the lesion was observed (8).

The estimated number of TMV particles per cell, averaged over all cells in the infected area of *N. glutinosa*, is about 10^3 (40) or two to four orders of magnitude lower than in a comparable systemic infection, where the number of particles per cell is estimated to be between 10^5 and 6 x 10^7 (20).

EARLY CYTOPATHIC CHANGES

In the necrotic lesion the cytopathic events leading to collapse and necrosis start with changes in the chloroplast, consisting of swelling and distortion of the chlorophyll lamellae and swelling of starch grains, about eight hours after infection (62). Later, the cytoplasmic membranes disintegrate, the number of mitochondria increases, and the chloroplasts degenerate before necrosis takes place. At this time (about 40 hours after inoculation), the nucleus still preserves its integrity. Apparently some initial events in the cell lead to changes in membrane permeability followed by swelling and bursting of the protoplast and by tonoplast disruption. Changes in membrane permeability measured by depolarization of trans-membrane electropotential (PD) have been observed in cells around tobacco ringspot-induced necrotic lesions in cowpea. A gradient of PD became evident with very depolarized cells (-50 to -60mV) in the remaining live cells of the lesion, moderately depolarized cells

(-120 to -140mV), and normal-potential-bearing cells (-150 to -170 mV) approximately 5 mm to 10 mm away from the lesion (54).

Changes in permeability in leaves have also been determined by measuring electrolyte leakage by conductometry. Thus in *N. tabacum* cv. Xanthi-nc, statistically significant increased electrolyte leakage was evident about seven hours before the first local lesion symptoms became visible (63). On the other hand, plasmolysis of cells, which inhibits the increase in permeability, suppresses the necrotic hypersensitive response (9). Perhaps the cell membranes are protected by the osmotic agents, although as an alternative Otsuki *et al.* (43) suggested that plasmolysis leads to a disruption of cell-to-cell contact, needed for the development of the hypersensitive reaction. Damage to cell membranes could also be the result of lipid peroxidation, due to activation of lipoxygenases, as observed in cowpea leaves infected by CMV (24). Thus, the numbers of free radicals as determined by electron paramagnetic resonance, increased markedly within eight hours of inoculation. Treatment with scavengers of free radicals such as 2-mercaptoethylamine inhibited local lesion formation. The amount of unsaturated fatty acids decreased, while that of saturated fatty acids remained unchanged. An accumulation of stable free radicals was also observed in tobacco leaves with TMV-induced lesions (46). Apparently, lipid peroxidation plays an important role in the necrotic reaction.

The changes in permeability of membranes could allow contact between various enzymes and their substrates, resulting in products of phenol oxidation. These could kill the cells. Disturbances in membrane permeability have also been implicated as the primary occurrence in hypersensitive necrotic reactions caused by fungi and bacteria (19).

In chlorotic local lesions, induced by turnip mosaic virus in *C. quinoa*, disintegration of chloroplasts was not observed, although most palisade and spongy mesophyll cells contained chlorotic chloroplasts forming large aggregates, up to 20 chloroplasts, instead of being uniformly distributed along the cell periphery. These chloroplasts contained relatively few starch grains (27).

In a starch lesion reaction, such as TMV in cucumber cotyledons, swelling of chloroplasts is first observed 2 to 2.5 days after inoculation. When viewed 5 days after inoculation these chloroplasts contain large starch grains, but neither they nor the cells disintegrate or die (8). This accumulation of starch grains could be associated with a reduced permeability of chloroplast and cell membranes, thereby decreasing movement of carbohydrate.

It would be of interest to see if the primary event in a starch lesion is not associated with a decrease in 3', 5'-cyclic adenosine monophosphate (cAMP). The concentration of cAMP was found to be lower in clover systemically infected by clover yellow mosaic virus than in healthy clover. The reduced concentration of cAMP was associated with an accumulation of starch grains which could be reversed by external application of cAMP (58). This dissociation of starch grains was associated with alternations in chloroplast and cell membranes as intramembranous particles, blebbing and wrinkling of membranes. cAMP is apparently involved in the regulation of starch - glucose interconversion, as an effector molecule for adenosine diphosphate - glucose pyrophosphorylase, or by inducing changes in the structure and function of chloroplast and cell membranes, or a combination of the two. Thus, the lack of cAMP may reduce the transport of glucose through the membranes leading to accumulation of starch.

HYPOTHESES EXPLAINING LOCALIZATION

Death of cells and ultrastructural changes in advance of the infection

Death of cells This explanation is not satisfactory even in necrotic local lesion hosts and still less so in a chlorotic or starch-lesion host, as virus particles are found outside the necrotic area in viable cells (40). Furthermore, when plants with fully developed lesions are placed at temperatures above 30°C, the virus resumes its spread.

Structural changes in advance of the infection leading to cell death Around the core of necrotic cells of the lesion in tobacco cv. Samsun NN, there is a zone of cells that stain darkly. Seven days after inoculation with TMV, the mesophyll cells surrounding the lesion showed structural features suggestive of metabolic activity. Compared with cells of normal mesophyll, cells within this zone had smaller vacuoles, more cytoplasm and ribosomes, and a marked system of endoplasmic reticulum. Crystal-containing spherosomes were observed and it was suggested that these were peroxisomes (47), and that structural changes induced in advance of infection leading to cell collapse, may be instrumental in virus localization. These changes may progress to a level that causes collapse of a narrow ring of cells not yet invaded by virus, resulting in a barrier to further movement of virus. This may be the result of high peroxidase activity (52), although activation of peroxidase is also observed in plants kept at 100% relative humidity, thereby inhibiting the necrotic reaction (42). Furthermore, in *N. glutinosa* the peripheral tissue was found to consist of necrotic cells, intermingled with healthy cells and others showing various levels of disorganization (21).

Modifications of the vascular bundles were observed also in the zone surrounding TMV-induced necrotic lesions on *N. glutinosa*

(14). At a distance of about 1 mm from the lesion, sieve plate pores and the plasmodesmata connecting the sieve elements with their companion cells were occluded by a callose-like substance. Closer to the lesion, the tracheids were partially or totally blocked by plugs of an electron dense material, apparently a phenolic compound. Occlusion of tracheids may be involved in the wilting and collapse of cells in the lesion, by blocking water transport. The role of these plugs in virus localization is questionable, as some tracheids were still partially pervious at the time when lesion necrotization was almost complete.

Barrier substances

Several barrier substances in cells surrounding the local lesion have been implicated as possible factors preventing virus movement.

Calcium pectate and cell wall glycoprotein Weintraub and Ragetli (61) suggested that deposition of calcium pectate in the middle lamella of cells surrounding TMV lesions in *N. glutinosa* may be involved in localizing the infection.

From Pinto beans infected with TMV or TNV two new glycoproteins could be extracted. These are not present in uninfected controls (5). However, as these glycoproteins were also obtained from sham-inoculated plants, they seem to be a wound response induced by mechanical inoculation. Their site of deposition either in front of or behind the virus, is not yet known, and therefore their role in localizing the infection is unclear.

Phenolic compounds The necrotic response is accompanied by increased production of phenolic compounds derived from phenylalanine (17). It has been assumed that the stimulation of polyphenol oxidase (PPO) and peroxidases is responsible for the rapid accumulation of a toxic concentration of quinones, thereby localizing the infection and causing necrosis. However, several lines of evidence suggest (31) that PPO activation and accumulation of phenolic compounds may be the result of injury and decompartmentalization as a result of necrosis, and not the cause for localizing the infection. For example, transferring TMV-infected Xanthi nc tobaccos to 20°C, after previous exposure to 30°C (causing systemic infection), resulted in lower PPO activity at the time of lesion appearance, compared with that in uninoculated plants (6).

Enzymes of the phenylpropanoid pathway, such as phenylalanine ammonia lyase (PAL), cinnamic acid-4-hydroxylase, and catechol O-methyltransferase (OMT), are strongly stimulated in TMV-infected leaves of hypersensitive tobaccos (16). There seems to be a close parallel in time and amplitude between the appearance of necrosis and stimulation of these enzymes, and their stimulation increases with the size of the necrotic area (29). PAL and OMT could be involved in the metabolism of phenols and necrogenesis, or in the biosynthesis of lignin and lignification because PAL catalyzes the deamination of phenylalanine to trans-cinnamic acid, which is then transformed to phenolic lignin precursors; and OMT methylates caffeic and 5-hydroxyferrulic acid as intermediate stages of lignin biosynthesis (30).

A flavonoid has also been found in *Gomphrena globosa* infected by tomato bushy stunt. Its accumulation did not involve enhanced PAL activity, but its role in localization is at present unknown (45).

Lignification and suberin deposition As mentioned above, the activation of the phenylpropanoid pathway could be involved in lignification, and indeed lignin deposition has been observed at the borders of necrotic lesions induced by TMV U-5 in *N. glutinosa* and in the lumen of tracheids situated in proximity to them (13). Lignin formation seems to be associated with wound-healing mechanisms, when the wound undergoes necrosis. Abrasion itself causes an increase in lignin materials in bean leaves, although at a lower level than in leaves with localized necrotic lesions (26). Lignin may be involved with cell-wall thickening, postulated to be a barrier to spread of virus. Increased lignification resulting from the activation of the phenylpropanoid metabolism has also been suggested as one of the processes that reduce cell-to-cell spread of *Phytophthora infestans* in potato tuber tissue (15) and *Botrytis* in wheat leaves (37). However, in tomato bushy stunt-induced necrotic lesions in *Gomphrena globosa*, conspicuous thickenings of lignin were deposited outside the cells of the halo surrounding the necrotic area, but these did not follow the cell perimeter (3). Deposition of suberin was also observed around these lesions, as well as around TMV- and TNV-induced local lesions on bean (11). The time of occurrence and distribution of suberin in the latter case corresponded closely with that of lignin. Apparently, the modification of cell walls by the matrix materials lignin and suberin occurs as a general response to injury, amplified by necrotic lesion formation.

Paramural bodies, cell wall thickening, and callose deposition Blocking of plasmodesmata or rupture of plasmodesmatal connections by paramural bodies and callose deposition along cell walls have been studied in relation to virus localization. Thus, membrane-bound vesicular (paramural) bodies were observed between the plasma membrane and the cell wall at the periphery of TMV-induced lesions on Pinto bean leaves (53). Similar vesicles (paramural bodies) between the primary cell wall and the plasma membrane were observed also in cells of PVX-induced lesions on *Gomphrena globosa* (1). However, paramural bodies associated with cell wall protrusions have also been observed in systemic infections, as in barley infected with barley stripe mosaic (38), and in bean leaves infected with bean pod mottle virus (25). Their specific association with virus localization is therefore questionable. These paramural bodies may be involved in enzyme reactions related to cell wall modifications in response to virus infection in general. By similar reasoning it is doubtful if the cell wall thickening observed around necrotic lesions is responsible for localizing the infection, as cell wall thickening has also been found in systemic infections. Thus, cell wall thickenings, suggested as blocking plasmodesmata, were observed along the periphery of lesions incited by potato virus M on beans (59), or around TMV-induced lesions on Pinto beans (12). However, there have also been reports of cell wall overgrowths in systemically infected tissue, as for example, in Chinese cabbage leaves infected with cauliflower mosaic virus (4), beans infected with bean pod mottle virus (25), and barley infected with barley stripe mosaic virus (38). It is apparent that, at least in these systemic infections, cell wall thickenings are ineffective in preventing virus spread.

The deposition of callose has received much attention in recent years. Wu and colleagues observed callose deposition around TMV-induced lesions on Pinto beans (64). When such plants received a hot water treatment or UV irradiation which results in larger lesions, i.e. partial breakdown of localization, callose deposition decreased. Also keeping the plants in the dark, which inhibits callose deposition is well known to increase lesion size. Large depositions of callose were also observed in association with cell walls surrounding the necrotic zone of PVX-incited local lesions on *G. globosa* (1). These caused a protrusion of the arms of the plasmodesmata until finally the increased callose deposition ruptured plasmodesmatal connections. Membrane-bound bundles of virus particles embedded in callose were observed outside the protoplast. Allison and Shalla (1) suggested that the callose deposition and exclusion of virus particles from the protoplast may limit the spread of virus in a hypersensitive host.

However, heavy callose depositions were also observed in tobacco infected with TSW and with tobacco rattle virus that produce systemic necrosis where the virus does not remain localized (51). On the other hand, around tomato bushy stunt necrotic lesions on *G. globosa*, callose deposition was slight and erratic (44), and no callose deposition was observed around TMV-induced starch lesions on cucumber cotyledons where the infection remains localized without necrosis (8).

Furthermore, callose deposition observed around lesions did not prevent accumulation of ^{45}Ca, ^{32}P and ^{35}S-methionine in the center of the lesion, when the nuclides were applied to the leaf stalk (48).

In conclusion it seems that the various barrier substances, as well as ultrastructural changes in advance of the infection, could be responses to *necrotization* and not necessarily responsible for *localization* although they may work in the same direction, helping to slow down virus movement. To obtain better insight on localization, it is therefore necessary to select a system where the infection remains localized but without necrotization, so that any response can be related to the localizing process, without having to consider the possible association with the necrotic reaction. The starch lesion system, such as TMV in cucumber cotyledons, seems to be the most suitable, as here the infection remains localized without necrotization. No callose deposition was observed around the starch lesion (33), and no ultrastructural changes were observed at the periphery of the lesion (8). Thus, except for the fact that the cells at the periphery of the lesion contained virus, they were otherwise normal. These virus-containing cells were connected through plasmodesmata with normal non-infected cells. Apparently, in this starch lesion, host localization of TMV is not due to ultrastructural changes in advance of the infection, or to blocking of plasmodesmata. It was suggested that these cells at the periphery of the lesion do not support virus synthesis (31).

Localization due to reduced virus multiplication - a substance(s) that inhibits virus multiplication

Metabolic inhibitors and ultraviolet irradiation Indications of a host-mediated process that suppresses virus multiplication in a starch lesion host were obtained in studies using antimetabolites (36, 49) and UV-irradiation (33). As seen from Fig. 1 virus concentration in cucumber cotyledons increased 11 times when cotyledons were injected with 10 µg/ml actinomycin D one day after inoculation with TMV. When the antibiotic was applied at later intervals, the effect on virus concentration decreased; it was negligible when applied on the third day or later after inoculation. Comparable results were obtained with chloramphenicol which apparently does not enter the host cell nucleus where TMV-RNA is thought to be synthesized. Similarly, short wave UV-irradiation (2537Å) significantly increased virus titers, but not when applied at later intervals. Actinomycin D and chloramphenicol also stimulated the biosynthesis of TNV in *C. amaranticolor* (10). Highest stimulation occurred when the antibiotics were applied from 12 hours before to 10 hours after inoculation. 2-Thiouracil, assumed to interfere with host RNA metabolism, also suppressed the localization of cowpea chlorotic mottle virus (28). Treatments within 24 hours after inoculation increased both lesion size and sap infectivity very markedly.

In ultrastructural studies of UV-irradiated cucumber cotyledons inoculated with TMV, it was observed that the central cells of the starch lesion contained high concentrations of TMV and even crystals, in comparison with non-irradiated controls (Fig. 2) (Bahat and Loebenstein, unpublished results). This marked increase in virus content in the central cells of the lesion after irradiation, to-

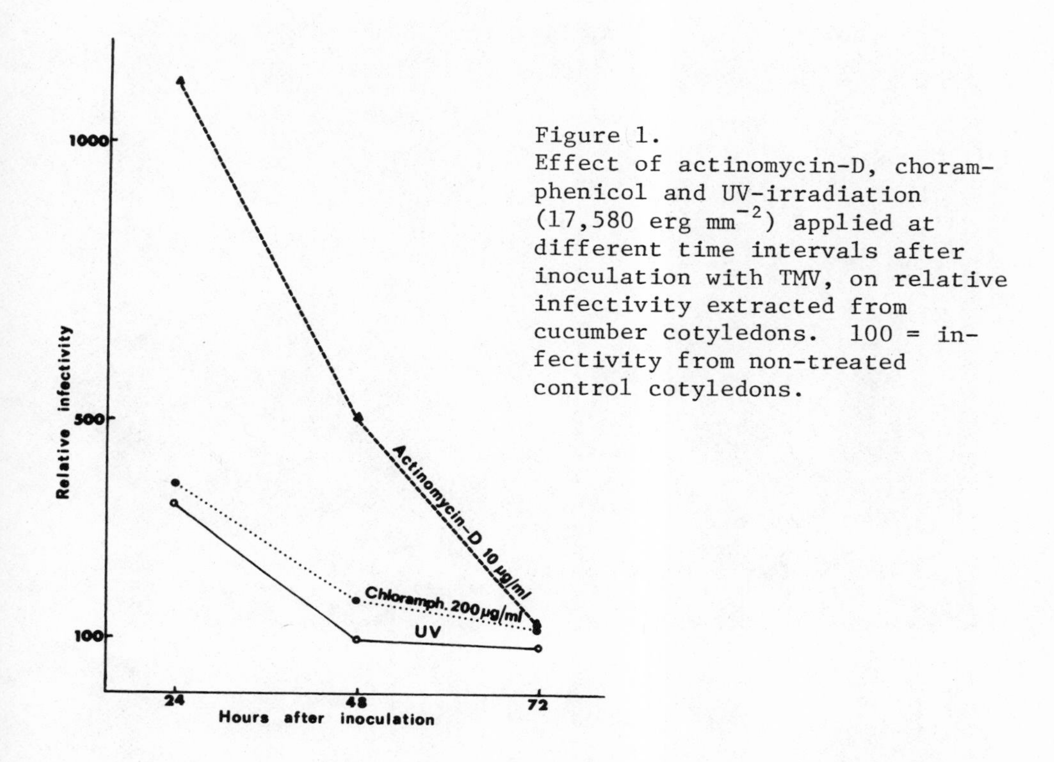

Figure 1.
Effect of actinomycin-D, choramphenicol and UV-irradiation (17,580 erg mm^{-2}) applied at different time intervals after inoculation with TMV, on relative infectivity extracted from cucumber cotyledons. 100 = infectivity from non-treated control cotyledons.

gether with the absence of ultrastructural changes at the periphery of the starch lesion mentioned above, suggest that at least in a starch lesion host the localizing mechanism operates mainly via inhibition of virus multiplication. This is an active process which requires new m-RNA synthesis and, therefore, treatments which inhibit transcription suppress localization and lead to increased virus multiplication in such hosts. The virus itself activates this mechanism.

Activation of the localizing mechanism by polyanions

Injection of polycarboxylates, with maleic acid as the anionic component of the polymer, into leaves of Samsun NN tobacco induced resistance to TMV (56). In the resistant tissue both the size and number of lesions decreased significantly. The resistance developed gradually after application of the inducer. Interference was completely suppressed in plants kept at 40°C before and after injection of ethylene maleic anhydride co-polymer (EMA-31), while actinomycin D given together with polyanion resulted in partial suppression. In the antiviral state, good correlations were observed between resistance and a specific ribosomal fraction (R_2), recovered by a two-

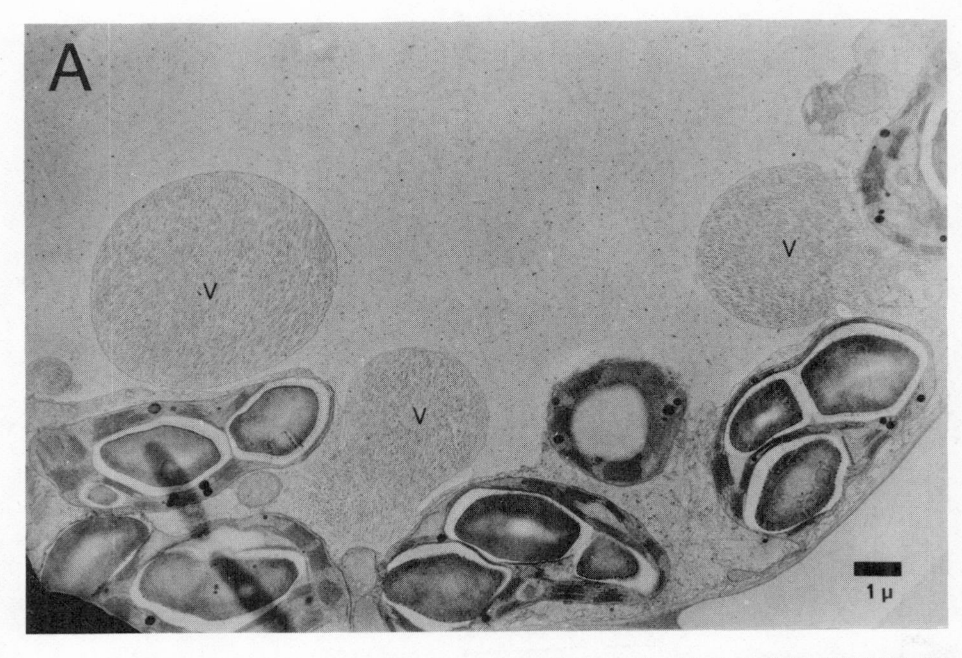

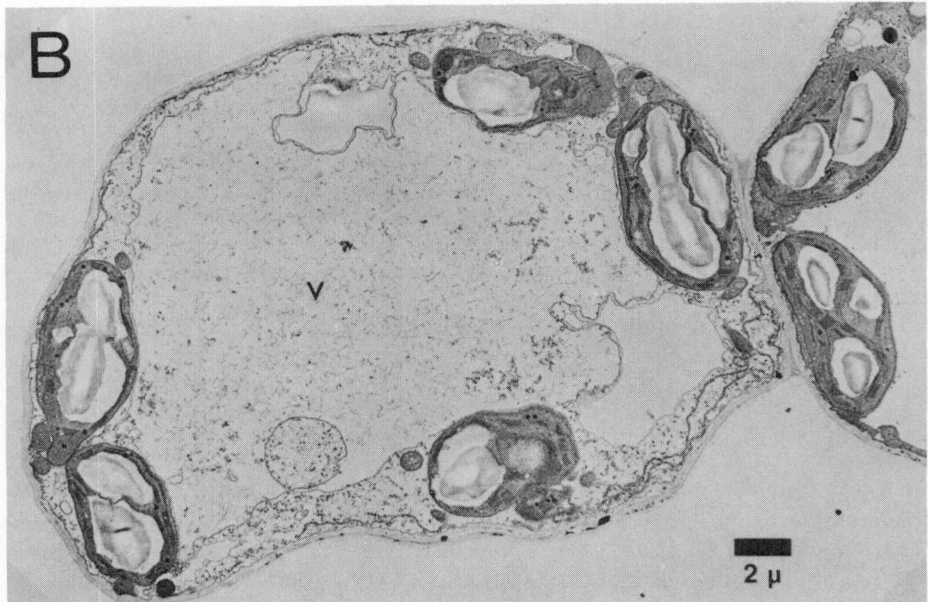

Figure 2 Crystals in cell from UV-irradiated cucumber cotyledon (A) compared to cell from non-irradiated TMV-inoculated cotyledon (B)

step procedure (57). This fraction increased considerably during the development of resistance. Actinomycin D and high temperatures, which suppressed development of interference, also abolished differences between R_2 from EMA-31-treated tissue and from control tissue. No differences were observed between EMA-31-treated and control tissue from Samsun, or in Samsun NN injected with polyanions that did not induce interference. It seems that *in vitro* concentrated preparations of R_2 form an unstable aggregate. It was suggested that these "sticky" ribosomes are involved in the resistant state.

Interestingly, activation of resistance with EMA-31 has so far been obtained only in hosts responding to infection with necrotic local lesions and not in starch lesion hosts. In the former, resistance was activated only in directly treated tissue, and not by applying the polyanion to other tissues (systemic effect).

Induction of resistance, *i.e.* activation of the mechanism responsible for localizing a virus, was observed also with other polyanions as poly I, Poly C (55) and polyacrylic acid (18). Polyacrylic acid induced resistance to TMV and TNV in tobacco cv. Xanthi nc leaves and was associated with the appearance of three additional soluble proteins. These three proteins were similar to those produced in TMV-infected leaves of cv. Xanthi plants that also are resistant to infection (60).

Substances that inhibit virus multiplication in infected and resistant tissue Antiviral or interfering agents have been extracted from both infected and uninfected resistant tissue. However, their connection with localization or induced resistance is still an open question, especially as they have been tested as inhibitors of infection applied during inoculation (by mixing with the test virus) and not as inhibitors of virus multiplication which should be applied to the infected cell 5-12 hours (or later) after inoculation (32). An antiviral factor (AVF) developing in local and systemically infected plants was reported by Sela and co-workers (50). The active principle extracted from leaves of *N. glutinosa* 48 hours after inoculation with TMV was first considered to be RNA, but later to be a phosphoglycoprotein with a molecular weight of about 22,000 (41). Preparations of the precursor of AVF could be made from non-infected *Nicotiana* leaves, either from those carrying the N-gene or from Samsun tobacco from which the N-gene is absent. These pre-AVF preparations could be activated to give antiviral activity by phosphorylation with crude enzyme preparations. Interestingly, however, active enzyme preparations could be obtained both from infected and non-infected plants carrying the N-gene though in intact plants TMV infection was necessary to obtain AVF (2). The activation required ATP, cAMP and cGMP. However, as AVF was tested mainly as an inhibitor of infection, its association with localizing the infection through inhibition of virus multiplication has yet to be shown.

Also, as AVF is obtained from infected tissue, the possibility that it is a product of the necrotization process cannot be ruled out yet.

Inhibitory substances have also been extracted from resistant uninfected tissues (35). A crude extract from uninfected apical halves of leaves of *Datura stramonium* previously inoculated on their basal halves with TMV or TNV, consistently interfered with infection by TMV (when mixed in the inoculum) to a much greater extent than did juice from the apical halves of control leaves. After partial purification, the active protein-like substance resembled interferon in some of its properties. It was stable at pH 2.5 and in perchloric acid, precipitated by zinc acetate, acetone and ammonium sulphate, sensitive to trypsin, and had an apparent molecular weight of approximately 28,000. However, as testing for activity was done by mixing the substance with the virus, the effect on virus multiplication and its role in localization and induced resistance have yet to be determined.

For a reliable evaluation of substances as inhibitors of virus multiplication, it is necessary to have a test system in which cells that have been infected synchronously can be treated uniformly at different times after inoculation with the test substance. The protoplast system seems to be the most suitable as it is possible to inoculate simultaneously all cells and get "one-step growth curves", in contrast to intact plants where only a small percentage of the cells become infected during the primary inoculation. It is also impossible to apply the test material uniformly to every cell in the intact plant. The possibility that a substance is released from the infected-resistant cell (into the medium) can also be tested in a protoplast system, whereas in the intact plant such a material diffuses into numerous surrounding cells and into the vascular system. The use of the protoplast system for both production (release into the medium) and assay of such substances is obvious.

This conceptual approach to using the protoplast system was, however, hampered by the finding of Japanese researchers that in isolated protoplasts from resistant plants, where infection is localized in the intact plant, virus multiplies to the same extent as in protoplasts from susceptible tobacco cultivars in which virus spreads systemically (43). This discrepancy between high virus titer in protoplasts from resistant plants and low virus content per infected cell in the intact plant, was found by us to be due to the presence of 2,4 dichlorophenoxyacetic acid (2,4-D) in the protoplast incubation medium. This growth regulating substance markedly enhanced TMV multiplication in protoplasts from two local lesion responding (resistant) cultivars while it reduced virus multiplication in protoplasts from a systemically reacting cultivar (susceptible). In the absence of 2,4-D the virus titer in protoplasts from the resistant variety was 10-20% of that in susceptible protoplasts (34).

Table 1 Effect of inhibitory substance(s) in incubation medium from resistant protoplasts on virus multiplication in resistant and susceptible protoplasts

Incubation medium (IM)	Test protoplasts	Hours of incubation of test protoplasts	Infectivity[1]		
			Protoplasts in incubation medium	Protoplasts in control medium[2]	% Inhibition
NN protoplasts (resistant)	NN	48	3.2	8.4	62
		72	11.7	23.4	50
		96	13.6	32.5	58
	Samsun	48	5.9	14.3	59
		72	9.8	25.1	61
		96	17.2	52.2	66
Samsun (susceptible)	NN	72	22.6	17.9	(-26)
	Samsun	72	45.0	41.9	(-11)

1 Average number of local lesions per 10^6 protoplasts on one half-leaf of *N. glutinosa*, calibrated to standard TMV solution (2.5 μg/ml) about 60 lesions per half-leaf

2 Medium in which uninoculated protoplasts had been suspended

Data from 3 - 4 experiments

Results for a substance(s) inhibiting virus multiplication released into the medium from TMV-infected protoplasts of a resistant variety are summarized in Table 1. The substance(s) inhibited virus multiplication in protoplasts from both resistant (Samsun NN) and susceptible (Samsun) plants when applied up to 18 hours after inoculation of the test protoplasts. It was not produced in protoplasts from susceptible plants nor from non-inoculated protoplasts of the resistant cultivar. This seems to be the first report on a substance(s) inhibiting the multiplication of a virus released from infected cells of a cultivar where the infection in the intact plant is localized. It is also evident that the protoplast system is most suitable for evaluating such substance(s).

CONCLUSIONS

Ultrastructural changes and various barrier substances associated with necrotic lesions could well be a response of the host to necrotization, but not necessarily be responsible for localizing the infection although they may help in slowing down virus movement. It is suggested that localization is an active process of the host-which is induced by the virus and probably by other agents whereby multiplication of the virus is inhibited. Preliminary evidence has been obtained that infected protoplasts from a local lesion responding cultivar, produce a substance(s) that inhibits virus multiplication.

REFERENCES

1. ALLISON, A.V. & SHALLA, T.A. (1974). The ultrastructure of local lesions induced by potato virus X : a sequence of cytological events in the course of infection. *Phytopathology* 64, 784 - 793.

2. ANTIGNUS, Y., SELA, I. & HARPAZ, I. (1977). Further studies on the biology of an antiviral factor (AVF) from virus-infected plants and its association with the N-gene of *Nicotiana* species. *Journal of General Virology* 35, 107 - 116.

3. APPIANO, A., PENNAZIO, S., D'AGOSTINO, G. & REDOLFI, P. (1977). Fine structure of necrotic local lesions induced by tomato bushy stunt virus in *Gomphrena globosa* leaves. *Physiological Plant Pathology* 11, 327 - 332.

4. BASSI, M., FAVALI, M.A. & CONTI, G.G. (1974). Cell wall protrusions induced by cauliflower mosaic virus in Chinese cabbage leaves : a cytochemical and autoradiographic study. *Virology* 60, 353 - 358.

5. BROWN, R.G. & KIMMINS, W.C. (1973). Hypersensitive resistance. Isolation and characterization of glycoproteins from plants with localized infections. *Canadian Journal of Botany* 51, 1917 - 1922.

6. CABANNE, F., SCALLA, R., MARTIN, C. (1968). Activité de la polyphénoloxydase au cours de la réaction d'hypersensibilité chez *Nicotiana Xanthi* n.c. infecté par le virus de la mosaique du tabac. *Annales de Physiologie végétale* 10, 199 - 208.

7. CHEO, P.C. (1970). Subliminal infection of cotton by tobacco mosaic virus. *Phytopathology* 60, 41 - 46.

8. COHEN, J. & LOEBENSTEIN, G. (1975). An electron microscope study of starch lesions in cucumber cotyledons infected with tobacco mosaic virus. *Phytopathology* 65, 32 - 39.

9. COUTTS, R.H.A. (1978). Suppression of virus induced local lesions in plasmolysed leaf tissue. *Plant Science Letters* 12, 77 - 85.

10. FACCIOLI, G. & RUBIES-AUTONELL, C. (1976). Action of antimetabolites on the biosynthesis of tobacco necrosis virus in locally-infected cells of *Chenopodium amaranticolor*. *Phytopathologische Zeitschrift* 87, 48 - 56.

11. FAULKNER, G. & KIMMINS, W.C. (1975). Staining reactions of the tissue bordering lesions induced by wounding, tobacco mosaic virus and tobacco necrosis virus in bean. *Phytopathology* 65, 1396 - 1400.

12. FAULKNER, G. & KIMMINS, W.C. (1978). Fine structure of tissue bordering lesions induced by wounding and virus infection. *Canadian Journal of Botany* 56, 2990 - 2999.

13. FAVALI, M.A., BASSI, M. & CONTI, G.G. (1974). Morphological cytochemical and autoradiographic studies of local lesions induced by the U5 strain of tobacco mosaic virus in *Nicotiana glutinosa* L. *Rivista di Patologia Vegetale Series IV* 10, 207 - 218.

14. FAVALI, M.A., CONTI, G.G. & BASSI, M. (1978). Modifications of the vascular bundle ultrastructure in the "Resistant Zone" around necrotic lesions induced by tobacco mosaic virus. *Physiological Plant Pathology* 13, 247 - 251.

15. FRIEND, J., REYNOLDS, S.B. & AVEGARD, A. (1973). Phenylalanine ammonia-lyase, chlorogenic acid and lignin in potato tuber tissue inoculated with *Phytophthora infestans*. *Physiolog-*

ical Plant Pathology 3, 495 - 507.

16. FRITIG, B., GOSSE, J., LEGRAND, M. & HIRTH, L. (1973). Changes in phenylalanine ammonia-lyase during the hypersensitive reaction of tobacco to TMV. *Virology* 55, 371 - 379.

17. FRITIG, B., LEGRAND, M. & HIRTH, L. (1972). Changes in the metabolism of phenolic compounds during the hypersensitive reaction of tobacco to TMV. *Virology* 47, 845 - 848.

18. GIANINAZZI, S. & KASSANIS, B. (1974). Virus resistance induced in plants by polyacrylic acid. *Journal of General Virology* 23, 1 - 9.

19. GOODMAN, R.N. (1968). The hypersensitive reaction in tobacco : a reflection of changes in host cell permeability. *Phytopathology* 58, 872 - 873.

20. HARRISON, B.D. (1955). Studies on virus multiplication in inoculated leaves. Ph.D. Thesis, London University.

21. HAYASHI, T. & MATSUI, C. (1965). Fine structure of lesion periphery produced by tobacco mosaic virus. *Phytopathology* 55, 387 - 392.

22. HOLMES, F.O. (1929). Local lesions in tobacco mosaic. *Botanical Gazette* 87, 39 - 55.

23. HOLMES, F.O. (1946). A comparison of the experimental host ranges of tobacco-etch and tobacco-mosaic viruses. *Phytopathology* 36, 643 - 659.

24. KATO, S. & MISAWA, T. (1976). Lipid peroxidation during the appearance of hypersensitive reaction in cowpea leaves infected with cucumber mosaic virus. *Annals of the Phytopathological Society of Japan* 42, 472 - 480.

25. KIM, K.S. & FULTON, J.P. (1973). Plant virus-induced cell wall overgrowth and associated membrane elaborations. *Journal of Ultrastructure Research* 45, 328 - 342.

26. KIMMINS, W.C. & WUDDAH, D. (1977). Hypersensitive resistance : determination of lignin in leaves with a localized virus infection. *Phytopathology* 67, 1012 - 1016.

27. KITAJIMA, E.W. & COSTA, A.S. (1973). Aggregates of chloroplasts in local lesions induced in *Chenopodium quinoa* Wild by turnip mosaic virus. *Journal of General Virology* 20, 413 - 416.

28. KUHN, C.W. (1971). Cowpea chlorotic mottle virus local lesion area and infectivity increased by 2-thiouracil. *Virology* 43, 101 - 109.

29. LEGRAND, M., FRITIG, B. & HIRTH, L. (1976). Enzymes of the phenylpropanoid pathway and the necrotic reaction of hypersensitive tobacco to tobacco mosaic virus. *Phytochemistry* 15, 1353 - 1359.

30. LEGRAND, M., FRITIG, B. & HIRTH, L. (1978). O-diphenol O-methyltransferases of healthy and tobacco-mosaic-virus-infected hypersensitive tobacco. *Planta* 144, 101 - 108.

31. LOEBENSTEIN, G. (1972a). Localization and induced resistance in virus-infected plants. *Annual Review of Phytopathology* 10, 177 - 208.

32. LOEBENSTEIN, G. (1972b). Inhibition, interference and acquired resistance during infection. In : *Principles and Techniques in Plant Virology*. Ed. C.I. Kado and H.O. Agrawal. pp. 32 - 61, Van Nostrand Reinhold Co. New York.

33. LOEBENSTEIN, G., CHAZAN, R. & EISENBERG, M. (1970). Partial suppression of the localizing mechanism to tobacco mosaic virus by UV irradiation. *Virology* 41, 373 - 376.

34. LOEBENSTEIN, G., GERA, A., BARNETT, A., SHABTAI, S. & COHEN, J. (1980). Effect of 2,4-dichlorophenoxyacetic acid on multiplication of tobacco mosaic virus in protoplasts from local-lesion and systemic-responding tobaccos. *Virology* 100, 110 - 115.

35. LOEBENSTEIN, G., RABINA, S. & VAN PRAAGH, T. (1966). Induced interference phenomena in virus infections. In : *Viruses in Plants*. Ed. A.B.R. Beemster and J. Dijkstra. pp. 151 - 157. North Holland Publishing Co. Amsterdam.

36. LOEBENSTEIN, G., SELA, B. & VAN PRAAGH, T. (1969). Increase of tobacco mosaic local lesion size and virus multiplication in hypersensitive hosts in the presence of actinomycin D. *Virology* 37, 42 - 48.

37. MAULE, A.J. & RIDE, J.P. (1976). Ammonia-lyase and O-methyltransferase activities related to lignification in wheat leaves infected with *Botrytis*. *Phytochemistry* 15, 1661 - 1664.

38. McMULLEN, C.R., GARDNER, W.S. & MYERS, G.A. (1977). Ultrastructure of cell-wall thickenings and paramural bodies induced by barley stripe mosaic virus. *Phytopathology* 67, 462-467.

39. MILNE, R.G. (1966a). Electron microscopy of tobacco mosaic virus in leaves of *Chenopodium amaranticolor*. *Virology* 28, 520 - 526.

40. MILNE, R.G. (1966b). Electron microscopy of tobacco mosaic virus in leaves of *Nicotiana glutinosa*. *Virology* 28, 527 - 532.

41. MOZES, R., ANTIGNUS, Y., SELA, I. & HARPAZ, I. (1978). The chemical nature of an antiviral factor (AVF) from virus-infected plants. *Journal of General Virology* 38, 241 - 249.

42. NIENHAUS, F. & HOOGEN, H. (1970). Stoffwechsel-physiologische Veranderungen in der Pflanze nach Virusinfektion unter Einfluss von Wundreiz. II. Enzymaktivitatsbestimmungen nach Acrylamidgelelcktrophorese. *Phytopathologische Zeitschrift* 69, 38 - 48.

43. OTSUKI, A., SHIMOMURA, T. & TAKEBE, I. (1972). Tobacco mosaic virus multiplication and expression of the N-gene in necrotic responding tobacco varieties. *Virology* 50, 45 - 50.

44. PENNAZIO, S., D'AGOSTINO, G., APPIANO, A. & REDOLFI, P. (1978). Ultrastructure and histochemistry of the resistant tissue surrounding lesions of tomato bushy stunt virus in *Gomphrena globosa* leaves. *Physiological Plant Pathology* 13, 165 - 171.

45. PENNAZIO, S., APPIANO, A. & REDOLFI, P. (1979). Changes occurring in *Gomphrena globosa* leaves in advance of the appearance of tomato bushy stunt virus necrotic local lesions. *Physiological Plant Pathology* 15, 177 - 182.

46. REUNOV, A.V. (1976). The use of paramagnetic resonance in plant virology. (in Russian) *Virus Diseases of Plants in the Far East* 25 (128), 71.

47. ROSS, A.F. & ISRAEL, H.W. (1970). Use of heat treatments in the study of acquired resistance to tobacco mosaic virus in hypersensitive tobacco. *Phytopathology* 60, 755 - 770.

48. SCHUSTER, G. & FLEMMING, M. (1976). Studies on the formation of diffusion barriers in hypersensitive hosts of tobacco mosaic virus and the role of necrotization in the formation of diffusion barriers as well as in the localization of virus infections. *Phytopathologische Zeitschrift* 87, 345 - 352.

49. SELA, B., LOEBENSTEIN, G. & VAN PRAAGH, T. (1969). Increase of tobacco mosaic virus multiplication and lesion size in hypersensitive hosts in the presence of chloramphenicol. *Virology* 39, 260 - 264.

50. SELA, I., HARPAZ, I. & BIRK, Y. (1966). Identification of the active component of an antiviral factor isolated from virus-infected plants. *Virology* 28, 71 - 78.

51. SHIMOMURA, T. & DIJKSTRA, J. (1975). The occurrence of callose during the process of local lesion formation. *Netherlands Journal of Plant Pathology* 81, 107 - 121.

52. SIMONS, T.J. & ROSS, A.F. (1971). Metabolic changes associated with systemic induced resistance to tobacco mosaic virus in Samsun NN tobacco. *Phytopathology* 61, 293 - 300.

53. SPENCER, D.F. & KIMMINS, W.C. (1971). Ultrastructure of tobacco mosaic virus lesions and surrounding tissue in *Phaseolus vulgaris* var. Pinto. *Canadian Journal of Botany* 49, 417 - 421.

54. STACK, J.P. & TATTAR, T.A. (1978). Measurement of transmembrane electropotentials of *Vigna sinensis* leaf cells infected with tobacco ringspot virus. *Physiological Plant Pathology* 12, 173 - 178.

55. STEIN, A. & LOEBENSTEIN, G. (1970). Induction of resistance to tobacco mosaic virus by poly I . poly C in plants. *Nature* 226, 363 - 364.

56. STEIN, A. & LOEBENSTEIN, G. (1972). Induced interference by synthetic polyanions with the infection of tobacco mosaic virus. *Phytopathology* 62, 1461 - 1466.

57. STEIN, A., LOEBENSTEIN, G. & SPIEGEL, S. (1979). Further studies of induced interference by a synthetic polyanion of infection by tobacco mosaic virus. *Physiological Plant Pathology* 15, 241 - 255.

58. TU, J.C. (1979). Alterations in chloroplast and cell membranes associated with cAMP-induced dissociation of starch grains in clover yellow mosaic virus infected clover. *Canadian Journal of Botany* 57, 360 - 369.

59. TU, J.C. & KIRUKI, C. (1971). Electron microscopy of cell wall thickening in local lesions of potato virus-M infected red kidney bean. *Phytopathology* 61, 862 - 868.

60. VAN LOON, L.C. & VAN KAMMEN, A. (1970). Polyacrylamide disc-electrophoresis of soluble leaf proteins from *Nicotiana tabacum* var. "Samsun" and "Samsun NN". II Changes in protein constitution after infection with tobacco mosaic virus. *Virology* 40, 199 - 211.

61. WEINTRAUB, M. & RAGETLI, H.W.J. (1961). Cell wall composition of leaves with a localized virus infection. *Phytopathology* 51, 215 - 219.

62. WEINTRAUB, M. & RAGETLI, H.W.J. (1964). An electron microscope study of tobacco mosaic virus lesions in *Nicotiana glutinosa* L. *Journal of Cell Biology* 23, 499 - 509.

63. WESTSTEIJN, E.A. (1978). Permeability changes in the hypersensitive reaction of *Nicotiana tabacum* cv. Xanthi n.c. after infection with tobacco mosaic virus. *Physiological Plant Pathology* 13, 253 - 258.

64. WU, J.H. & DIMITMAN, J.E. (1970). Leaf structure and callose formation as determinants of TMV movement in bean leaves as revealed by UV irradiation studies. *Virology* 40, 820 - 827.

THE PROTECTIVE EFFECTS OF SYSTEMIC VIRUS INFECTION

ROBERT W. FULTON

Department of Plant Pathology
University of Wisconsin
Madison, Wisconsin, U.S.A.

INTRODUCTION

Observations made some 40 to 50 years ago provided the background information on the cross protection phenomenon. Briefly stated, cross protection is the insusceptibility of a virus-infected plant to infection by other strains of the same virus. There are numerous exceptions to this statement; some of the terminology is confusing and some inappropriate. I shall review some of the experimental evidence on how cross protection is manifested and recount some of the terminology that has been used. Then I will discuss features of the cross protection reaction which suggest that an active defense mechanism is involved.

HISTORICAL BACKGROUND

In 1929 McKinney (29) observed that tobacco infected with a strain of tobacco mosaic virus (TMV) causing a green mosaic did not develop additional symptoms when inoculated with a strain of TMV causing a bright yellow mosaic. Essentially the same observation was made in 1931 by Thung (49): ordinary TMV protected against subsequent infection by a "white moasic" strain. Thung used the terms "acquired virus immunity", "antagonistic action" and "domination" to describe his observations.

Salaman (43) found that a mild strain of potato virus X (PVX) protected tobacco or *Datura* against infection by a severe strain. He later showed (44) that the same protection was obtained with a mild strain of potato virus Y against a severe strain. Salaman used the term "acquired immunity" and suggested that "protective inoculation" might have practical value.

Each of these early reports is based on the protection brought about by infection with a relatively mild strain of virus against a more severe strain. Experimentally, this is an easy system to use although it is not the only system possible.

A somewhat different system is provided by tobacco ringspot and other viruses. These induce local and systemic necrotic lines and rings from which tobacco as it grows "recovers" and develops new leaves that are essentially free from symptoms. In working with this virus, Price (35) applied the term "acquired immunity" to the development of normal-appearing leaves on tobacco infected with tobacco ringspot virus. Since these leaves did not develop additional symptoms when they were reinoculated, they resembled leaves infected by a mild strain of virus in being immune to re-infection. Price further showed that recovered leaves were protected against symptom development by some but not all other strains of tobacco ringspot virus. Recovered leaves did contain virus but at lower concentration than in leaves with ringspot symptoms (37).

Price's (35) use of the term "acquired immunity" as pertaining to the development of recovered leaves was criticized by Salaman (45), Valleau (50) and others on several grounds. They were reluctant to call leaf tissue immune which contained virus, although it did not display symptoms. Similarly, there were objections to the term "recovery" (50) on the basis that such leaves did display some slight effects of virus infection. The term "chronically diseased" was suggested (31), but seems inappropriate because it does not suggest the principle feature of the ringspot syndrome, the production of seemingly healthy leaves subsequent to initial acute systemic necrotic disease. "Acquired tolerance" (47) probably describes the syndrome; nevertheless, the term "recovery", i.e. from symptoms, is expressive and commonly used.

Virus diseases of the ringspot type provide unique opportunities for experimental manipulation. Even slight symptom development induced by a challenge inoculation can be detected readily on recovered leaves. Moreover, these leaves are ideal for reciprocal tests in which the ability of each of two viruses to protect against the other can be tested readily.

It may be appropriate here to point out that there are slightly different interpretations of the term "cross protection". One interpretation reserves the term for experimental manipulations in which each of two virus isolates is shown to protect against the other. A second interpretation uses the term to describe protection by one isolate against a different isolate. We use the term in the latter sense because "cross" was first applied to the protection by one strain against another ("cross immunization", 25) and it has been widely used in this sense. The term is applicable to systems in which only one isolate can be used as a challenge. Such systems may involve a virus strain producing only necrotic local lesions and not systemically invading its hosts. There are numerous systems in which it is not possible to test both strains against each other.

A system that has often been used in investigating cross protection is that described by Kunkel (25) involving strains of TMV and *Nicotiana sylvestris*. In this species, some strains of TMV cause prominent mosaic symptoms, while others cause distinct local necrotic lesions with no systemic symptoms. Kunkel found that mosaic leaves were protected against strains causing necrotic lesions (exceptions to this will be discussed in the section on superinfection). Kunkel showed further that leaves heavily inoculated with a mosaic-type strain became solidly immune within four or five days to infection with a necrotic lesion strain. Immunity seemed to be closely confined to tissues invaded by the protecting virus. The protection induced by mosaic type strains was specific; infection by cucumber mosaic or tobacco ringspot viruses did not confer protection against necrotic lesion strains of TMV. "Attenuated" (i.e. mild) strains of TMV protected against common or aucuba strains. Kunkel used the term "cross immunization" in referring to the immunizing effect of one strain against the other.

The TMV-*N. sylvestris* system was used by Sadasivan (42) in somewhat different, quantitative tests involving cross protection. He showed that when unrelated viruses were mixed with strains of TMV that caused lesions on *N. sylvestris*, the number of lesions were fewer than the number caused by unmixed inoculum. TMV strains causing mosaic symptoms in *N. sylvestris*, however, had a much greater effect in reducing lesion numbers. This effect was directly proportional to the amount of the mosaic (immunizing) strain present in the mixture with the local lesion-inducing strain.

Sadasivan (42) concluded that the effect of the mosaic strain was due to competition for a limited number of infection sites. Successful infection by the mosaic-type strain precluded the development of a necrotic lesion. This effect may be somewhat different from that resulting from prior multiplication of virus within the cells which become unable to support multiplication of a second strain. The term "interference" has been applied to the effects obtained with mixed inoculum and it seems appropriate. Price (38), however, restricted use of the term to reduction of infection by unrelated viruses, but this has not been generally accepted.

The term "premunity" was proposed by Quanjer (49) to describe the immunizing effect of previous infection by a plant virus against related strains.

CROSS PROTECTION AND VIRUS TAXONOMY

The concept that infection by one strain of virus rendered a plant immune from superinfection by other strains, besides being easily taught and apparently logical, has important implications for virus taxonomy. If strains of the same virus protect against each other, then cross protection between a known and an unknown strain is evidence of a relationship between the strains. This test has been widely used, but it has been recognized that failure to protect had no significance as far as relationships between strains are involved. Considerable evidence, even from the earliest observations, has demonstrated that some strains, which on other bases were closely related, failed to cross protect. McKinney (29), for example, noted that a "mild dark green" mosaic virus failed to protect tobacco against infection by the yellow strain of TMV, although both appeared to be TMV strains. Price (37) found that a yellow strain of tobacco ringspot virus protected against the common strain, but that the latter did not protect against the yellow strain. Matthews (28) found that some strains of potato virus X (PVX) did not completely protect *Datura stramonium* against any ringspot strain of PVX. The failure to protect was associated with strain-specific antigens not possessed by the ringspot strains. Matthews suggested that the degree of relationship among strains might affect cross protection reactions.

Other data, however, do not support close correlation between serological similarity and cross protection. Bawden and Kassanis (3) found that tobacco veinal necrosis virus failed to cross protect against several other potato virus strains known to be closely related to it serologically. There are various other reports of superinfection occurring when the viruses involved appeared, otherwise, to be closely related (2, 19). We have shown (15) variations among strains of tobacco streak virus in their ability to cross protect, but these could not be correlated with serological differences.

WHY DOES SUPERINFECTION OCCUR ?

Any theory that seeks to explain the basis of cross protection must also accommodate the observations on superinfection, which occurs commonly and may be more frequent than is apparent. Bercks (7), working with strains of potato virus X in tobacco, found that the challenge strain could often be detected serologically although it had not produced symptoms. He suggested that in the absence of symptoms induced by a challenge virus, cross protection should be verified by other methods which would demonstrate the presence or absence of the challenge virus. More recently, Cassels and Herrick (9) have demonstrated that a mild strain of TMV in tomato suppressed the development of a severe strain used as a challenge, but that severe strain antigen could be detected in challenged plants. These observations suggest that superinfection may occur commonly, but be undetected. Questions of terminology are raised. Is cross protection to be considered the absence of symptoms attributable to a challenge virus, or does the term necessitate a demonstration that the challenge virus does not replicate ?

An observation which, at the time, appeared to contradict previous observations was that *Nicotiana sylvestris* infected by TMV strains causing mosaic was not completely protected against superinfection by strains causing necrotic local lesions (14). If the challenge inoculum was sufficiently concentrated, lesions appeared in mosaic leaves, but only in the dark green tissue. The concentration of protecting virus in this tissue was much lower than in the light green tissue, and this was presumed to be related to the susceptibility to superinfection.

These results resemble those of Bercks (7) and Cassels and Herrick (9) who demonstrated superinfection serologically. The chief difference seems to be that in the TMV-*N. sylvestris* system the presumably low degree of multiplication by the challenge strain results in necrotic symptoms. Of several strains of TMV used with this system, the two which caused necrosis of healthy *N. sylvestris* with the least multiplication were the most efficient in superinfecting mosaic *N. sylvestris* (14).

These observations, as well as others, indicate that cross protection often does not represent complete immunity; superinfection may occur, but when it does, symptoms may be fewer, milder, or inapparent. Thus the concept arose that complete immunity is not a criterion for cross protection. Bennett (6), for example, pointed out that varying degrees of protection occur, and other authors have considered protection to be relative rather than absolute.

If cross protection is relative, then what is the degree of protection necessary for demonstrating a strain relationship ? The question is of importance because there have been numerous reports that virus infection may reduce susceptibility to unrelated viruses. Price (35) reported that tobacco mosaic virus symptoms were less severe in tobacco leaves recovered from tobacco ringspot virus symptoms. McKinney (30), Oshima (34), Nitzany and Cohen (33), Marrou and Migliori (27), and Prochazkova (39) found that systemic infection of *N. sylvestris* by cucumber mosaic or any of several other viruses reduced the numbers of lesions produced by challenge inoculation with unrelated, necrotic lesion-inducing strains of TMV. The opposite reaction, synergism, may occur with some combinations of viruses. Bennett (5) found that when tomatoes recovered from dodder latent mosaic virus were inoculated with tobacco etch virus or with TMV, dodder latent mosaic virus symptoms reappeared and persisted. Other examples of such synergism could be cited and seem to be related to changes in the level of concentration of one of the two viruses in the plant.

In spite of suppressive effects of unrelated viruses, there is usually a distinct difference between the high level of specific protection provided by one virus against a related strain and the distinctly lower level of non-specific protection.

THEORIES OF THE NATURE OF CROSS PROTECTION

Numerous theories have been proposed to explain the phenomenon of cross protection. To be complete, a theory should also explain superinfection by related strains as well as explain the non-specific reductions in susceptibility that unrelated viruses may induce. It may be appropriate to inquire how well the theories fit experimental observations.

Köhler and Hauschild (24) postulated that the protecting virus pre-empted a specific precursor in plant cells. The unavailability of the precursor prevented multiplication of the challenge virus. This theory assumes that there is a specific precursor for each virus, for which there is no evidence. A similar theory postulated that specific sites might be occupied by a protecting virus and unavailable to the challenge virus. While it is difficult to accommodate specificity in these theories, there must be a limited pool of material capable of being diverted to virus replication in a cell, and any diversion could result in non-specific depression of replication by a challenge virus.

Gibbs (17) postulated that the RNA of a challenge virus might become irreversibly but ineffectually bound to the replicase of the virus already in the cell. Binding might occur if recognition sites on the two viral RNAs were similar, but, if they were not identical, the binding might not lead to replication. Challenge viruses unrelated to the protecting virus would not bind to replicase already present and would thus not be prevented from multiplying. While replicase specificity may be a factor in cross protection, it is not easy to explain superinfection, because it usually occurs to a much less extent than the same amount of virus would cause in a healthy plant.

Ross (41) pointed out that a cell supporting virus synthesis would be flooded with mRNA. More ribosomes than normal would be complexed and hence unavailable for an introduced virus. This mechanism could apply equally well to related or unrelated viruses, and may well explain the effect of an unrelated virus in reducing susceptibility. It is difficult, however, to accommodate the demonstrable evidence of specificity of cross protection by this mechanism.

Kavanau (23) suggested that challenge virus might aggregate with the protecting virus, and so fail to replicate. This may occur, but cytological evidence indicates that there are various degrees of the tendency of viruses to aggregate within a plant cell.

Another hypothesis proposed by Ross (41) suggests that the basic mechanism is the ability of replicase of the protecting virus to recognize and copy the RNA of the challenge virus. No new replicase would be produced and templates of the protecting virus would so far outnumber those of the challenge virus that any effective new template would have little effect. The challenge virus would not be able to make much progress unless it could code for a replicase with greater affinity for its own RNA than for the RNA already present.

This theory could account for superinfection occurring, with limited replication of the challenge virus on the basis of degrees of affinity of a replicase for an introduced challenge virus. It is less easy to accommodate this theory to the effects of one virus in suppressing infection by an unrelated virus. This and the effects observed with mixed inoculum may, however, be due to different mechanisms. The various manifestations of virus interaction that have been grouped under the term "cross protection" may involve more than one mechanism.

An anomalous interaction between two distinct viruses was described by Bawden and Kassanis (3). Tobacco severe etch virus (SEV) inoculated to tobacco with, or after, potato virus Y (PVY) or *Hyocyamus* virus 3 (HV3) replaced them. Tobacco infected with SEV could not be infected by HV3. This was for many years an isolated example. It was eventually found that there is a slight degree of serological relationship between SEV and the other two viruses, all members of the potyvirus group.

More recently it was shown that tomato ringspot virus protects against elm mosaic (cherry leaf roll) virus (53), although they are serologically unrelated (12). In the reverse test, elm mosaic virus did not protect against tomato ringspot virus. Both viruses are typical nepovirses.

Partly in an effort to formulate a theory that might account for the anomalous interactions, as well as cross protection between related strains, DeZoeten and Fulton (10) suggested that uncoating of a mechanically inoculated virus occurs exterior to the cytoplasm of a cell, in ectodesmata or plasma membranes. It seemed unlikely that a challenge virus would find an environment within a cell conducive to uncoating while newly formed nucleic acid was being coated. Thus incoming nucleic acid of a challenge virus would meet an environment filled with free protein subunits of a protecting virus. If these "recognized" the challenge nucleic acid it would be coated, and no replication would occur.

All of the theories outlined here are based on the effect of the presence of the protecting virus within the cells inoculated, or of materials involved in replication of the virus. Specificity, or lack of it, is attributed to specific recognition of nucleic acid by replicase, or by protein subunits. What evidence is there then that an active resistance of the plant is involved ?

EVIDENCE FOR INVOLVEMENT OF ACTIVE DEFENSE IN CROSS PROTECTION

Some early observations on interference and cross protection suggest that host-mediated factors are involved, although they were not always so interpreted. One who did so was Thompson (48) who mixed unrelated viruses causing local lesions with TMV and compared lesion counts with controls. Lesion counts were reduced. Heating TMV (crude sap) to 60° for 10 minutes much reduced the interfering effect, but did not much reduce infectivity of the TMV. Dialysis of the infectious sap had a similar effect. TMV partially purified by three successive $(NH_4)_2SO_4$ precipitations, then adjusted to the same infectivity as the sap also lost much of its interfering property. Thompson suggested that a factor in addition to infectious particles was involved in interference. These observations, however, do not in themselves constitute evidence of specific antiviral material having been formed as a result of virus infection. There are numerous alterations in the chemical composition of leaves resulting from virus infection and some of the new materials might interfere with infection non-specifically.

There are, however, other indications of a restriction in virus multiplication in infected plants. The dark green areas of mosaic, TMV-infected leaves contain much less virus than the light green areas (14, 18). Atkinson and Matthews (1) found that the boundaries between light and dark green areas were stable. The protoplasmic connections appeared normal between cells containing high concentrations of virus rods and cells containing low concentrations. They postulated that some diffusible agent must be involved in the formation of these clusters of cells in which virus multiplication is suppressed.

A similar situation seems to exist with cucumber mosaic virus (CMV) (26). Virus concentration normally is low in the dark green areas of infected leaf tissue. A large proportion of protoplasts derived from these areas did not fluoresce when they were treated with antiserum to CMV or CMV coat protein and counterstained with fluorescent antibody. If infected plants were kept at 30°, however, green areas were invaded by CMV and the extractible infectivity equalled that of the light green areas. Incubation of protoplasts derived from dark green areas at 18° and 25° did not increase infectivity, which was interpreted as indicating a persistently resistant state. Similar results were obtained by Shalla and Petersen (46) with TMV. They found that only up to 20% of the protoplasts from epidermal cells of heavily inoculated tobacco leaves contained serologically demonstrable TMV. These leaves were, however, completely resistant to superinfection. Shalla and Petersen suggested that a diffusible substance is produced at infection centers which inhibits replication of other strains, but not other viruses.

The dark green areas of mosaic leaves may be analogous with the symptom-free leaves which develop following systemic necrosis caused by various ringspot viruses. The concentration of virus is lower in these leaves than in leaves with symptoms (37), although this is not invariably the case (13). With strains which superinfect such leaves, the amount of infection is usually much lower than in healthy leaves (15). Some factor evidently limits virus multiplication in these leaves.

The dark green tissue of TMV-infected leaves apparently contains a substantial number of cells which contain no virus at all. Murakishi and Carlson (32) found that when explants of bits of dark green tissue regenerated into plants, about half did not contain virus. These plants were not susceptible to infection by TMV two months after initiation, but became susceptible two and one-half months after initiation. The dark green tissue was not clonally derived (8), a conclusion also reached by Atkinson and Matthews (1). Thus it is unlikely to have arisen from isolated somatic mutations early in leaf development. Carlson and Murakishi (8) postulated a diffusible inhibitor of TMV multiplication as being involved in suppressing virus multiplication in the dark green tissue.

Foglein *et al.* (11) reported that when protoplasts were prepared from various tissues of TMV-infected plants in which virus synthesis had slowed or stopped, RNA synthesis was renewed. Presumably separation of infected protoplasts from the remaining tissue removed some restraining factor.

If a diffusible inhibitor of TMV multiplication is produced by a TMV-infected plant, it is not clear why the dark green areas permit superinfection. Nor is it clear why inoculation of dark green areas with a mosaic strain does not prevent subsequent infection by a local lesion-inducing strain (14) as it does in healthy leaves.

Kassanis *et al.* (20) suggested a relationship between the hypothetical restraining factor and the b proteins. They found that tobacco infected with potato virus Y, cucumber mosaic virus, potato virus X, potato aucuba mosaic virus, or alfalfa mosaic virus showed varying degrees of resistance to TMV. Resistance was correlated with the appearance of three new proteins, the b proteins as were induced by polyacrylic acid (16) (the association of these proteins with systemic acquired resistance is covered in this volume by Gianinazzi, Loebenstein, Harrison, Van Loon and others). Kassanis and White (21) showed that actinomycin D inhibited the development of resistance to TMV induced by earlier infection by potato virus Y or by polyacrylic acid. Other data, however, do not support the idea that the b proteins directly affect virus multiplication. Kassanis and White (22) observed that while polyacrylic acid decreased yield of TMV in tobacco protoplasts, the effect differed from the increased resistance to virus infection in tobacco leaves treated with polyacrylic acid. The b proteins produced in polyacrylic acid-treated leaves had no effect on virus yield in protoplasts.

Similar data have been presented by van Loon (51) and van Loon and Dijkstra (52) who described the appearance of new proteins in plants infected with mosaic as well as necrosis-inducing viruses. They also described a specificity for systemic induced resistance in Samsun NN tobacco induced by either TMV or tobacco necrosis virus. The resistance was more effective against challenge inoculation with the same virus than with the other virus.

CONCLUSIONS

Apparently host factors may play a decisive role in the cross protection phenomenon. Wenzel (54) suggested that products coded for by a virus may repress its synthesis and, with lower effectiveness, the synthesis of other viruses. Obviously there exists some control of virus synthesis; it would have a selective advantage in preventing the extermination of a host by a virus. Wenzel also suggested that synergism, that is stimulation of multiplication of one virus by another, occurs because the "helper" virus somehow renders repressors of the other virus non-functional.

It seems possible that part of the difficulty in attempting to explain cross protection is in looking for a single explanation for what may be a set of different reactions. Some may be relatively weak and non-specific, others may be marked and highly specific. The weak reactions may be attributed to ribosome unavailability or to other disruption of normal host functions by the first virus to infect. Interference may well be a phenomenon different from this and involve only competition for a limited number of infection sites. The specificity in cross protection is difficult to explain; an intermediary such as replicase that is capable of exhibiting the proper specificity seems an attractive hypothesis. Coupled with these factors is the suppression of virus multiplication in the host by mechanisms largely unknown at present, but which may play a major role in cross protection.

REFERENCES

1. ATKINSON, P.H. & MATTHEWS, R.E.F. (1970). On the origin of dark green tissue in tobacco leaves infected with tobacco mosaic virus. *Virology* 40, 344 - 356.

2. BALD, J.G. (1948). Potato virus X : effectiveness of acquired immunity in older and younger leaves. *Austral. Council Sci. Ind. Res. J.*, 20, 247 - 251.

3. BAWDEN, F.C. & KASSANIS, B. (1945). The suppression of one plant virus by another. *Ann. Appl. Biol.* 32, 52 - 57.

4. BAWDEN, F.C. & KASSANIS, B. (1951). Serologically related strains of potato virus Y that are not mutually antagonistic in plants. *Ann. Appl. Biol.* 38, 402 - 410.

5. BENNETT, C.W. (1949). Recovery of plants from dodder latent mosaic. *Phytopathology* 39, 637 - 646.

6. BENNETT, C.W. (1951). Interference phenomena between plant viruses. *Ann. Rev. Microbiol.* 5, 295 - 308.

7. BERCKS, R. (1948). Serologische Beiträge zur Frage der Abwehr von Zweitinfektionen bei X-Viren. *Phytopath. Z.* 25, 54 - 61.

8. CARLSON, P.S. & MURAKISHI, H.H. (1978). Evidence on the clonal versus non-clonal origin of dark green islands in virus-infected tobacco leaves. *Plant Sci. Let.* 13, 377 - 382.

9. CASSELS, A.C. & HERRICK, C.C. (1977). Cross protection between mild and severe strains of tobacco mosaic virus in doubly inoculated tomato plants. *Virology* 78, 253 - 260.

10. DEZOETEN, G.A. & FULTON, R.W. (1975). Understanding generates possibilities. *Phytopathology* 65, 221 - 222.

11. FOGLEIN, F.J., KALPAGAM, C., BATES, D.C., PREMECZ, G., NYITRAI, A. & FARKAS, G.L. (1975). Viral RNA synthesis is renewed in protoplasts isolated from TMV-infected Xanthi tobacco

leaves in an advanced stage of virus infection. *Virology* 67, 74 - 79.

12. FULTON, J.P. & FULTON, R.W. (1970). A comparison of some properties of elm mosaic and tomato ringspot viruses. *Phytopathology* 60, 114 - 115.

13. FULTON, R.W. (1949). Virus concentration in plants acquiring tolerance to tobacco streak. *Phytopathology* 39, 231 - 243.

14. FULTON, R.W. (1951). Superinfection by strains of tobacco mosaic virus. *Phytopathology* 41, 579 - 592.

15. FULTON, R.W. (1978). Superinfection by strains of tobacco streak virus. *Virology* 85, 1 - 8.

16. GIANINAZZI, S. & KASSANIS, B. (1974). Virus resistance induced in plants by polyacrylic acid. *J. Gen. Virol.* 23, 1 - 9.

17. GIBBS, A. (1969). Plant virus classification. *Adv. Virus Res.* 14, 263 - 328.

18. HOLMES, F.O. (1928). Accuracy in quantitative work with tobacco mosaic virus. *Bot. Gaz.* 86, 66 - 81.

19. HOLMES, F.O. (1934). A masked strain of tobacco mosaic virus. *Phytopathology* 24, 845 - 873.

20. KASSANIS, B., GIANINAZZI, S. & WHITE, R.F. (1974). A possible explanation of the resistance of virus-infected tobacco plants to second infection. *J. Gen. Virol.* 23, 11 - 16.

21. KASSANIS, B. & WHITE, R.F. (1974). Inhibition of acquired resistance to tobacco mosaic virus by actinomycin D. *J. Gen. Virol.* 25, 323 - 324.

22. KASSANIS, B. & WHITE, R.F. (1978). Effect of polyacrylic acid and b proteins on TMV multiplication in tobacco protoplasts. *Phytopath. Z.* 91, 269 - 272.

23. KAVANAU, J.L. (1949). On correlation of the phenomena associated with chromosomes, foreign proteins and viruses. III. Virus associated phenomena. *Amer. Nat.* 83, 113 - 138.

24. KÖHLER, E. & HAUSCHILD, I. (1947). Betrachtungen und versuche zum Problem der "erwarbenen Immunität" gegen Virus-

infektionen bei Pflanzen. *Zuchter* 17/18, 97 - 105.

25. KUNKEL, L.L. (1934). Studies on acquired immunity with tobacco and aucuba mosaics. *Phytopathology* 24, 437 - 466.

26. LOEBENSTEIN, G., COHEN, J., SHABTAI, S., COUTTS, R.H.A. & WOOD, K.R. (1977). Distribution of cucumber mosaic virus in systemically infected tobacco leaves. *Virology* 81, 117 - 125.

27. MARROU, J. & MIGLIORI, A. (1971). Interference entre les virus de la mosaique du concombre et de la mosaique du tabac. II. Influence d'une infection prealable des feuilles de tabac "Xanthi" n.c. par le VMC sur l'installation du VMT. *Ann. Phytopathol.* 3, 431 - 437.

28. MATTHEWS, R.E.F. (1949). Criteria of relationship between plant virus strains. *Nature* 163, 175.

29. MCKINNEY, H.H. (1929). Mosaic diseases in the Canary Islands, West Africa and Gibraltar. *J. Agric. Res.* 39, 557 - 578.

30. MCKINNEY, H.H. (1941). Virus-antagonism tests and their limitations for establishing relationships between mutants, and non-relationships between distinct viruses. *Am. J. Bot.* 28, 770 - 778.

31. MCKINNEY, H.H. & CLAYTON, E.E. (1944). Acute and chronic symptoms in the tobacco ringspot disease. *Phytopathology.* 34, 60 - 76.

32. MURAKISHI, H.H. & CARLSON, P.S. (1976). Regeneration of virus-infected tobacco leaves. *Phytopathology* 66, 931 - 932.

33. NITZANY, F.E. & COHEN, S. (1960). A case of interference between alfalfa mosaic virus and cucumber mosaic virus. *Virology* 11, 771 - 773.

34. OSHIMA, N. (1955). Interference between two unrelated plant viruses. *Jubilee Publ. Comm., 60th B'days Prof. Y. Tochimai & Prof. T. Fukushi.* 230 - 234.

35. PRICE, W.C. (1932). Acquired immunity to ringspot in Nicotiana. *Contrib. Boyce Thompson Inst.* 4, 359 - 403.

36. PRICE, W.C. (1936). Specificity of acquired immunity from tobacco ringspot diseases. *Phytopathology* 26, 665 - 675.

37. PRICE, W.C. (1936). Virus concentration in relation to acquired immunity from tobacco ringspot. *Phytopathology* 26, 503 - 529.

38. PRICE, W.C. (1964). Strains, mutation, acquired immunity, and interference. In : *Plant Virology*, Ed. by M.K. Corbett and H.D. Sisler, 527 pp. University of Florida Press.

39. PROCHAZKOVA, Z. (1970). Interaction of cucumber mosaic virus and potato virus Y with tobacco mosaic virus. *Biol. Plant.* (Praha) 12, 297 - 304.

40. QUANJER, H.M. (1946). "Premunity". *Phytopathology* 35, 892.

41. ROSS, A.F. (1974). Interaction of viruses in the host. *Acta Hortic.* 36, 247 - 260d.

42. SADASIVAN, T.S. (1940). A quantitative study of the interaction of viruses in plants. *Ann. Appl. Biol.* 27, 359 - 367.

43. SALAMAN, R.N. (1933). Protective inoculation against a plant virus. *Nature* 131, 468.

44. SALAMAN, R.N. (1937). Acquired immunity against the 'Y' potato virus. *Nature* 139, 924.

45. SALAMAN, R.N. (1938). The potato virus "X" : its strains and reactions. *Phil. Trans. Roy. Soc. (B)* 229, 137 - 217.

46. SHALLA, T.A. & PETERSEN, L.J. (1978). Studies on the mechanism of viral cross protection. *Phytopathology* 68, 1681 - 1683.

47. SMITH, K.M. (1939). The present status of plant virus research. *Biol. Rev.* 8, 136 - 179.

48. THOMPSON, A.D. (1960). Possible occurrence of interference systems in virus-infected plants. *Nature* 187, 761 - 762.

49. THUNG, T.H. (1931). Semtstof en plantencel bij enkele virusziekten van de tabaksplant. *Z. Ned. Indisch Natuurwetensch. Congr. Bandoeng, Java,* 450 - 463.

50. VALLEAU, W.D. (1941). Experimental production of symptoms in so-called recovered ringspot tobacco plants and its bearing on acquired immunity. *Phytopathology* 31, 522 - 533.

51. VAN LOON, L.C. (1975). Polyacrylamide disc electrophoresis of the soluble leaf proteins from *Nicotiana tabacum* var. Samsum and Samsun NN. IV. Similarity of qualitative changes of specific proteins after infection with different viruses and their relationship to acquired resistance. *Virology* 67, 566 - 575.

52. VAN LOON, L.C. & DIJKSTRA, J. (1976). Virus-specific expression of systemic acquired resistance in tobacco mosaic virus- and tobacco necrosis virus-infected 'Samsum NN' and'Samsun' tobacco. *Neth. J. Plant Path.* 82, 231 - 237.

53. VARNEY, E.H. & MOORE, J.D. (1952). Strain of tomato ringspot virus from American elm. *Phytopathology* 42, 476 - 477. (Abstr.).

54. WENZEL, G. (1971). Vergleichende Untersuchungen des Interferenzverhaltens phytopathogener Viren auf Tabak. *Phytopath. Z.* 71, 147 - 162.

REGULATION OF CHANGES IN PROTEINS AND ENZYMES ASSOCIATED WITH ACTIVE DEFENCE AGAINST VIRUS INFECTION

L.C. VAN LOON

Department of Plant Physiology
Agricultural University, Arboretumlaan 4
6703 BD Wageningen, The Netherlands

INTRODUCTION

Plant viruses possess only a limited amount of genetic information. Apart from the gene(s) for coat protein and one or two large polypeptides presumed to be involved in virus replication, information for only one to a few specific disease-inducing proteins can be present. This conclusion is reinforced by the existence of viroids, the smallest pathogens known, which contain as few as 359 nucleotides. No specific viral protein(s) involved in pathogenesis have as yet been identified, neither by *in vitro* translation of viral nucleic acids in cell-free protein-synthesizing systems, nor by *in vivo* analysis of changes in protein profiles of host plants upon infection. Contrary to metabolic changes after infection with fungi or bacteria, those resulting from virus infection appear to be solely reactions of the host plant and connected primarily with the type and severity of the symptoms produced: essentially all of these changes are also found after infection with fungi and bacteria that induce similar symptoms. Apparently, the ability of the host plant to react in the particular way it does, is present in a "cryptic" form all the time, but is evoked only by the triggering action of the infecting virus. The genetic information underlying symptom expression may never be expressed during normal plant development, or, if it is, upon virus infection it is expressed in an untimely or uncoordinated way, no longer subject to common regulatory controls, thus giving rise to the particular symptoms characteristic of the disease.

During active defence, information to counteract multiplication and/or spread of the virus is expressed. Most often, active defence takes the form of a hypersensitive reaction in which the virus is effectively localized in a narrow ring of tissue around a necrotic lesion at the site of virus penetration. Initially, at least, the virus is not physically impeded from spreading by the barriers of oxidized phenols which form in necrotizing tissue, because a) viruses are able to escape from the local lesion site by appropriate changes in environmental conditions, e.g. tobacco mosaic virus (TMV) escapes from local lesions on *N* gene-containing tobacco plants by a shift in temperature from 20 to 30°; b) virus may be found up to 1.5 mm from the necrotic area in cells that otherwise appear healthy; and c) local necroses, induced by hot water treatment, of cells involved in virus multiplication in systemically infected plants do not stop systemic spread of the virus. Furthermore, both localization without necrosis, e.g. of bean yellow mosaic virus (BYMV) in chlorotic lesions on *Tetragonia expansa*, or of TMV in starch lesions on cucumber cotyledons, and necrosis without localization, e.g. of tobacco rattle virus (TRV) or of potato virus Y^n (PVY^n) on tobacco, occur. Thus, necrosis is not a prerequisite for virus localization, and active defence can operate in its absence. Although it might be expected that virus localization depends on host RNA and protein synthesis, specific inhibitors such as actinomycin D or chloramphenicol have been found at most only to weaken the localizing mechanism without suppressing it completely. Virus multiplication is unrestricted, however, in protoplasts from hypersensitively reacting tobacco, tomato, and cowpea cultivars. Under these conditions, no hypersensitive necrosis ensues and virus multiplication is similar to that in protoplasts from susceptible cultivars. These observations support the concept that the localizing mechanism is operative in neighbouring, not yet infected cells rather than in the cells in which the virus is actively multiplying.

PHYSIOLOGICAL CHARACTERISTICS OF THE HYPERSENSITIVE REACTION

The hypersensitive reaction is non-specific in that basically similar necrotic lesions may be formed after infection with viral, fungal, or bacterial pathogens. Although host plant varieties with genetically determined differential sensitivity to one or more virus strains will show the hypersensitive reaction only after infection with specific strains, the reaction itself is always similar.

Once local lesions start to develop, many changes in enzyme activities become apparent. Notable among these are increases in oxidative and hydrolytic enzymes. Respiration is increased concomitantly with a shift in glucose catabolism towards the pentose phosphate pathway and a stimulation of glucose-6-phosphate dehydrogenase and 6-phosphogluconate dehydrogenase. In addition, aromatic biosynthesis is strongly activated (Table 1).

Table 1 Comparison of changes in enzye activities

	Hypersensitive reaction	Natural senescence	Artificial ageing	Wounding or injury
Respiratory activity	+	-	+	+
Glucose-6-phosphate dehydrogenase	+		+	+
6-Phosphogluconate dehydrogenase	+		+	+
Aromatic biosynthesis	+			
Phenylalanine ammonia-lyase	+	-	+	+
Cinnamic acid 4-hydroxylase	+			
Caffeic acid o-methyl-transferase	+			
Polyphenoloxidase	+	-	+	+
Peroxidase	+	+	+	+
Catalase	+	+		
Ribonuclease	+	+	+	+
Acid phosphatase	+	0	+	+
Protease	+	-	+	+

\+ increase - decrease 0 unchanged

Enzymes selected have been determined for at least two different virus - host plant combinations, by two different research groups, or both, or are otherwise well established

Most of these changes are similar to those occurring in non-infected plants during artificial ageing as in the accelerated senescence of detached leaves, or after wounding, and differ from alterations which occur during natural senescence. Only ribonucl-ease and peroxidase are strongly increased both during natural senescence and during artificial ageing or after wounding. Where-as acid phosphatase remains constant and protease, polyphenoloxid-ase, and phenylalanine ammonia-lyase (PAL) decrease during normal leaf development, all four enzymes increase upon detachment and after wounding. Leaf detachment or injury also induces a large in-crease in respiration and stimulates activities of the pentose phosphate pathway enzymes. Of the activities listed in Table 1, only polyphenoloxidase, peroxidase, and ribonuclease (RNase) have been shown to increase significantly in plants developing systemic symptoms without necrosis, albeit later and to a lesser extent than in hypersensitively reacting hosts. Such increases in peroxidase activity are positively correlated with symptom severity. The enzyme changes characteristic of the hypersensitive response show a pattern resembling those typical of artificial ageing, wounding, stress and injury.

Although quantitatively the enzyme changes more resemble those of wounding than of ageing, it could be that localization in necrotic lesions starts as a localized accelerated senescence, the resulting necrosis being responsible for the wounding response. One senescence-specific peroxidase isoenzyme appears prematurely in young bean leaves infected with southern bean mosaic virus (SBMV) (16) and all quantitative changes in peroxidase isoenzymes in TMV-infected Samsun NN tobacco are similar to those occurring during natural senescence (51). However, in both SBMV-infected bean leaves and TMV-infected tobacco leaves additional qualitative changes occur, which are specific for injury, and characteristic of virus-induced necrosis, respectively. Whereas detachment of tobacco leaves gives rise to accumulation of an ageing-specific nuclease, local lesion formation increases only an "injury-specific" nuclease under conditions when no senescence of the tissue is apparent (60). Therefore, the observed changes in enzyme activities show characteristics of either injury, or a combination of injury and senescence.

Does accelerated senescence or local tissue injury contribute to virus localization ? It is commonly assumed that a reduction in the number of lesions formed is a measure of increased resistance. However, lesion number only measures the events leading to the establishment of infection and merely reflects the susceptibility of the tissue in terms of available infectable sites. If after the successful penetration of the virus the host plant is able to limit effectively virus multiplication and spread, the defence mechanism is activated separately around each infection center. The number of infectable sites may vary but is irrelevant with regard to the effectiveness of the plant reaction. Upper leaves of tobacco plants that react hypersensitively to TMV remain symptomless and free of virus irrespective of the virus dose and, hence, of the number of lesions developing on inoculated lower leaves. On the contrary, a single infection site on a systemically reacting tobacco cultivar will lead to extensive virus multiplication and invasion of the younger leaves, which develop mosaic and severe malformations. For virus localization to be effective, once the virus has penetrated, multiplication and spread must be limited. Such limitation is reliably reflected by lesion size, because no virus can be recovered from tissue more than a few cell layers beyond the lesion edge.

The hypersensitive reaction is indeed characterized by strongly reduced spread of virus, even before macroscopically visible symptoms are apparent (35). Further activation of the defence mechanism, recognized as a consequence of hypersensitive reactions induced by viruses and other pathogens, occurs in the form of acquired resistance. In particular, systemic acquired resistance is characterized primarily by a decrease in lesion size, at least in tobacco (40, 51) and in bean (40). Additional effects on lesion number, noted in tobacco, bean, and other plants, may be ascribed to the failure to recognize very small lesions. That lesion number and size may be affected differently is illustrated by the fact that lesion number in tobacco increases, whereas lesion size decreases progressively with leaf age (12, 51). Also, in *Nicotiana glutinosa* systemic resistance is reflected by a decrease in lesion size, but by an increase in number (18).

All the enzymes which increase during the hypersensitive reaction strongly increase in non-infected leaves upon detachment. However, such artificial ageing increases both TMV multiplication and lesion size in tobacco and bean leaves (32, 51). Hence, at least, the accelerated senescence of detached leaves, together with the accompanying enzyme changes, does not contribute to virus localization. On the other hand, additional wounding by punching out leaf discs, followed by incubation of the discs under conditions similar to those for detached leaves, reduces lesion enlargement even beyond that observed on intact plants (Table 2). Unfortunately, no further data are available on relative rates of lesion enlargement in wounded leaves, possibly because damaged leaves are not suitable material for virus inoculation and lesion identification. If responses to mechanical, physical, or chemical wounding would indeed affect virus localization, their effects must be limited to the immediate vicinity of the wounds. Ross (40) was

Table 2 Relative lesion diameter on TMV-infected Samsum NN tobacco leaves under different physiological conditions

Condition	Lesion diameter
Intact plant	100
Detached leaf	123 ± 9
Leaf discs 2.5 cm diameter	71 ± 5

unable to influence the rate of lesion enlargement in tobacco or bean by mechanical abrasion, chemical or heat injury of adjacent leaves, although heat injury, at least, substantially increases polyphenoloxidase and peroxidase activity in adjacent leaves (26). Polyphenoloxidase also increases following inoculation with water or trimming of plants. Such treatments either do not affect, or increase lesion diameter (51). These results together with observations that polyphenoloxidase activity does not increase during the hypersensitive reaction of bean to alfalfa mosaic virus (AMV) (56) and appears to be absent from *Chenopodium amaranticolor* reacting hypersensitively to tobacco necrosis virus (TNV) (15), corroborate the earlier conclusion by Van Loon and Geelen (54) that stimulation of polyphenoloxidase is a secondary phenomenon, not related to lesion formation or to virus localization.

REGULATION OF ENZYME CHANGES AND LESION SIZE BY PLANT HORMONES

Since pathogenesis caused by viruses reflects aberrations of plant growth and development, hormones as natural regulators of these processes may be involved. This concept also applies to the hypersensitive reaction characterized by accelerated ageing, stress and wounding. Endogenous levels of auxins, gibberellins and cytokinins are high in rapidly growing leaves and gradually decline during maturation and senescence. Generally, treatment with these hormones alone or in combination delays senescence. In contrast, exogenous abscisic acid (ABA) or ethylene usually promotes senescence.

Auxins are known to counteract increases in ribonuclease and acid phosphatase (14, 30), but may either stimulate or decrease peroxidase activity (22, 53) (Table 3). Effects of auxins on the localization of TMV in hypersensitively reacting tobacco are complex and may in part result from auxin-induced ethylene production (53). However, since application of the natural auxins indole acetic acid and phenyl acetic acid decreases lesion enlargement, whereas rejuvenation of the leaves would be associated with an increase in lesion size, auxin regulation of leaf physiological age does not appear to be a controlling factor in lesion enlargement. No data are available for gibberellins.

Cytokinins are the most powerful inhibitors of senescence in most plant species and counteract all associated enzyme changes. Their effect is especially marked in detached leaves which, following severing from the stalk, are devoid of root-produced endogenous cytokinins and, consequently, show accelerated ageing. Detachment of an inoculated leaf stimulates lesion enlargement in TMV-infected tobacco (Table 2) and this effect is selectively counteracted by kinetin (53). In intact plants cytokinins have

Table 3 Influence of plant hormones on enzyme activities

	Auxin	Gibberellin	Cytokinin	Abscisic acid	Ethylene
Respiratory activity	+		–		+
Glucose-6-phosphate dehydrogenase					
6-Phosphogluconate dehydrogenase					
Aromatic biosynthesis					
Phenylalanine ammonia-lyase					+
Cinnamic acid 4-hydroxylase					+
Caffeic acid o-methyl-transferase					
Polyphenoloxidase	V		+		+
Peroxidase	V		–		+
Catalase					+
Ribonuclease	–		–	+	+
Acid phosphatase	–		–	+	+
Protease			+		+

+ increase – decrease V variable

been reported to influence lesion size and virus multiplication in various ways, depending on the physiological condition of the leaves, the concentration used, and the time of application with respect to inoculation. Since cytokinin content appears to increase during the hypersensitive reaction, Balazs *et al.* (5) suggested that cytokinins play a role in the inhibition of lesion spread. However, increasing the cytokinin content before appearance of lesions by spraying with a range of concentrations of kinetin six hours after inoculation did not significantly influence final size of lesions. Such observations suggest that cytokinins do not play a role in regulating the rate of lesion enlargement in leaves of intact plants.

Although the main function of ABA in leaves seems to be the promotion of stomatal closure in response to water stress, ABA accumulates in leaves during autumn and may accelerate senescence when applied exogenously. ABA can stimulate ribonuclease and acid phosphatase activity (14) (Table 3). In TMV-infected tobacco, ABA increases lesion number, an effect associated with a physiologically older stage of the leaf. However, it also enhances virus multiplication and lesion size (7) in contrast to what happens in physiologically older leaves. Endogenous ABA progressively increases in tobacco as TMV lesions develop (4); this would sustain lesion growth rather than limit lesion size. Inoculation of lower leaves of *N. glutinosa* with TMV results in a reduction in the ABA levels of upper leaves, and when challenge inoculated, these leaves develop slightly, though significantly smaller lesions (18).

Young growing tissues evolve more ethylene than do older tissues. However, ethylene increases considerably during specific developmental processes such as fruit ripening and leaf abscission. Although, like ABA, exogenously applied ethylene promotes leaf senescence, it is questionable whether it plays any role in its regulation.

Thus, the plant hormones implicated in the regulation of leaf senescence influence the rate of lesion growth, reflected in differences in final size, either marginally or in contrast to their activity in retarding or promoting senescence. Hence, leaf senescence, while increasing susceptibility to infection, does not appear to play a role in regulating lesion spread. Since ABA increases lesion size while stimulating RNase and acid phosphatase, whereas auxins decrease lesion size while counteracting the stimulation, these enzymes also do not contribute to virus localization. For RNase this conclusion is strongly supported by observations that a) water inoculation increases RNase activity severalfold, but when such leaves are challenge inoculated with virus two days later, lesion size is not significantly different from that in leaves not inoculated previously and b) upon detachment RNase activity strongly increases, as does lesion size (3, 51).

Whereas ethylene decreases during natural senescence or artificial ageing, it strongly increases upon wounding. It is produced as a typical wound response by undamaged cells adjacent to disrupted tissue and commonly mediates many of the wound reactions: it considerably stimulates respiration and may increase the activities of, among other enzymes, RNase, peroxidase, polyphenoloxidase, catalase, acid phosphatase, and the key enzymes of aromatic biosynthesis, PAL and cinnamic acid 4-hydroxylase (CAH) (1, 61) (Table 3). When Balazs *et al.* (8) compared ethylene production of Xanthi-nc. tobacco during the hypersensitive reaction to TMV and systemic infection by cucumber mosaic virus (CMV), they observed that ethylene production was markedly increased during local lesion formation at the very beginning of tissue necrosis, whereas it did not change perceptibly in the systemic infection. Similar results with other host plants were reported by Nakagaki *et al.* (33) and by Pritchard and Ross (37). Nakagaki and co-workers concluded that ethylene does not inhibit TMV synthesis, but is connected with necrosis of cells invaded by the virus. However, ethylene synthesis increases prior to lesion appearance and remains elevated after the necrotic process ends (37). In local infections that result in chlorotic, non-necrotic lesions, ethylene evolution may increase similarly (21). A possible involvement of ethylene in virus localization is suggested by observations that treating selected tobacco leaves with ethylene prior to inoculation of other leaves with TMV, will reduce lesion size in the latter (41), thus mimicking acquired resistance. Since ethylene can regulate increases in many enzymes that are stimulated during the hypersensitive reaction, as well as inhibit lesion spread in a way comparable to acquired resistance, its role in inducing systemic resistance is obviously important in understanding virus localization.

BIOCHEMICAL CHARACTERISTICS OF THE HYPERSENSITIVE REACTION

Systemic induced resistance is acquired when, following primary local infection, non-inoculated leaves develop the ability to react to a challenge inoculation with smaller and/or fewer lesions (12, 39). Since tobacco leaves with and without induced resistance respond similarly to treatments that affect lesion size, Ross (40) postulated that the mechanism of virus localization is the same under both conditions. In resistant leaves, however, the localizing mechanism would become operative earlier than in control leaves and/or operate at a higher level through stimulation or enhancement of the processes normally functioning to limit lesion size. Acquired resistance is not virus-specific; almost any virus, as well as fungi and bacteria, that induce necrotic lesions in tobacco will induce systemic effects that are, effectively, indistinguishable from those induced by TMV (39). Acquired resistance thus appears to be part of the general non-specific defence mechanism of plants to pathogens that induce localized infections. Hence, changes causally involved in the expression of systemic resistance should be both relatively non-specific, and function likewise to limit lesion size after a primary inoculation.

Enzymes of aromatic biosynthesis

Challenge inoculation of systemically resistant tobacco leaves with TMV causes an earlier and greater decrease in the concentration of phenolic compounds around the time of lesion formation than does inoculation of control leaves (44). At the same time, PAL activity abruptly increases in a narrow ring of tissue around the developing lesion (19), enhancing the capacity for phenol biosynthesis in living cells bordering lesions at a time critical with respect to the induction in such cells of a mechanism that eventually limits lesion size (44). Although the increase in PAL activity was reported by Simons and Ross (44) to be greater in systemically resistant than in control leaves, Fritig *et al.* (19) found PAL levels to be of the same order of magnitude in both. However, when calculated on the basis of cell numbers, in later stages of infection when necrosis becomes progressively more extensive in leaves without than in those with induced resistance, PAL activity remained higher in the latter.

Studies by Fritig *et al.* (20) and Legrand *et al.* (28) have shown that PAL, CAH, and caffeic acid-o-methyltransferase are sequentially stimulated and that these stimulations agree well with concomitant increases in the levels of all the phenylpropanoids and with an increased rate of incorporation of labelled phenylalanine into these compounds. Although the maximum concentrations of methoxylated compounds are reached rather late, the onset of stimulation of the phenylpropanoid pathway appears to be an early event; both PAL activity and the biosynthesis of scopoletin are already enhanced before the first local lesions are visible (20). Scopoletin and its glucoside, scopolin, accumulate around the developing necrotic center, their fluorescence indicating a zone of stimulated cells which eventually will become necrotic. Changes in the activities of the enzymes of aromatic biosynthesis can be used as biochemical markers of the necrotic reaction; symptoms of necrosis and changes in enzyme activity develop at the same time and the increases in enzyme activities are proportional to the number of developing necrotic lesions. Also, the greater the increase in enzyme activity, the larger is the final lesion size. However, if the reaction of Samsun NN tobacco to the common strain of TMV is compared with that to other strains of TMV that cause much smaller lesions, the increases in enzyme activities on a per cell basis, are progressively larger when final lesion size remains smaller. This inverse proportionality between enzyme stimulation and lesion size likewise holds for systemically resistant leaves (25). Since with smaller local lesions both total peroxidase activity and the average stimulation per cell are higher, smaller lesions are associated also with a higher capacity for phenylpropanoid turnover.

Whereas increased synthesis of aromatic compounds and their eventual oxidation seem primarily related to necrogenesis, it is possible that some of the metabolic intermediates promote virus localization, for instance by stimulating neighbouring cells to synthesize anti-viral substances. Cells stimulated early during lesion development will become necrotic in one or two days, but cells further away from the infection center will be stimulated when lesion spread is already slowing down. Such cells have far more time before they become necrotic to synthesize specific compounds that must attain a certain level to be effective in counteracting multiplication or spread of the virus. Legrand *et al.* (29) have suggested that stimulation of o-diphenol-o-methyltransferases and the corresponding changes in the concentration of the naturally occurring o-diphenolic substrates are related to increased synthesis of lignin-like compounds. These products could be involved in the thickening of walls of cells surrounding the local lesion; this is associated with the cessation of lesion spread and the development of resistance to further spread of the virus. Such lignification has been implicated in limitation of lesions in TMV U5-infected *N. glutinosa* (17).

Any treatment that abolishes the hypersensitive reaction such as a temperature shift, administration of antimetabolites, preparation of isolated mesophyll cells or protoplasts, and osmotic shock, at the same time suppresses increases in enzyme activities. Even when protoplasts are prepared from pre-infected leaves already carrying lesions, the stimulated PAL activity decreases abruptly and the increase disappears completely during the rather short time, six to eight hours, required for maceration of the tissue. Since the isolation of the protoplasts involves incubation in solutions of high osmotic strength, hypertonic osmotic pressure appears to be the determining factor for the non-occurrence of the hypersensitive response. Under such conditions, plasmodesmatal connexions between neighbouring cells are broken. This disconnexion could impede the cell to cell movement of any factor synthesized in infected cells which would activate cells in advance of infection. During protoplast isolation, the hypothetical factor causing necrosis would diffuse from infected cells into the maceration medium and, hence, activation is lost. Alternatively, suppression of the synthesis of this eliciting factor could equally well account for the non-occurrence or rapid disappearance of PAL stimulation (27), especially since Premecz *et al.* (36) have shown that hypertonic osmotic pressure causes striking changes in the regulation of RNA and protein synthesis.

Peroxidase activity

Systemically resistant leaves do not have increased respiration, but after challenge inoculation, Simons and Ross (43) detected earlier increases in the activities of a number of respiratory enzymes and found that the maximal activities were reached earlier and were higher in leaves with than in those without induced resistance. However, as lesions appeared sooner and in greater numbers in resistant leaves, this pattern of metabolic activity appeared to be directly related to the amount of necrosis developing in the early stage of infection. Only peroxidase and catalase activities increased in parallel with the development of resistance and remained at high levels. Van Loon and Geelen (54) similarly found that peroxidase levels were much higher and that activation of this enzyme occurred earlier in leaves with acquired resistance. Such leaves also showed high catalase but normal polyphenoloxidase activity. In contrast, in soybean leaves reacting hypersensitively to cowpea chlorotic mottle virus, development of acquired resistance appears to be associated with increased polyphenoloxidase rather than peroxidase activity (10).

In tobacco peroxidase activity progressively increases with leaf age. In young, expanding leaves TMV lesion size is not correlated with peroxidase activity, but from mature, full-grown to senescent leaves, lesion size decreases inversely with the logarithm of enzyme activity. Artificial ageing, trimming of plants, root inundation, drought or salt stress all stimulate leaf peroxidase activity to different extents. However, when such leaves are inoculated, lesion size is invariably increased. Likewise, injection of leaves with tobacco or horseradish peroxidase either has no effect on or causes enlargement of lesions. Different isoenzymes are affected differently by the treatments, but no correlation between changes in individual peroxidase isoenzymes and effects on lesion size could be established (51). The increased peroxidase activity associated with acquired resistance is not due to "injury-sensitive" isoperoxidases but to those whose activity increases during senescence. Moreover, the changes in response to challenge inoculation are the same, whether the lower leaves were intact or had been mechanically injured and/or infected. It thus appears that the reduction in final lesion size characteristic of systemically induced resistance is due to factors other than peroxidase activity (11). Similarly, no correlation between peroxidase activity and lesion size was noted by Stein and Loebenstein (47) after treatment of tobacco plants with polyanions. Initial leaf peroxidase activity thus seems to reflect a physiological state rather than to be directly responsible for regulating the rate of lesion enlargement.

On the other hand, the increase in peroxidase activity during lesion development may not merely reflect tissue necrosis. Compartmentation of peroxidase and its phenolic substrates will be lost as a result of cellular disruption. Together with the dramatic increase in the levels of phenols and the concomitant stimulation of polyphenoloxidase and peroxidase, as well as the reduction in catalase activity (54), this will greatly enhance the capacity of the cells to oxidize phenols and produce lignin-like barrier substances. As pointed out by Weststeijn (57), however, the supply of phenolic substrates and the content of o-diphenol-oxidizing enzymes in young will exceed those in old leaves; nevertheless, the rate of lesion enlargement in young is greater than in old leaves. Weststeijn (57) therefore proposed that peroxidase may be instrumental in producing an "inducing" agent, which, after movement to neighbouring cells, elicits the resistance mechanism. While there are reports that new isoenzymes, apparently unconnected with senescence, appear in Pinto beans after infection with SBMV (16), AMV, or TMV (45), in *N. glutinosa* after infection with TMV or TNV (45), and in tobacco after infection with TMV (48, 54), this apparently did not occur with other viral infections where either no new isoenzymes were observed or their appearance was related to premature senescence (9, 34, 59). A possible involvement of peroxidase can, however, be inferred from genetic analysis of peroxidase and symptom type in red clovers susceptible and resistant to BYMV. Although the genes determining hypersensitive and systemic necrosis were not identical with those for peroxidase, a closely linked specific cathodic isoenzyme appeared to act as a modifier contributing to virus localization and tissue necrosis (42).

Pathogenesis-related proteins

Systemically resistant leaves are further characterized by the presence of specific pathogenesis-related proteins (PRs). The appearance after virus infection of newly induced, specific PRs was originally demonstrated in tobacco and in *N. glutinosa* by Van Loon and Van Kammen (55), and quickly confirmed by Gianinazzi *et al.* (24). Similar proteins have since been recognized in virus-infected cowpea leaves (13). After infection of tobacco with TMV, PRs appear in the inoculated leaves at the onset of necrosis and increase in amount with time. From about seven days on, they also become detectable in the non-inoculated parts where virus is absent and where systemically induced resistance is expressed upon challenge inoculation.

None of these proteins appears as a result of natural or artificial ageing, mechanical or chemical wounding, or water, drought or salt stress. However, in Xanthi-nc tobacco they can be induced artificially by injection of leaves with polyacrylic or salicylic acid. Only salicylic acid is effective in inducing PRs in the tobacco cultivar Samsun NN (2). These observations prove that PRs are host-specific. Specific staining of electrophoretic patterns has shown that these proteins are not isozymic forms of 30 enzymes known to increase in activity upon TMV infection, nor do they contain carbohydrate, lipid, or RNA (48).

The ten new proteins identified in hypersensitively reacting Samsun NN tobacco were provisionally numbered IV to I and N to S in order of decreasing electrophoretic mobility (Fig. 1). They

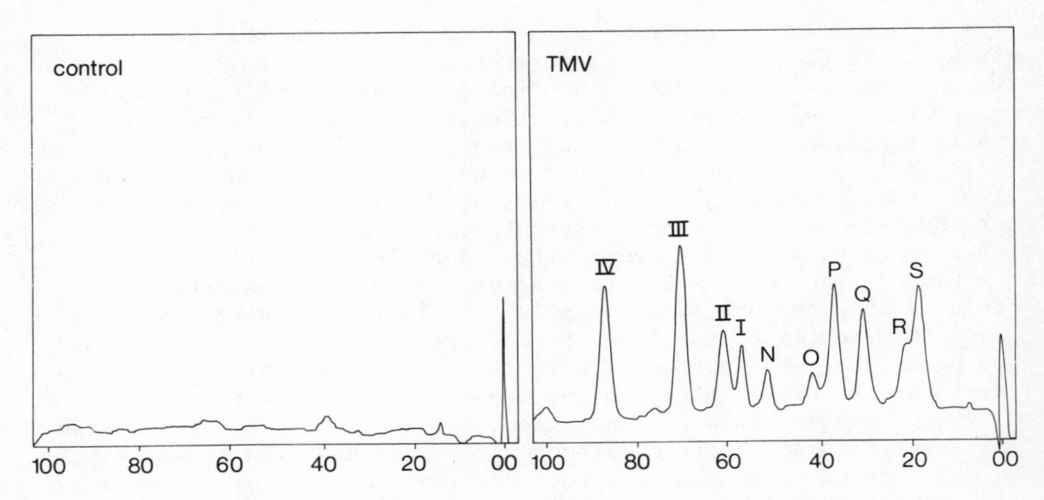

Figure 1 Densitometer tracings of electrophoretic patterns in 10% polyacrylamide gels of pH 3-soluble proteins from water-inoculated and 7-day TMV-infected Samsun NN tobacco leaves after prior treatment of the total soluble protein extracts with 12.5 μg trypsin per ml for 24 hr at 2°C.

are partly similar to those induced in *N. sylvestris* by TNV and differ from those induced in the *N* gene-carrying *N. glutinosa*. Although hypersensitivity to all strains of TMV in tobacco depends on the presence of the *N* gene, *N. glutinosa* expresses different proteins. On the other hand, the Samsun variety, that lacks the *N* gene, expresses the same proteins as Samsun NN. Hence, the PRs in tobacco are not related to the *N* gene. The lines Aurea S and Aurea N (46) contain an additional PR migrating between bands IV and III. The same band has been identified in the cultivar White Burley in England by Antoniw and White (2), but is absent from White Burley from Wageningen. It has been found also in Judy's Pride Burley and its hybrids by Gianinazzi *et al.* (23). This additional PR thus seems to be an independent marker present in part of the Burley tobacco population.

Tobacco PRs are preferentially extracted at low pH. They are soluble proteins and differ from the low pH-extractable proteins of non-infected leaves by their resistance to proteases. When total soluble leaf protein extracted at pH 8 is incubated either at room temperature under conditions where endogenous tobacco proteases are active, or at 2° in the presence of trypsin or chymotrypsin, pH 3-soluble proteins from non-infected plants are completely degraded, whereas those from infected plants are not

(Fig. 1). It seems unlikely that PRs are themselves stable end products of proteolysis. They are generated neither by repeated freezing and thawing of non-infected leaves, nor by treatment of protein extracts from non-infected leaves with proteases or detergents. Furthermore, their appearance can be inhibited by both actinomycin D and cycloheximide. Since protease activity is strongly increased in hypersensitively reacting tobacco, the PRs appear well adapted to function in an environment characterized by accelerated proteolysis such as occurs in ageing and dying cells. Although there is a distinct correlation between occurrence of these proteins and acquired resistance to further TMV infection, it has not been shown whether PRs limit spread of lesions or merely reflect metabolic changes accompanying localization.

During lesion development, the necrotic reaction and virus localization can be abolished by raising the temperature from 20 to 30°. At 30° the *N* gene is not expressed and the plants will develop systemic mosaic symptoms. If Samsun NN plants are kept at 20°, the amount of PRs keeps increasing even from 7 to 14 days after inoculation when lesion spread has virtually ceased. In contrast, when plants infected for seven days were transferred to 30°, further PR increase was completely suppressed. After six days at 30° the amounts of bands I, II, and IV had decreased, while the intensity of band III was unchanged. Virus synthesis rapidly resumed as shown by both an increase in the amount of extractable virus and by the appearance of free TMV coat protein in the electrophoretic patterns (49). These results demonstrate that the presence of PRs is not sufficient to prevent systemic spread of the virus. Likewise, if necrosis is caused by systemically spreading viruses such as TRV or PVY^n, PRs accumulate to high levels (50) but spread of these viruses does not appear to be impeded. If PRs function in limiting multiplication or spread of some viruses such as TMV, they apparently do so effectively only under conditions where a hypersensitive reaction has been initiated. Moreover, because PRs keep increasing at 20° for at least 14 days, their function may be related to their continuous accumulation, the amounts observed being merely the result of rather than the condition for their action.

Ethylene

Ethylene synthesis starts earlier in leaves with than in those without induced resistance. Pritchard and Ross (37) found ethylene synthesis to be increased about three hours before lesions were visible in control leaves and eight hours earlier in systemically resistant leaves. In both cases, ethylene quantitatively paralleled the amount of necrosis and reached a maximum when lesion growth abated. As treatment of tobacco leaves with ethylene prior to inoculation enhanced the capacity of the leaves to localize TMV, and because the amounts produced on infection seemed sufficient to assure its diffusion to several layers of cells in advance of virus movement, the early emanation of ethylene could well induce changes that either alter the reaction of those cells to the virus or modify virus movement into and out of the cells (37).

When Samsun NN tobacco infected with TMV is placed in an ethylene atmosphere, the leaves yellow, senesce prematurely and local lesion spread may be enhanced, a phenomenon which led Pritchard and Ross (37) to conclude that the ethylene synthesized during lesion development contributes to continued lesion growth. This conclusion agrees with observations that placing inoculated leaf discs in an atmosphere of ethylene or on solutions of 2-chloroethylphosphonic acid (ethephon) that decomposes into ethylene after uptake, increases lesion size and also stimulates TMV multiplication in systemically reacting tobacco cultivars (6, 33). However, when infected leaves on intact plants are sprayed with ethephon 12 hours after inoculation final lesion size is reduced in a concentration-dependent manner. After spraying with ethephon, there is a burst of ethylene which, in timing, is similar to that which occurs naturally in inoculated leaves. This natural burst is mimicked even better, both in time and spatially, by pricking leaves with needles dipped in ethephon solution. Under these conditions, small reddish-brown rings, resembling virus-induced local lesions developed around the punctures within a few hours. During the following days these rings slightly enlarged and turned dark brown whereas the centers became flaccid and whitish.

When the pH 3-extractable proteins from leaves with ethephon-induced local lesions were analysed by gel electrophoresis, changes qualitatively identical to those resulting from TMV infection were evident. Ethephon induced all ten PRs, whereas no such changes took place after pricking with water or after chemical injury (52). Quantitatively, however, there were differences in the relative amounts of individual PRs induced by ethephon as opposed to TMV, notably a deficiency in band II. It may be significant that this band is always comparatively low in non-inoculated, systemically resistant leaves from TMV-induced plants.

ROLE OF ETHYLENE IN THE HYPERSENSITIVE REACTION

Independently of the infecting virus, a hypersensitive reaction in tobacco is thus accompanied by an activation of the phenylpropanoid pathway, increased peroxidase activity, induction of PRs, and stimulated synthesis of ethylene. Since the same changes occur when plants develop systemic necrosis upon infection with TRV or PVY^n, none of these biochemical characteristics appears specific to the local lesion type of reaction to virus infection. However, systemic necrosis may be considered to be a hypersensitive reaction which is initiated too late to stop virus spread and hence, develops into a necrosis all along the path of spread of the virus. Nevertheless, peroxidase activity may increase similarly during non-necrotic infection, although this is not accompanied by new isoenzymes and occurs only late when symptoms are already apparent (54). Small amounts of PRs, such as are present in systemically resistant leaves, are also found relatively late in infection with some mosaic-type inducing viruses, e.g. potato virus Y^o and CMV. However, when challenge inoculated with TMV or TNV, such leaves invariably exhibit enhanced virus localization (50). The development of chlorotic lesions caused by BYMV on *T. expansa* is likewise accompanied by increased ethylene production (21). Such plants when exposed to 1% CO_2, which competitively inhibits ethylene action, failed to develop chlorotic lesions and showed symptomless infection. However, when treated with 1% ethylene, BYMV-infected plants developed necrotic spots within the chlorotic lesions. This effect was best shown in the older, mature leaves. Necrosis could also be induced by spraying inoculated leaves with 0.1% ethephon twice daily for four days (4).

In a hypersensitive reaction, stimulation of PAL, increase in peroxidase activity, induction of PRs, and increased emanation of ethylene are highest in the immediate vicinity of the lesions. PAL and ethylene production fall off to control levels within a few mm from the lesion center, whereas increased peroxidase activity and PRs are not confined to this zone. When lesion development progresses, there is a radial spread of stimulated activities preceding the virus and a slowing down when localization is complete (28). Stimulation is thus propagated in a wave-like fashion from the lesion center, reaching cells further away only later when cells stimulated earlier have become necrotic. Of the different alterations, stimulation of peroxidase and induction of PRs become evident only after appearance of lesions whereas stimulation of PAL and a short, large burst of ethylene occur just before lesions become macroscopically visible. Maximal ethylene production is usually attained between one and two days after inoculation and is proportional to the number of infection sites. Its production remains elevated for the rest of the incubation period during which lesions slowly enlarge (Fig. 2).

Since ethylene was found to induce PRs, and because cells at the lesion edge keep producing ethylene during lesion development, ethylene was investigated as a trigger for the other changes observed. Pricking of ethephon solution into leaves increased peroxidase activity even more than did infection with TMV. Moreover, under both conditions, the changes in the anodic and the

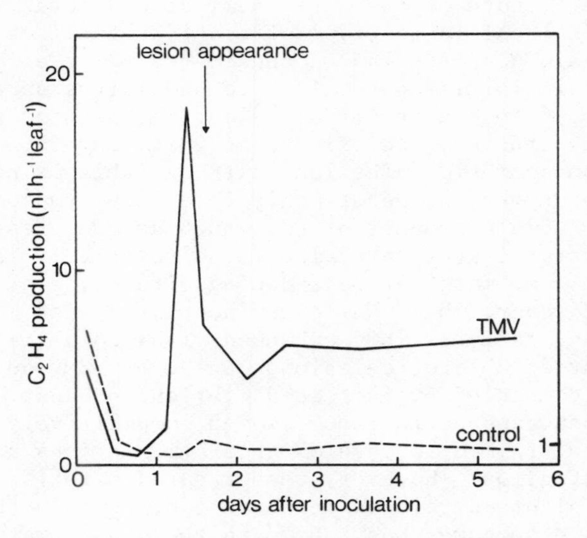

Figure 2 Ethylene production of water-inoculated and TMV-infected Samsun NN tobacco leaves.

cathodic isoenzymes were identical, as was the appearance of the new anodic isoenzyme (52). Thus, ethylene appears to be responsible not only for the induction of PRs but also for the alterations in peroxidase activity characteristic of TMV infection. Whereas pricking with solutions of mineral acids leads to only whitish spots on the leaves, ethephon caused reddish-brown pigmentation around the pricking site. As such browning is considered to result from the oxidation of phenolic compounds, ethylene may affect phenylpropanoid metabolism as does infection with TMV. Stimulation of PAL and CAH by ethylene are well documented (1). Otherwise, accumulation of scopoletin as a result of ethephon treatment of tobacco has been observed by Reuveni and Cohen (38) to occur in stems but not in leaves. However, synthesis of scopolin or scopoletin in leaves, followed by rapid translocation to the stem does not seem inconceivable (38). Although stimulation of the enzymes of aromatic biosynthesis and accumulation of phenols have not yet been demonstrated in ethephon-treated tobacco leaves, the similarity of lesion morphology after pricking with solutions of ethephon and infection with TMV suggest a similar regulation of phenylpropanoid metabolism.

Changes characteristic of systemically resistant leaves are induced also in untreated leaves of ethephon-pricked plants. Peroxidase activity is increased and after seven days, PRs are present in low concentrations. When such leaves are inoculated with TMV, enlargement of lesions is strongly decreased, up to 50%, quantitatively similar to that caused by systemic resistance induced by TMV (52). Furthermore, TMV and ethephon are equally effective in inducing higher systemic resistance in upper than in lower leaves. No systemic resistance is induced at 30°. When ethephon was applied to plants kept at 30°, no new proteins were induced and when after seven days plants were returned to 20° and challenge inoculated, limitation of lesions was not enhanced.

Local application of ethephon thus mimics most, if not all, of the changes associated with the hypersensitive reaction in tobacco, including the induction of systemically acquired resistance. Since the formation of local lesions in response to TMV is itself accompanied by a burst of ethylene (Fig. 2), it appears that this ethylene is responsible for at least the increase in peroxidase activity and the induction of both new proteins and systemic resistance. The involvement of ethylene does not seem to be confined to tobacco. In Pinto beans infected with necrosis-inducing viruses such as TNV or cowpea mosaic virus, at least four proteins extractable at low pH appeared to be greatly stimulated, if not newly induced. Pricking bean leaves with an ethephon solution caused identical alterations in the protein pattern, whereas pricking with water resulted only in the appearance of a different band.

COMPARISON OF ETHYLENE AND SALICYLIC ACID INDUCED RESISTANCE

Injection with salicylic acid has been reported to induce at least three PRs and acquired resistance in Samsun NN tobacco leaves (2). However, salicylic acid did not enhance ethylene production, nor did it increase peroxidase activity. When introduced by injection or by uptake, through the petiole, it induced PRs IV, III, and I in relatively large amounts, whereas II and the more slowly migrating components N to S were almost absent. When inoculated with TMV, salicylic acid-treated leaves developed lesions that remained substantially smaller than those which developed on water-injected leaves. No systemic effect occurred, however; PRs could not be detected in upper leaves of salicylic acid-injected plants, and lesion size was not reduced upon challenge inoculation. Thus, the effect of salicylic acid is localized. Since salicylic acid induces localized acquired resistance without increasing peroxidase activity, these observations a) further confirm that increased peroxidase activity is not a prerequisite for induced resistance and b) indicate the strong association between the reduction in spread of TMV lesions in tobacco with the presence of the new protein components.

Since both PRs and acquired resistance are induced by salicylic acid without an increase in ethylene production, it may seem questionable whether ethylene can be causally involved in their induction. However, salicylic acid is not considered to be a natural constituent of tobacco leaves. Its effects may be explained by assuming that it is structurally related to a compound, presumably phenolic, that functions as the local inducer under natural conditions. It seems unlikely that this compound is present in sufficient amounts in uninfected tobacco leaves. If so, it could be liberated by treatments such as mechanical wounding, but such treatments induce neither PRs nor acquired resistance. The compound thus has to be induced by infection. The stimulation of aromatic biosynthesis, the oxidation of phenols, or both might be instrumental in this respect. Since under natural conditions ethylene itself does induce PRs and acquired resistance, and may also mediate the increase in PAL activity, the following course of events may be envisaged.

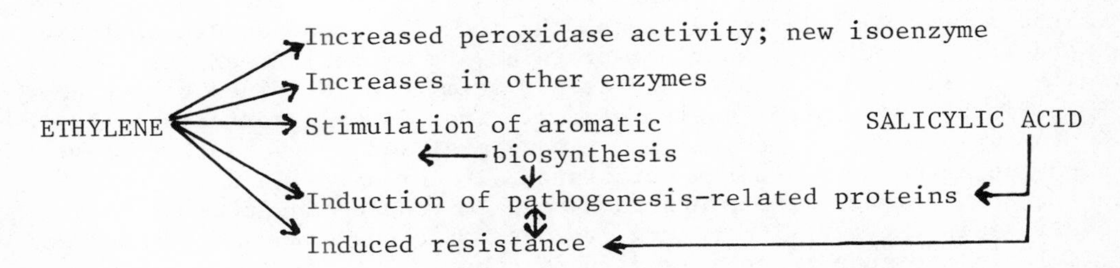

Ethylene is directly responsible for the increase in peroxidase activity, for the appearance of its new isoenzyme, and probably for increases in other enzymes. It likewise stimulates aromatic biosynthesis and through the action of a specific product of phenylpropanoid metabolism induces PRs, which may induce resistance. Salicylic acid mimics the last two steps. However, independent effects on PR induction and development of acquired resistance cannot be ruled out.

Increased ethylene synthesis occurs only in cells surrounding the developing lesions. As a gas, it will diffuse at most a few cell layers in advance of infection before it escapes from the tissue. It is these cell layers that respond by increased enzyme activities by induction of PRs and restriction of further virus multiplication and spread. In protoplasts, the severing of the connection between the cells releases them from the action of ethylene, which, being insoluble in water, evanesces. If ethylene acts as a trigger, it is clear that in preparation of protoplasts, the stimulus for increased PAL synthesis will rapidly disappear, the hypersensitive reaction will be inhibited and virus multiplication will resume.

Ethylene itself does not induce necrosis in tobacco even when individual leaves are gassed with as high as 300 µl/l ethylene for 40 hours (37). There is a burst of ethylene, similar in time to the one in hypersensitively reacting tobacco, during induction by CMV in cucumber cotyledons of chlorotic lesions in which the virus is not localized (31). Thus, increased production of ethylene before the appearance of necrotic lesions in TMV-infected tobacco does not appear to be responsible for the induction of hypersensitivity. Together with increased cell leakage (58), however, it is so far the earliest change which distinguishes hypersensitivity from a systemic reaction. The activation of ethylene synthesis thus marks a crucial event in pathogenesis, the resulting local accumulation of ethylene emanation being a causative factor in redirecting plant metabolism during the hypersensitive reaction.

REFERENCES

1. ABELES, F.B. (1973). *Ethylene in Plant Biology*. Academic Press, New York.

2. ANTONIW, J.F. & WHITE, R.F. (1980). The effects of aspirin and polyacrylic acid on soluble leaf proteins and resistance to virus infection in five cultivars of tobacco. *Phytopathologische Zeitschrift* (In press).

3. BAGI, G. & FARKAS, G.L. (1967). On the nature of increase in ribonuclease activity in mechanically damaged tobacco leaf tissue. *Phytochemistry* 6, 161 - 169.

4. BAILISS, K.W., BALAZS, E. & KIRÁLY, Z. (1977). The role of ethylene and abscisic acid in TMV-induced symptoms in tobacco. *Acta Phytopathologica Academiae Scientiarum Hungaricae* 12, 133 - 140.

5. BALAZS, E., BARNA, B. & KIRÁLY, Z. (1977). Heat-induced local lesions with high peroxidase activity in a systemic host of TMV. *Acta Phytopathologica Academiae Scientiarum Hungaricae* 12, 151 - 156.

6. BALAZS, E. & GABORJANYI, R. (1974). Ethrel induced leaf senescence and increased TMV susceptibility in tobacco. *Zeitschrift für Pflanzenkrankheiten und Pflanzenschutz* 81, 389 - 393.

7. BALAZS, E., GABORJANYI, R. & KIRÁLY, Z. (1973). Leaf senescence and increased virus susceptibility in tobacco: the effect of abscisic acid. *Physiological Plant Pathology* 3, 341 - 346.

8. BALAZS, E., GABORJANYI, R., TOTH, A. & KIRÁLY, Z. (1969). Ethylene production in Xanthi tobacco after systemic and local virus infections. *Acta Phytopathologica Academiae Scientiarum Hungaricae* 4, 355 - 358.

9. BATES, D.C. & CHANT, S.R. (1970). Alterations in peroxidase activity and peroxidase isozymes in virus-infected plants. *Annals of Applied Biology* 65, 105 - 110.

10. BATRA, G.K. & KUHN, C.W. (1975). Polyphenoloxidase and peroxidase activities associated with acquired resistance and its inhibition by 2-thiouracil in virus-infected soybean. *Physiological Plant Pathology* 5, 239 - 248.

11. BIRECKA, H., CATALFAMO, J.L. & URBAN, P. (1975). Cell wall and protoplast isoperoxidases in tobacco plants in relation to mechanical injury and infection with tobacco mosaic virus. *Plant Physiology* 55, 611 - 619.

12. BOZARTH, R.F. & ROSS, A.F. (1964). Systemic resistance induced by localized virus infections: extent of changes in uninfected plant parts. *Virology* 24, 446 - 455.

13. COUTTS, R.H.A. (1978). Alterations in the soluble protein patterns of tobacco and cowpea leaves following inoculation with tobacco necrosis virus. *Plant Science Letters* 12, 189 - 197.

14. DE LEO, P. & SACHER, J.A. (1970). Control of ribonuclease and acid phosphatase by auxin and abscisic acid during senescence of *Rhoeo* leaf sections. *Plant Physiology* 46, 806 - 811.

15. FACCIOLI, G. (1979). Relation of peroxidase, catalase and polyphenoloxidase to acquired resistance in plants of *Chenopodium amaranticolor* locally infected by tobacco necrosis virus. *Phytopathologische Zeitschrift* 95, 237 - 249.

16. FARKAS, G.L. & STAHMANN, M.A. (1966). On the nature of changes in peroxidase isoenzymes in bean leaves infected by southern bean mosaic virus. *Phytopathology* 56, 669 - 677.

17. FAVALI, M.A., BASSI, M. & CONTI, G.G. (1974). Morphological, cytochemical and autoradiographic studies of local lesions induced by the U5 strain of tobacco mosaic virus in *Nicotiana glutinosa* L. *Rivista di Patologia Vegetale* S. IV, 10, 207 - 218.

18. FRASER, R.S.S., LOUGHLIN, S.A.R. & WHENHAM, R.J. (1979). Acquired systemic susceptibility to infection by tobacco mosaic virus in *Nicotiana glutinosa* L. *Journal of General Virology* 43, 131 - 141.

19. FRITIG, B., GOSSE, J., LEGRAND, M. & HIRTH, L. (1973). Changes in phenylalanine ammonia-lyase during the hypersensitive reaction of tobacco to TMV. *Virology* 55, 371 - 379.

20. FRITIG, B., LEGRAND, M. & HIRTH, L. (1972). Changes in the metabolism of phenolic compounds during the hypersensitive reaction of tobacco to TMV. *Virology* 47, 845 - 848.

21. GABORJANYI, R., BALAZS, E. & KIRÁLY, Z. (1971). Ethylene production, tissue senescence and local virus infections. *Acta Phytopathologica Academiae Scientiarum Hungaricae* 6, 51 - 55.

22. GALSTON, A.W., LAVEE, S. & SIEGEL, B.Z. (1968). The induction and repression of peroxidase isozymes by 3-indole acetic acid. In : *Biochemistry and Physiology of Plant Growth Substances*, Ed. by F. Wightman & G. Setterfield, pp. 455 - 472. The Runge Press Ltd., Ottawa.

23. GIANINAZZI, S., AHL, P. & CORNU, A. (1980). b-Protein variation in virus-infected intraspecific tobacco hybrids. *Acta Phytopathologica Academiae Scientiarum Hungaricae*, in press.

24. GIANINAZZI, S., MARTIN, C. & VALLÉE, J.-C. (1970). Hypersensibilité aux virus, température et protéines solubles chez le *Nicotiana* Xanthi n.c. Apparition de nouvelles macromolécules lors de la répression de la synthèse virale. *Compte rendu de l'Académie des Sciences de Paris* 270 D, 2383 - 2386.

25. HIRTH, L., LEGRAND, M. & FRITIG, B. (1977). Quelques aspects biochimiques des réactions de défense de tabacs hypersensibles infectés par le VMT. In : *Travaux dédiés à G. Viennot-Bourgin*, Ed. by G. Viennot-Bourgin, pp. 133 - 142. Société Française de Phytopathologie, Paris.

26. KÖNIG, D. & NIENHAUS, F. (1970). Stoffwechselphysiologische Veränderungen in der Pflanze nach Virusinfektion unter Einfluss von Wundreiz 1. Polyphenoloxidase-, Peroxydase- und Cytochromoxydaseaktivität. *Phytopathologische Zeitschrift* 68, 193 - 205.

27. KOPP, M., GEOFFROY, P. & FRITIG, B. (1979). Phenylalanine ammonia-lyase levels in protoplasts isolated from hypersensitive tobacco pre-infected with tobacco mosaic virus. *Planta* 146, 451 - 457.

28. LEGRAND, M., FRITIG, B. & HIRTH, L. (1976). Enzymes of the phenylpropanoid pathway and the necrotic reaction of hypersensitive tobacco to tobacco mosaic virus. *Phytochemistry* 15, 1353 - 1359.

29. LEGRAND, M., FRITIG, B. & HIRTH, L. (1978). *O*-diphenol *o*-methyltransferases of healthy and tobacco-mosaic-virus-infected hypersensitive tobacco. *Planta* 144, 101 - 108.

30. LONTAI, I., VAN LOON, L.C. & BRUINSMA, J. (1972). Effects of auxin on the activity of RNA-hydrolysing enzymes from senescing and ageing barley leaves. *Zeitschrift für Pflanzenphysiologie* 67, 146 - 154.

31. MARCO, S. & LEVY, D. (1979). Involvement of ethylene in the development of cucumber mosaic virus-induced chlorotic lesions in cucumber cotyledons. *Physiological Plant Pathology* 14, 235 - 244.

32. NAKAGAKI, Y. & HIRAI, T. (1971). Effect of detached leaf treatment on tobacco mosaic virus multiplication in tobacco and bean leaves. *Phytopathology* 61, 22 - 27.

33. NAKAGAKI, Y., HIRAI, T. & STAHMANN, M.A. (1970). Ethylene production by detached leaves infected with tobacco mosaic virus. *Virology* 40, 1 - 9.

34. NOVACKY, A. & HAMPTON, R.E. (1968). Peroxidase isozymes in virus-infected plants. *Phytopathology* 58, 301 - 305.

35. OHASHI, Y. & SHIMOMURA, T. (1971). Necrotic lesion induced by heat treatment on leaves of systemic host infected with tobacco mosaic virus. *Annals of the Phytopathological Society of Japan* 37, 22 - 28.

36. PREMECZ, G., RUZICSKA, P., OLAH, T. & FARKAS, G.L. (1978). Effect of "osmotic stress" on protein and nucleic acid synthesis in isolated tobacco protoplasts. *Planta* 141, 33 - 36.

37. PRITCHARD, D.W. & ROSS, A.F. (1975). The relationship of ethylene to formation of tobacco mosaic virus lesions in hypersensitive responding tobacco leaves with and without induced resistance. *Virology* 64, 295 - 307.

38. REUVENI, M. & COHEN, Y. (1978). Growth retardation and changes in phenolic compounds, with special reference to scopoletin, in mildewed and ethylene-treated tobacco plants. *Physiological Plant Pathology* 12, 179 - 189.

39. ROSS, A.F. (1961). Systemic acquired resistance induced by localized virus infections in plants. *Virology* 14, 340 - 358.

40. ROSS, A.F. (1966). Systemic effects of local lesion formation. In : *Viruses of Plants*, Ed. by A.B.R. Beemster & J. Dijkstra, pp. 127 - 150, North-Holland Publishing Company, Amsterdam.

41. ROSS, A.F. & PRITCHARD, D.W. (1972). Local and systemic effects of ethylene on tobacco mosaic virus lesions in tobacco. *Phytopathology* 62, 786.

42. SHEEN, S.J. & DIACHUN, S. (1978). Peroxidases of red clovers resistant and susceptible to bean yellow mosaic virus. *Acta Phytopathologica Academiae Scientiarum Hungaricae* 13, 21 - 28.

43. SIMONS, T.J. & ROSS, A.F. (1971). Metabolic changes associated with systemic induced resistance to tobacco mosaic virus in Samsun NN tobacco. *Phytopathology* 61, 293 - 300.

44. SIMONS, T.J. & ROSS, A.F. (1971). Changes in phenol metabolism associated with induced systemic resistance to tobacco mosaic virus in Samsun NN tobacco. *Phytopathology* 61, 1261 - 1265.

45. SOLYMOSY, F., SZIRMAI, J., BECZNER, L. & FARKAS, G.L. (1967). Changes in peroxidase-isozyme patterns induced by virus infection. *Virology* 32, 117 - 121.

46. SPURR, H.W. & BURK, L.G. (1977). Aurea N, a burley tobacco with a single dominant gene for hypersensitivity to TMV. *Proceedings of the American Phytopathological Society* 4, 142, No. 278.

47. STEIN, A. & LOEBENSTEIN, G. (1976). Peroxidase activity in tobacco plants with polyanion-induced interference to tobacco mosaic virus. *Phytopathology* 66, 1192 - 1194.

48. VAN LOON, L.C. (1972). *Pathogenese en Symptoomexpressie in Viruszieke tabak; een onderzoek naar veranderingen in oplosbare eiwitten.* Thesis. Laboratorium voor Virologie, Landbouwhogeschool, Wageningen.

49. VAN LOON, L.C. (1975). Polyacrylamide disk electrophoresis of the soluble leaf proteins from *Nicotiana tabacum* var. "Samsun" and "Samsun NN" III. Influence of temperature and virus strain on changes induced by tobacco mosaic virus. *Physiological Plant Pathology* 6, 289 - 300.

50. VAN LOON, L.C. (1975). Polyacrylamide disc electrophoresis of the soluble leaf proteins from *Nicotiana tabacum* var. "Samsun" and "Samsun NN" IV. Similarity of qualitative changes of specific proteins after infection with different viruses and their relationship to acquired resistance. *Virology* 67, 566 - 575.

51. VAN LOON, L.C. (1976). Systemic acquired resistance, peroxidase activity and lesion size in tobacco reacting hypersensitively to tobacco mosaic virus. *Physiological Plant Pathology* 8, 231 - 242.

52. VAN LOON, L.C. (1977). Induction by 2-chloroethylphosphonic acid of viral-like lesions, associated proteins, and systemic resistance in tobacco. *Virology* 80, 417 - 420.

53. VAN LOON, L.C. (1979). Effects of auxin on the localization of tobacco mosaic virus in hypersensitively reacting tobacco. *Physiological Plant Pathology* 14, 213 - 226.

54. VAN LOON, L.C. & GEELEN, J.L.M.C. (1971). The relation of polyphenoloxidase and peroxidase to symptom expression in tobacco var. "Samsun NN" after infection with tobacco mosaic virus. *Acta Phytopathologica Academiae Scientiarum Hungaricae* 6, 9 - 20.

55. VAN LOON, L.C. & VAN KAMMEN, A. (1970). Polyacrylamide disc electrophoresis of the soluble leaf proteins from *Nicotiana tabacum* var. "Samsun" and "Samsun NN" II. Changes in protein constitution after infection with tobacco mosaic virus. *Virology* 40, 199 - 211.

56. VEGETTI, G., CONTI, G.G. & PESCI, P. (1975). Changes in phenylalanine ammonia-lyase, peroxidase and polyphenoloxidase during the development of local necrotic lesions in Pinto bean leaves infected with alfalfa mosaic virus. *Phytopathologische Zeitschrift* 84, 153 - 171.

57. WESTSTEIJN, E.A. (1976). Peroxidase activity in leaves of *Nicotiana tabacum* var. *Xanthi* nc. before and after infection with tobacco mosaic virus. *Physiological Plant Pathology* 8, 63 - 71.

58. WESTSTEIJN, E.A. (1978). Permeability changes in the hypersensitive reaction of *Nicotiana tabacum* cv. Xanthi nc. after infection with tobacco mosaic virus. *Physiological Plant Pathology* 13, 253 - 258.

59. WOOD, K.R. (1971). Peroxidase isoenzymes in leaves of cucumber (*Cucumis sativus* L.) cultivars systemically infected with the *W* strain of cucumber mosaic virus. *Physiological Plant Pathology* 1, 133 - 139.

60. WYEN, N.V., UDVARDY, J., ERDEI, S. & FARKAS, G.L. (1972). The level of a relatively purine-specific ribonuclease increases in virus-infected hypersensitive or mechanically injured tobacco leaves. *Virology* 48, 337 - 341.

61. YANG, S.F. & PRATT, H.K. (1978). The physiology of ethylene in wounded plant tissues. In : *Biochemistry of Wounded Plant Tissues*, Ed. by G. Kahl, pp. 595 - 622. Walter de Gruyter & Co., Berlin.

ANTIVIRAL AGENTS AND INDUCERS OF VIRUS RESISTANCE : ANALOGIES WITH INTERFERON

S. GIANINAZZI

Station d'Amélioration des Plantes
INRA, B.V. 1540
21034 Dijon Cedex, France

*'En biologie, comme en justice,
il y a des moments où l'accumulation
des preuves indirectes est si forte
que même les présomptions méritent
d'être prises en considération'*

Jacqueline De Maeyer (16)

INTRODUCTION

The possibility that the active defence mechanisms against viruses in plants may be analogous to the interferon system in animals is not a new hypothesis. In 1940, Price (43) compared the 'apparent recovery' of virus-infected plants to acquired immunity in animals, and later Loebenstein (37) in 1963 suggested that acquired resistance to viruses may be caused by an interferon-like mechanism. More recently, we reported the appearance of new soluble proteins in plant cells that become resistant to viruses and pointed out the possibility of an analogy between these and interferon in animals (19, 22, 23).

Although research for the molecular basis of active mechanisms of resistance in plants has stimulated many studies, it would be beyond the scope of the present chapter to review exhaustively all the literature covering this subject. Emphasis is placed rather more on those studies which have helped to clarify to some extent the mechanisms of active resistance in plants and which seem to strengthen the possibility of an analogy between these and the interferon system in animals. A large number of substances exist that can affect the development of virus diseases in plants and these may be divided into two main groups a) antiviral agents, b) inducers of virus resistance.

ANTIVIRAL AGENTS

Substances acting as antiviral agents can originate from such widely different sources as plants, microorganisms, insects and molluscs,or be synthetic compounds. On the basis of their mode of action, these antiviral agents can be grouped into a) inhibitors of virus infection *in vitro* and *in vivo* and b) inhibitors of virus replication.

Inhibitors of virus infection in vitro *and* in vivo

Inhibitors of virus infection *in vitro* can act in various ways by either complexing, chemically modifying and/or denaturing, dissociating (hydrolysing) or precipitating viruses. For example, it has been known for a long time that the *in vitro* enzymatic oxidation of phenolic compounds such as catechol or tyrosine, with quinone formation, can be responsible for oxidative virus inactivation (7, 27, 38). In fact, lack of or reduction in infectivity often observed in crude inocula coming from macerated tissue is usually attributed to this. Certain naturally occurring compounds such as tannins, naphthoquinone derivatives or saponins can also act as virus inactivators (44) and since these compounds are widespread in certain plant families, they may affect virus transmission more often than is realized.

Table 1 Characteristics of *in vivo* inhibitors of virus infection

Isolated from very different sources

Active when applied to leaf tissues before or at the time of virus inoculation but having very little or no effect when added shortly after virus inoculation

Lack of virus specificity but some host specificity

Inhibitors of virus infection *in vivo* prevent the virus from establishing itself; their properties are summarized in Table 1. Several of these inhibitors are in plant saps, for example that of *Chenopodium album*. As can be seen in Table 2, *C. album* sap strongly inhibits infection by three viruses on four plants, two coming from the Leguminosae, one from the Solanaceae and one from the Amaranthaceae. It is interesting to note that this sap is least active on the plant *Gomphrena globosa* coming from the same order as *C. album*, that is, the Centrosperma. The inhibitory substance in *C. album* sap is proteinaceous in nature (51) as are the potent inhibitors isolated from pokeweed (*Phytolacca americana*) (63) and carnation (*Dianthus caryophyllus*)(44), and as are, probably, many from other plant species. The inhibitor from carnation has been rigorously purified and characterized (45) and its main features are summarized in Table 3.

Table 2 Effect of *Chenopodium album* sap on the infection of four species of plants by three viruses (51)

Virus	Test plant	Inhibition %			
		Final dilution of *C. album* sap			
		$2x10^{-1}$	$2x10^{-2}$	$2x10^{-3}$	$2x10^{-4}$
Tobacco necrosis	*Phaseolus vulgaris*	100	98	88	60
Tobacco necrosis	*Vigna sinensis*	96	92	76	32
Tobacco mosaic	*Nicotiana glutinosa*	94	63	21	12
Potato virus X	*Gomphrena globosa*	85	59	0	0

These inhibitors of virus infection have been reported to be effective in hypersensitive as well as in systemic hosts when mixed with the virus suspension before inoculation. They do not appear, however, to be able to protect the leaf against subsequent mechanical inoculation of virus when they are absorbed by the leaf via its petiole (44). The non-infective virus-inhibitor mixtures can be easily fractionated, for example by centrifugation or chromatography, to give virus that is fully infective (45).

Table 3 Properties of inhibitor from carnation (45)

Proteinaceous, containing 16 amino acids

Molecular weight 14,000 (133 amino acid residues)

Lysine ϵ-NH_2group is functional

Positive (+) net charge, up to pH 7.8

100% inhibition of 0.06% TMV on *N. glutinosa* by 0.6 µg/ml

Thermal inactivation : 80^oC (10 min 0.2M-PO_4, pH 6.6)

Concentration in sap : 7 mg/l (uncorrected)

In vivo inhibitors have a very broad action spectrum with little or no virus specificity. For example, the carnation inhibitor is active against at least 17 plant viruses while the pokeweed inhibitor can be active even against animal viruses such as the influenza virus (56). In contrast, some host specificity seems to exist in the sense that these inhibitors are largely ineffective in preventing infection of the species from which they are obtained, or of closely related species (Table 2).

The fact that the effectiveness of the inhibitor varies with the host plant suggests that they may act either directly through the host or that they may affect the interaction between the virus and its host. Direct effect through the host seems unlikely because they cannot prevent virus infection when absorbed through the leaf petiole (44) or when infection involves zoospores of *Olpidium brassicae* (55). It seems most likely, therefore, that in the presence of the inhibitor some essential process of infection is prevented. One such process is the attachment of the virus to the host cell, and Ragetli and Weintraub (45) have proposed that the inhibitors could act on the host through cationic and anionic sites on the leaf surface. Those that are polyanions, such as polyaspartic acid, would compete directly with the negatively charged virus nucleic acid for the cationic receptor sites on cells. The inhibitor isolated from carnation, would initially or primarily interact via its ϵ-amino groups with anionic host sites and subsequently interact through its carboxyl groups with cationic receptor sites. Differences in sensitivity to inhibitors of different plant species could subsequently be explained by differences in the distribution or number of cationic and anionic receptor sites on leaf surfaces.

In the plant kingdom certain taxonomic groups are notable for having proteinaceous inhibitors of virus infection. However, while members of the Centrosperma, for example, are particularly rich in potent inhibitors, they are also among those plants that are least sensitive to virus inhibitors. It seems probable therefore that in nature these proteinaceous virus inhibitors do not play a vital role in controlling virus infection, and that if they represent conservation of certain genes during evolution, then the retention on the latter is not essential for the survival of many species.

Inhibitors of virus replication

This second main group of antiviral agents prevents multiplication of viruses *in vivo*. Many chemicals have been reported to act as inhibitors of virus replication in plants, including several synthetic analogues of RNA and DNA bases or nucleosides and antibiotics most of which can also be active against animal viruses (29, 40). If these substances affect virus multiplication and not infection itself then they should be active when applied after inoculation.

Commoner and Mercer (11) reported that the pyrimidine base analogue 2-thiouracil (TU) was a powerful inhibitor of tobacco mosaic virus (TMV) replication in tobacco leaf discs, and that this inhibition was reversed by an excess of uracil but not by other pyrimidine bases. However, contradictory results have since been obtained, including some which even claim that TU favours the biosynthesis and infectivity of viruses. Results are also contradictory for another inhibitor of virus replication, actinomycin D (AMD), which inhibits DNA - dependent RNA synthesis. Several authors have pointed out that, in fact, either AMD or TU are only effective when applied at or very shortly after the time of virus inoculation; Dawson and Schlegel (14) determined the possible sequence of inhibition of TMV replication by these and other inhibitors using a system in which TMV multiplies synchronously in tobacco leaves (Table 4).

Table 4 Sequence of TMV replication established by Dawson and Schlegel (14) using virus inhibitors

INFECTION → AMD sensitive step → 3°C sensitive step → TU sensitive step → 12°C sensitive step → CX sensitive step → TMV-RNA → VIRION

AMD - actinomycin D TU - 2-thiouracil
CX - cycloheximide

Multiplication of several viruses has been shown to be sensitive to AMD and as Rottier and co-workers (48) stated 'it is tempting to speculate, therefore, that inhibition of multiplication by AMD is a common feature of all plant RNA viruses and by consequence, that the involvement of an early nuclear function is a general phenomenon'. The involvement of a specific part of the host DNA as one of the first steps in the multiplication of viruses may play a decisive role in the determination of host range, that is host specificity.

While such observations are very useful for understanding the molecular biology of virus replication and could provide us with a way for determining the basis of the antiviral activity of different substances, their application in the field seems to be less obvious. In fact, these compounds are generally unsuitable for use on crops both because of their high cost and toxicity to animals. Furthermore, many are not specific to steps in virus multiplication and can also disturb metabolism of the host plant. For example, application of TU *via* the root not only interfers with host-RNA metabolism (42) but can also prevent the development of acquired resistance occurring in plants after a primary virus infection (6).

The work carried out on virus inhibitors has shown how difficult it is to find a chemotherapeutic treatment that is effective in controlling the spread of plant viruses. However, certain of these inhibitors seem to have an interest for *in vitro* systems. Callus tissue from potato virus Y-infected tobacco plants has been obtained virus-free by growing it on medium containing TU (31), and treatment with virazole (a nucleoside analogue) during the regeneration process *in vitro* of plants from potato virus X-infected protoplasts resulted in subsequent elimination of PVX from most of the regenerated tobacco plants (50). This therapeutic potential of virus inhibitors for certain *in vitro* propagation systems could become of economic importance with the increase in the production of *in vitro* plants.

The question now arises as to whether all these inhibitors of virus establishment or replication act on the processes of virus infection in a way analogous to the interferon system in animals. Interferons, which are also protein in nature, are produced in tissue after virus infection or by other inducers of a non-viral nature. They generally show host specificity and are always most active in the host species producing them. In contrast, inhibitors of virus infection tend to lack host specificity and they always show less or no activity in the plant species from which they are isolated. Inhibitors of virus replication are synthetic or natural compounds isolated from widely different sources and not a product of virus infection. Their mode of action does not involve host synthesis of new cellular messenger RNA and protein as does the interferon system. For these reasons, therefore, any analogy between the action of antiviral agents in plants and the interferon system in animals is excluded.

INDUCERS OF VIRUS RESISTANCE

This group concerns all those substances that induce an antiviral state in the host by acting through AMD-sensitive mechanisms involving synthesis of new proteins by the host. These substances can be natural or synthetic molecules and they do not generally affect virus infectivity when mixed directly with the virus *in vitro*. Certain so-called virus inhibitors act in a way different from that described in the previous section and for this reason it seems more appropriate to include them in the category of inducers. An example is the proteinaceous compound isolated from *Boerhaavia diffusa* (60) which is active against different viruses; the development of resistance in treated or non-treated upper and lower *Nicotiana glutinosa* leaves and its reversal by AMD suggests that the transcription mechanism of the host cell may be involved (61).

Polyanions, which can induce interferons in animals (15), are also inducers of virus resistance in plants (9, 19, 52, 53). Polyacrylic (PAA) and polymetacrylic acid, active in animals (17), can give complete resistance to TMV two to three days after their injection into tobacco leaves (19). PAA, which does not affect virus infectivity *in vitro*, has been reported to induce resistance in several plants from widely different families and in different plant-parasite combinations, including bacterial and fungal infections (1, 4, 9, 18, 19, 36). Polyanion structure seems to be a prerequisite for virus-induced resistance, non-charged macromolecules such as polyacrylamide being ineffective. Resistance develops independently of the way in which PAA is administered to plants; it can be injected, sprayed on to the leaves, absorbed through the petiole or taken up by the root system. When absorbed by tobacco roots, PAA decreased the number of TMV lesions developing on leaves by 42 - 92% depending on the concentration and the molecular size of the PAA used, the lower molecular weight being more effective (33).

Several observations on PAA-induced resistance suggest that host metabolic processes are involved. The PAA effect can be reversibly inhibited by high temperature as can the hypersensitive reaction to TMV (19). If tobacco with TMV necrotic local lesions is transferred to a higher temperature (e.g. 32°C) the mechanism of resistance of the hypersensitive host breaks down, virus multiplication starts again and the infection becomes systemic (30). A similar phenomenon occurs with PAA, injected leaves becoming susceptible to TMV when transferred to a higher temperature (Fig. 1). Furthermore, this suppression of either the hypersensitive reaction or the PAA effect by the higher temperature is reversible. A PAA-injected leaf, transferred to 32°C and not infected with TMV at this higher temperature, becomes resistant again to TMV when it is returned to a lower temperature (19). This observation that PAA retains its ability to induce resistance for a long time may be due to its being poorly or not at all metabolized within the plant, as is also the case in animals (15). The fact that TMV is able to multiply at 32°C in the presence of an amount of PAA which at lower temperatures induces resistance, indicates that PAA does not directly prevent virus multiplication.

Another similarity to the hypersensitive reaction is the induction of PAA-injected leaves of new soluble leaf proteins (Fig. 2) (19). These are the same as those called b-proteins (22, 57) or new protein components (59) detected in tissue surrounding necrotic local lesions, that is in tissue where the virus is localized and which is resistant to a second infection (39, 46). These observations first stimulated the hypothesis that the presence of new proteins in a plant tissue that has acquired the ability to stop virus spread may be analogous to the interferon system in animals (19, 22). The b-proteins disappear in PAA-injected or TMV-infected hypersensitive tobacco leaves transferred from 20°C to 32°C (19), that is under conditions where the virus in cells around the necroses starts to

Figure 1 Leaves of *Nicotiana tabacum* var. Xanthi-nc. :

a) TMV-infected on the left side 3 days after transfer from 20°C to 32°C : all the infected part becomes necrotic on returning to 20°C after 3 days at 32°C, indicating that virus multiplication has occurred at the higher temperature

b) injected on the left side with polyacrylic acid (mol. wt. 3500, 50 µl/ml) at 20°C, 3 days later transferred to 32°C and infected with TMV after 3 days at this higher temperature : all the infected side becomes necrotic on returning to 20°C as in (a), indicating that polyacrylic acid does not protect the plant at the higher temperature

c) as in (b) but infected with TMV 3 days after returning from 32°C to 20°C : no necroses appear indicating that inhibition by the higher temperature of the polyacrylic acid effect can be reversed

multiply actively again and where there is breakdown of the mechanisms of induced resistance of the host plant.

White (62) similarly induced virus resistance and b-proteins by injecting benzoic acid, salicylic acid or aspirin, all negatively charged molecules, into tobacco leaves. Ross (47) suggested that a chemical messenger may be involved in inducing resistance in non-infected leaves of infected plants. It could be that certain poly-anions and negatively charged compounds have their effect in mimicking such a chemical messenger in cells. This hypothesis is strengthened by the analogy that exists between the hypersensitive reaction to viruses and the mode of action of PAA.

Both the PAA effect and b-protein synthesis are inhibited by AMD (20, 32), as is resistance induced by a primary infection with a virus (31). A similar inhibition of resistance and interferon production is known to occur in animal tissues (53). This implies that resistance and b-protein production are dependent on DNA coded information of the cell and underlines the possibility that these proteins induced in virus-resistant tissue may be analogous to interferon in animals. It is interesting, in the light of this hypothesis, to compare the properties known at present for b-proteins with those described for interferon in animals.

b-proteins, as interferon proteins, are both small enough to remain in solution even during prolonged ultracentrifugation but large enough to remain within dialysis tubings.

Interferons in their native form are glycoproteins while b-proteins do not seem to be sensitive to the periodic acid Schiff stain after separation on electrophoretic gels. However, it has recently been shown that mouse interferon can be chemically or enzymatically transformed into a carbohydrate-free protein, called aglycointerferon or interferoid, which is as active as the native glycosylated interferon. This carbohydrate-free interferoid has a molecular weight of about 15000 daltons (54). Molecular weight determinations of SDS-treated b-proteins give similar molecular weights of around 15000 daltons for the proteins that have been called b_1, b_1', b_2 and b_3 (5, 25, 26). Hedrick-Smith plots (28) of the electrophoretic mobilities of b-proteins from two different tobacco varieties on gels of different acrylamide concentration (Fig. 3) give lines for b_1, b_1', b_2 and b_3 the slopes of which are not significantly different from one another, independently of their origin. This confirms that these 4 proteins must have similar molecular weights and that each probably consists of a single polypeptide chain of similar size which are 'charge isomers' in the terminology of Hedrick and Smith (4, 5, 26). The protein b_4, however, appears to be different and molecular weight determinations give a value of about 29000 daltons (25).

b-proteins are preferentially extracted at low pH (2.8) and are stable on dialysis against buffers ranging from pH 2 - 10, as are

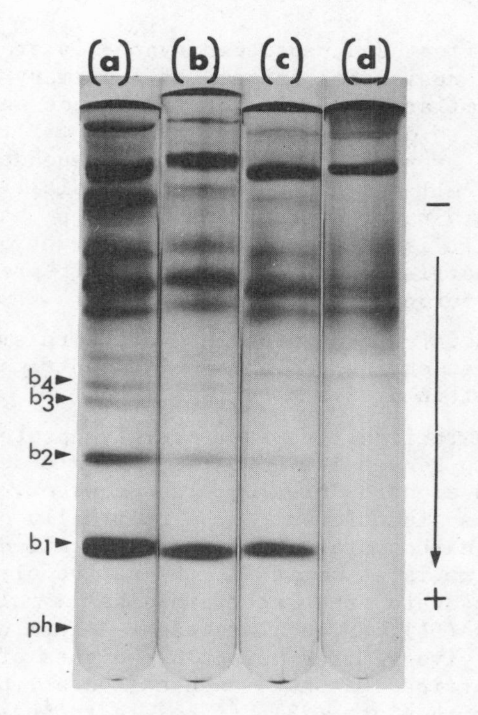

Figure 2 Electrophoretic separation of soluble leaf proteins from *N. tabacum* var. Xanthi-nc 3 days after either infection with (a) TMV, (b) *Pseudomonas syringae*, (c) injection with polyacrylic acid and (d) healthy. b_1 - b_4, new soluble leaf proteins ; ph, phenol front

interferons (25, 54). They are particularly rich in aromatic amino acids as compared to other soluble plant proteins (25). Amino acid analysis of b_1 has shown that 12% of its residues are aromatic with a relatively high content of tyrosine and tryptophan (5, 25). These amino acids have strongly hydrophobic side chains which could perhaps confer hydrophobic properties on the b-proteins, similar to those of interferon (54), and thus make them suitable for co-polymerization with other proteins (10).

In their native form b-proteins are host specific as are interferons and show a heterogeneity not only between species but also within a given species (Fig. 4 and 5). As can be seen in Fig. 5, three new protein bands b_1, b_2 and b_3 are induced in the hypersensitive *N. tabacum* variety Burley 49 by TMV, while four new bands, one of which is different (b_1'), are induced by *Thielaviopsis basicola* in the variety Judy's Pride Burley. With TMV infection the latter reacts sensitively and produces no b-proteins whilst with the fungal pathogen it gives necrotic local lesions with b-protein accumulation. Since both varieties produce b-proteins, but only one of them (Burley 49) is hypersensitive to TMV, the gene responsible for the hypersensitive reaction in tobacco to TMV, in this case the dominant gene N, cannot be directly responsible for the synthesis of the b-proteins. This is underlined by the fact that when the N gene from *N. glutinosa* was incorporated into *N. tabacum*, there was not an associated transfer of the gene for the b_1" protein characteristic of *N. glutinosa* (Fig. 4). However, it is possible that the genes specifying resistance to pathogens, such as the gene N for TMV in tobacco, could play a role in the regulation of the gene(s) coding for b-proteins. Gene(s) coding for these proteins can be transmitted sexually; for example, b_1' produced by Judy's Pride Burley is also found in the hybrids between the latter and Burley 49 after TMV infection (Fig. 5). Mendelian analysis of F_1 and F_2 generations of crosses between different tobacco varieties with their backcrosses indicates that production of the protein b_1' is probably under control of a single gene (P. Ahl, A. Cornu and S. Gianinazzi, unpublished data).

b-proteins, as well as interferons (54), seem to be able to move out of the cells producing them since they can be detected in intercellular leaf spaces (L.C. Van Loon, personal communication). They do not, however, appear to be able to move from one leaf to another through the vascular system of the plant. This has been shown in grafting experiments (Fig. 6); the variety Samsun NN, which does not produce the b_1' protein (21, 57), can be grafted on to a hybrid Samsun NN x Judy's Pride Burley root-stock which synthesizes this protein. When the leaves of the hybrid are infected with TMV, resistance is induced in the leaves of the Samsun NN graft but no b_1' can be detected in the latter, although the proteins b_1, b_2 and b_3 characteristic of this variety are present (57). On the other hand, when the hybrid is grafted on to a Samsun NN root-

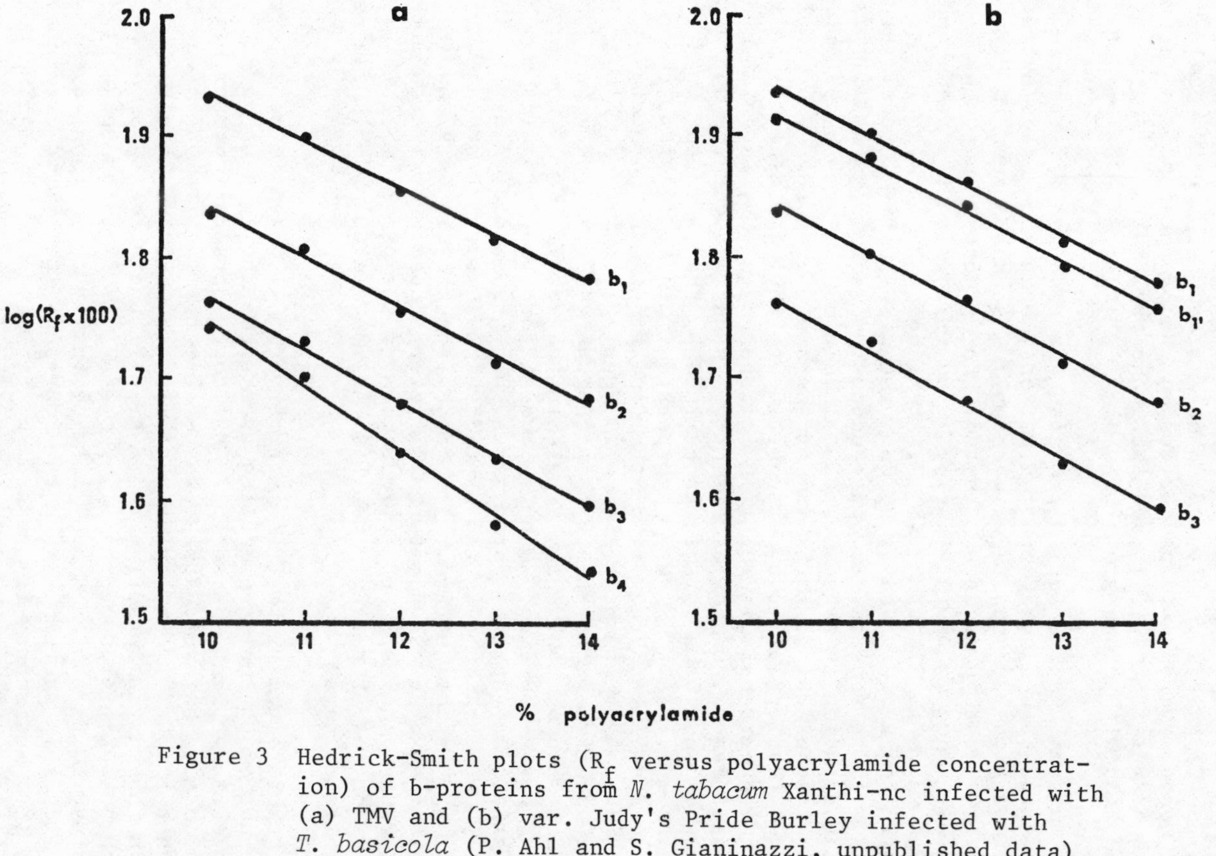

Figure 3 Hedrick-Smith plots (R_f versus polyacrylamide concentration) of b-proteins from *N. tabacum* Xanthi-nc infected with (a) TMV and (b) var. Judy's Pride Burley infected with *T. basicola* (P. Ahl and S. Gianinazzi, unpublished data)

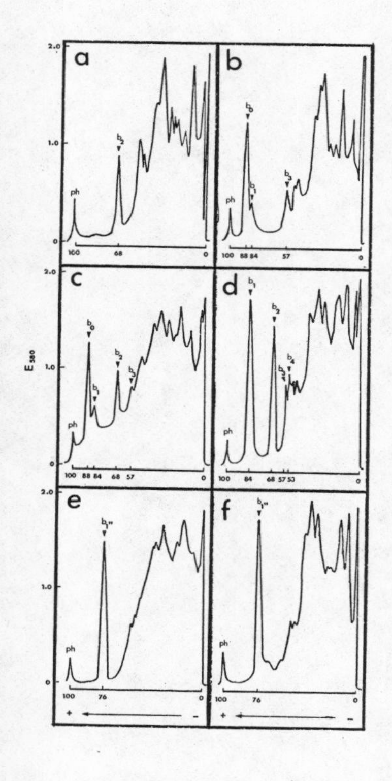

Figure 4.
b-protein variations between different *Nicotianas* species: (a) *N. tomentosiformis* treated with polyacrylic acid, (b) *N. sylvestris*, (c) amphidiploid *N. Sylvestris* X *N. tomentosiformis*, (d) *N. tabacum* var. Xanthi-nc (e) *N. glutinosa* and (f) *N. debneyi* infected with tobacco necrosis virus (P. Ahl and S. Gianinazzi, unpublished data).

stock, all four b-proteins including b_1 are produced in the graft. Such experiments provide experimental evidence for a mobile chemical messenger involved in inducing both resistance in non-infected leaves of infected plants, as Ross suggested (47), and b-protein production.

Virtually all types of viruses and a variety of stimuli can induce interferon in animal cells. This inducer non-specificity is also true of b-protein synthesis in plant tissues (Table 5) (35). Furthermore, in higher plants there is increasing evidence that new host-protein production associated with the hypersensitive type of resistance is a general phenomenon since it has been observed not only in tobacco but also in cowpea (12), cucumber (2), *Gomphrena globosa* (P. Redolfi and S. Pennazio, personal communication) and Pinto bean (unpublished data). Interferon induction in vertebrate animals is similarly known to occur in a wide range of species (54).

The amount of these resistance-related proteins in infected tobacco tissues increases with time even after the virus has stopped spreading. This is similar to that observed for interferon in infected animal tissues (54). By determining both the degree of

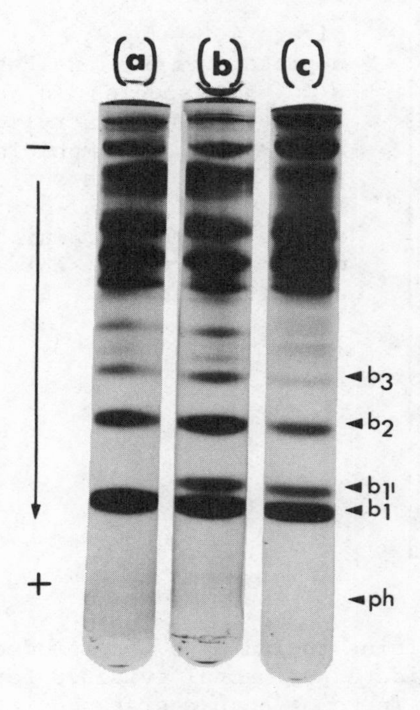

Figure 5 Differences in b-proteins from (a) *N. tabacum* var. Burley 49, (c) var. Judy's Pride Burley and (b) one of their hybrids, 7 days after infection with either (a,b) TMV or (c) *T. basicola*

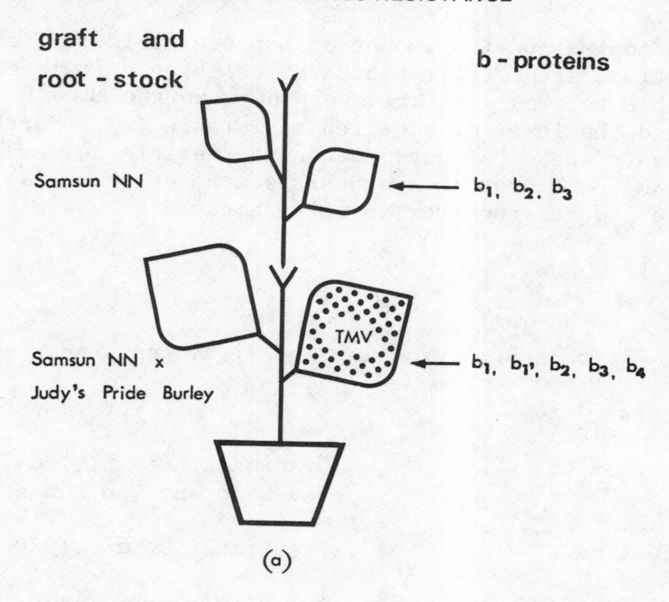

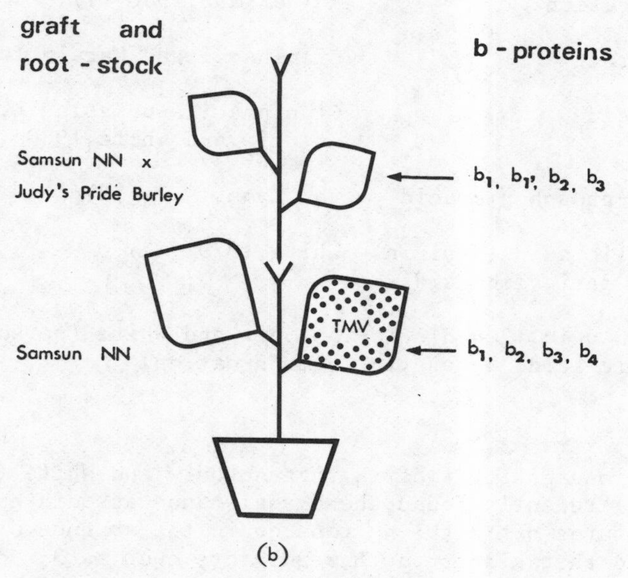

Figure 6 b-protein production in grafting experiments with *N. tabacum* : (a) var. Samsun NN grafted on to a hybrid (var. Samsun NN x var. Judy's Pride Burley) root-stock, (b) hybrids grafted on to a var. Samsun NN root-stock. b-proteins detected 2 weeks after inoculation of a root-stock leaf with TMV (S. Gianinazzi and P. Ahl, unpublished data)

resistance to TMV and the amount of b-proteins in leaf tissues at different times after their synthesis has been stimulated, it has been possible to show a relationship between the amount of these proteins and the level of acquired resistance (1). Furthermore in tobacco plants where PAA does not induce resistance, as for example in Samsun NN, no b-protein synthesis can be detected (4) although both can be induced together in this same plant by aspirin (62).

Table 5 Inducers of b-proteins and virus resistance in tobacco

Inducers	References
Virus infection	Gianinazzi *et al.*,1969, 1970 (22,23) Van Loon and Van Kammen,1970 (59)
Fungal infection	Gianinazzi *et al.*,1980 (26)
Bacterial infection	Ahl *et al.*,1980 (1)
Bacterial extract	Gianinazzi and Martin,1975 (20)
Polyacrylic acid	Gianinazzi and Kassanis,1974 (19) Antoniw and White,1980 (4)
2-Chloroethyl-phosphonic acid	Van Loon,1977 (58)
Acetyl-salicylic acid (aspirin) benzoic acid, salicylic acid	White,1979 (62)
3-indoleacetic acid, 2,4-dichlorophenoxyacetic acid, 6-benzylaminopurine	Antoniw and White (personal communication)

Another interesting fact is that Antoniw and White (personal communication) recently found these resistance-associated proteins in callus cultures of Xanthi-nc tobacco in the absence of virus. They suggested that plant growth regulators such as 3-indoleacetic acid (IAA), 2,4-dichlorophenoxyacetic acid and 6-benzylaminopurine, usually included in the callus culture media, could be responsible for the appearance of these proteins, since these three substances induced both b-proteins and resistance when injected into leaves of Xanthi-nc. This induction of b-proteins in tobacco callus may be at least partly responsible for the difficulty encountered in infecting certain healthy cultures and/or for the eradication of virus from infected cultures when the explant is infected with only a small amount of virus.

The antiviral effect of interferon has been shown both *in vivo* and *in vitro* using animal tissue cultures (54). Unfortunately, similar studies on the resistance-associated proteins of plants have been hampered up to the present by the lack of a suitable experimental system. In experiments using tobacco protoplasts, partially purified b-proteins did not induce resistance to TMV infection when they were added to the culture medium (34). However, we know that the hypersensitive reaction does not occur in protoplasts even when they are isolated from hypersensitive tobacco plants. Furthermore, Cassells and co-workers (8) showed that when protoplasts were isolated from PAA-injected resistant leaves, they lost their resistance and became sensitive to virus infection.

In spite of this lack of direct experimental evidence, it would be surprising if all the similarities that exist between the resistance-associated protein in plants and interferon in animals were purely fortuitous. The voluminous amount of work that has been done on the resistance-associated proteins in tobacco (b-proteins) indicates that these may be more than just biochemical markers of resistant tissues. They may be involved in the resistance mechanism itself. If this is true then the hypothesis that an interferon-like system exists in plants becomes plausible.

In the interferon system in animals the inducer molecule for virus resistance triggers induction mechanisms which activate the transcription of normally silent genetic information resulting in the production of new soluble proteins. These proteins, called interferoid in animals, are glycosylated and secreted as interferon. The binding of the interferon to a susceptible cell initiates a number of alterations in cells which could be responsible for their subsequent antiviral state. Some of these are direct surface alteractions but others require activation of another set of normally silent genetic information coding for interferon action(s), which function through the respective messenger RNA(s) and protein mediators. These various cellular alterations induced by interferons are measurable as antiviral activities, toxicity enhancement, effects on cell surfaces, regulation of interferon synthesis and immunomodulations (54).

Taking PAA, which is active in both plants and animals, as the inducer molecule for virus resistance, the different steps occurring during the expression of the active defence mechanisms in plants can now be considered by analogy with the interferon system in animals. In plants, modification of the genetic expression results in the synthesis of new soluble leaf proteins (b-proteins in tobacco), which may be analogous to interferons in animals (22, 19). The appearance of these proteins coincides with the active control of virus synthesis in leaf cells.

Some authors have suggested that in plants PAA can perhaps

protect against virus infection in more than one way, for example, by also affecting membrane properties with alterations in the turgidity (8) or elongation (18) of cells having antiviral activities. In fact, when b-protein synthesis is similarly induced in tobacco in tissues surrounding necrotic local lesions during the hypersensitive reaction, membrane alteration can be detected. For example, absorption of radioactive leucine-^{14}C or ^{32}P through the petiole of TMV-infected hypersensitive tobacco results in a non-uniform distribution of these molecules compared with the control, indicating that membrane permeability is altered in the infected leaf (24).

The molecular basis of the antiviral state in plants induced by compounds such as PAA is not known, although antiviral activities have been reported to exist in virus-infected plants. The precise chemical nature of these antiviral activites has not yet been established. The antiviral factor described by Sela and Applebaum (49) from tobacco, for example, is considered to be protein in nature (41). Its synthesis is linked to the presence of the dominant N-gene and its activity, as for the antiviral activities in the interferon system, is not sensitive to AMD (3).

CONCLUSIONS

There is now no doubt that active defence mechanisms against viruses exist in plants and that they can be stimulated in different ways, either after a primary infection by a pathogen or by injection or absorption of natural or synthetic compounds such as polyanions. The activation of such mechanisms results in plants becoming highly resistant (immunized in the terminology proposed by J. Kuć in this NATO-ASI) to a second infection. However, the molecular basis of such mechanisms is still not clear; strong evidence exists that they involve synthesis of new proteins in which host DNA plays an important role. These findings indicate that even if the active defence mechanisms in plants against viruses are not identical with the interferon system in animals, they are not completely different.

The discovery of chemical inducers such as PAA and aspirin represent a significant practical advance. This together with the eventual introduction of chemical inducers in tissue culture media could provide possibilities in the future for obtaining and maintaining valuable stock plants which are free from virus. The broad spectrum of the effect of PAA against viruses, fungi and bacteria underlines its possible practical importance.

In conclusion, it becomes clear from all these facts that if we are to increase our knowledge of the molecular basis of active defence mechanisms in plants, further studies on interactions between viral replication and host DNA function are essential. In this field of molecular genetics inducers of resistance and resistance associated proteins, such as the b-proteins in tobacco, could provide interesting tools.

REFERENCES

1. AHL, P., BENJAMA, A., SAMSON, R. & GIANINAZZI, S. (1981). Induction chez le Tabac par *Pseudomonas syringae* de nouvelles protéines (proteines 'b') associées au developpement d'une résistance non spécifique à une deuxième infection. *Phytopathologische Zeitschrift* (in press).

2. ANDEBRHAN, T., COUTTS, R.H.A., WAGIH, E.E. & WOOD, R.K.S. (1980). Induced resistance and changes in the soluble protein fraction of cucumber leaves locally infected with *Colletotrichum lagenarium* or tobacco necrosis virus. *Phytopathologische Zeitschrift* 98, 47 - 52.

3. ANTIGNUS, Y., SELA, I. & HARPAZ, I. (1977). Further studies on the biology of an antiviral factor (AVF) from virus-infected plants and its association with the N-gene of *Nicotiana* species. *Journal of General Virology* 35, 107 - 116.

4. ANTONIW, J.F. & WHITE, R.F. (1980). The effects of aspirin and polyacrylic acid on soluble leaf proteins and resistance to virus infection in five cultivars of tobacco. *Phytopathologische Zeitschrift* 93, 331 - 341.

5. ANTONIW, J.F., RITTER, C.E., PIERPOINT, W.S. & VAN LOON, L.C. (1980). Comparison of three pathogenesis-related proteins from plants of two cultivars of tobacco infected with TMV. *Journal of General Virology* 47, 79 - 87.

6. BATRA, G.K. & KUHN, C.W. (1975). Polyphenoloxidase and peroxidase activities associated with acquired resistance and its inhibition by 2-thiouracil in virus-infected soybean. *Physiological Plant Pathology* 5, 239 - 248.

7. BEST, R.J. (1937). On the presence of an 'oxidase' in the juice expressed from tomato plants infected with the virus of tomato spotted wilt. *Australian Journal of Experimental Biological and Medical Science* 15, 191 - 199.

8. CASSELLS, A.C., BARNETT, A. & BARLASS, M. (1978). The effect of polyacrylic acid treatment on the susceptibility of *Nicotiana tabacum* cv. Xanthi-nc to tobacco mosaic virus. *Physiological Plant Pathology* 13, 13 - 21.

9. CASSELLS, A.C. & FLYNN, T. (1978). Studies on polyacrylic acid induced resistance to viral and non-viral plant pathogens. *Pesticide Science* 9, 365 - 371.

10. CHOTHIA, C. & JANIN, J. (1975). Principles of protein-protein recognition. *Nature* 256, 705 - 708.

11. COMMONER, B. & MERCER, F.I. (1951). Inhibition of biosynthesis of tobacco mosaic virus by thiouracil. *Nature* 168, 113 - 114.

12. COUTTS, R.H.A. (1978). Alterations in the soluble protein patterns of tobacco and cowpea leaves following inoculation with tobacco necrosis virus. *Plant Science Letters* 12, 189 - 197.

13. DAWSON, W.O. & KUHN, C.W. (1972). Enhancement of cowpea chlorotic mottle virus biosynthesis and *in vivo* infectivity by 2-thiouracil. *Virology* 47, 21 - 29.

14. DAWSON, W.O. & SCHLEGEL, D.E. (1976). The sequence of inhibition of tobacco mosaic virus synthesis by actinomycin D, 2-thiouracil and cycloheximide in a synchronous infection. *Phytopathology* 66, 177 - 181.

15. DE CLERCQ, E., ECKSTEIN, F. & MERIGAN, T.C. (1970). Structural requirements for synthetic polyanions to act as interferon inducers. *Annals of the New York Academy of Sciences* 173, 444 - 461.

16. DE MAEYER, J. (1974). Interferon : 'du vent dans les voiles'. *La Recherche* 43, 5, 280 - 282.

17. DE SOMER, P., DE CLERCQ, E., BILLIAU, A., SCHONNE, E. & CLAESEN, M. (1968). Antiviral activity of polyacrylic and polymethacrylic acids. *Journal of Virology* 2, 878 - 885.

18. FERNANDEZ, T.F. & GABORJANYI, R. (1976). Reversion of dwarfing induced by virus infection : effect of polyacrylic and gibberellic acids. *Acta Phytopathologica Academiae Scientarium Hungaricae* 11, 271 - 275.

19. GIANINAZZI, S. & KASSANIS, B. (1974). Virus resistance induced in plants by polyacrylic acid. *Journal of General Virology* 23, 1 - 9.

20. GIANINAZZI, S. & MARTIN, C. (1975). A naturally occurring active factor inducing resistance to virus infection in plants. *Phytopathologische Zeitschrift* 83, 23 - 26.

21. GIANINAZZI, S., AHL, P. & CORNU, A. (1980). b-protein variation in virus-infected intraspecific tobacco hybrids. *Acta*

Phytopathologica Academiae Scientarium Hungaricae (in press).

22. GIANINAZZI, S., MARTIN, C. & VALLEE, J.C. (1970). Hypersensibilité aux virus, température et protéines solubles chez le *Nicotiana* Xanthi n.c. Apparition de nouvelles macromolécules lors de la répression de la synthese virale. *Comptes-Rendus Académie des Sciences, Paris* 270, 2283 - 2386.

23. GIANINAZZI, S., VALLEE, J.C. & MARTIN, C. (1969). Hypersensibilité aux virus, température et protéines solubles chez le *Nicotiana* Xanthi n.c. *Comptes-Rendus Acadêmie des Sciences, Paris* 268, 800 - 802.

24. GIANINAZZI, S., VALLEE, J.C. & MARTIN, C. (1972). Modification de la perméabilité au cours du phénomène d'hypersensibilité chez le *Nicotiana tabacum* var. Xanthi n.c. infecté avec le virus de la Mosaique du Tabac. *Comptes-Rendus Acadêmie des Sciences Paris* 275, 1383 - 1386.

25. GIANINAZZI, S., PRATT, H.M., SHEWRY, P.R. & MIFLIN, B.J. (1977). Partial purification and preliminary characterization of soluble leaf proteins specific to virus infected plants. *Journal of General Virology* 34, 345 - 351.

26. GIANINAZZI, S., AHL, P., CORNU, A., SCALLA, R. & CASSINI, R. (1980). First report of host b-protein appearance in response to a fungal infection in tobacco. *Physiological Plant Pathology* 16, 337 - 342.

27. HAMPTON, R.E. & FULTON, R.W. (1961). The relation of polyphenol oxidase to instability *in vitro* of prune drawf and sour cherry necrotic ringspot viruses. *Virology* 13, 44 - 52.

28. HEDRICK, J.L. & SMITH, A.J. (1968). Size and charge isomer separation and estimation of molecular weights of protein by disc electrophoresis. *Archives of Biochemistry and Biophysics* 126, 155 - 164.

29. HIRAI, T. (1979). Action of antiviral agents. In : *Plant Disease : An Advanced Treatise*, Ed. by J.G. Horsfall and E.B. Cowling. Vol. I, Academic Press, 285 - 306.

30. KASSANIS, B. (1952). Some effects of high temperature on the susceptibility of plants to infection with viruses. *Annals of Applied Biology* 39, 358 - 369.

31. KASSANIS, B. (1978). Forty years' research on plant viruses

at Rothamsted Experimental Station. *Rothamsted Report for* 1978, Part 2.

32. KASSANIS, B. & WHITE, R.F. (1974). Inhibition of acquired resistance to tobacco mosaic virus by actinomycin D. *Journal of General Virology* 25, 323 - 324.

33. KASSANIS, B. & WHITE, R.F. (1975). Polyacrylic acid-induced resistance to tobacco mosaic virus in tobacco cv. Xanthi. *Annals of Applied Biology* 79, 215 - 220.

34. KASSANIS, B. & WHITE, R.F. (1978). Effect of polyacrylic acid and b-proteins on TMV multiplication in tobacco protoplasts. *Phytopathologische Zeitschrift* 91, 269 - 272.

35. KASSANIS, B., GIANINAZZI, S. & WHITE, R.F. (1974). A possible explanation of the resistance of virus-infected tobacco plants to second infection. *Journal of General Virology* 23, 11 - 16.

36. KLUGE, S. & MARCINKA, K. (1979). The effects of polyacrylic acid and virazole on the replication and component formation of red clover mottle virus. *Acta Virologica* 23, 148 - 152.

37. LOEBENSTEIN, G. (1963). Further evidence of systemic resistance induced by localized necrotic virus infection in plants. *Phytopathology* 53, 306 - 308.

38. MARTIN, C. (1958). Etude de quelques déviations de métabolisme chez les plantes atteintes de maladies à virus. Thesis, University of Paris, France.

39. MARTIN, C. & GALLET, M. (1966). Nouvelles observations sur le phénomène d'hypersensibilité aux virus chez les végétaux. *Comptes-Rendus Academie des Sciences, Paris* 263, 1316 - 1318.

40. MAUGH, T.H., II. (1976). Chemotherapy : Antiviral agents come of age. *Science* 192, 128 - 132.

41. MOZES, R., ANTIGNUS, Y., SELA, I. & HARPAZ, I. (1978). The chemical nature of an antiviral factor (AVF) from virus infected plants. *Journal of General Virology* 38, 241 - 249.

42. PORTER, C.A. & WEINSTEIN, L.H. (1961). Incorporation of 2-thiouracil-S^{35} into RNA and acid soluble nucleotides of Varmor 48 tobacco. *Virology* 15, 504 - 506.

43. PRICE, W.C. (1940). Acquired immunity from plant virus dis-

eases. *The Quarterly Review of Biology* 15, 338 - 361.

44. RAGETLI, H.W.J. (1975). The mode of action of natural plant virus inhibitors. *Current Advances in Plant Science* 19, 321 - 334.

45. RAGETLI, J.P.H. & WEINTRAUB, M. (1974). The influence of inhibitors on the reaction of indicator plants. Contribution No. 339. Research Station, Agriculture Canada, Vancouver, B.C., 1 - 13.

46. ROSS, A.F. (1961a). Localized acquired resistance to plant virus infection in hypersensitive hosts. *Virology* 14, 329 - 339.

47. ROSS, A.F. (1961b). Systemic acquired resistance induced by localized virus infections in plants. *Virology* 14, 340 - 358.

48. ROTTIER, P.J.M., REZELMAN, G. & VAN KAMMEN, A. (1979). The inhibition of cowpea mosaic virus replication by actinomycin D. *Virology* 92, 299 - 309.

49. SELA, I. & APPELBAUM, S.W. (1962). Occurrence of an antiviral factor in virus-infected plants. *Virology* 17, 453 - 548.

50. SHEPARD, J.F. (1977). Regeneration of plants from protoplasts of potato virus X-infected tobacco leaves. II. Influence of virazole on the frequency of infection. *Virology* 78, 261 - 266.

51. SMOOKLER, M.M. (1971). Properties of inhibitors of plant virus infection occurring in the leaves of species in the Chenopodiales. *Annals of Applied Biology* 69, 157 - 168.

52. STEIN, A. & LOEBENSTEIN, G. (1972). Induced interference by synthetic polyanions with the infection of tobacco mosaic virus. *Phytopathology* 62, 1461 - 1466.

53. STEIN, A., LOEBENSTEIN, G. & SPIEGEL, S. (1979). Further studies of induced interference by a synthetic polyanion of infection by tobacco mosaic virus. *Physiological Plant Pathology* 15, 241 - 255.

54. STEWART II, W.E. (1979). The interferon system. Springer-Verlag, Wien and New York.

55. TEAKLE, D.S. & NIENHAUS, F. (1974). The effect of plant virus inhibitors on transmission of tobacco necrosis by *Olpidium brassicae*. *Phytopathologische Zeitschrift* 80, 1 - 8.

56. TOMLINSON, J.A., WALKER, V.M., FLEWETT, T.H. & BARCLAY, G.R. (1974). The inhibition of infection by cucumber mosaic virus and influenza virus by extracts from *Phytolacca americana*. *Journal of General Virology* 22, 225 - 232.

57. VAN LOON, L.C. (1976). Specific soluble leaf proteins in virus-infected tobacco plants are not normal constituents. *Journal of General Virology* 30, 375 - 379.

58. VAN LOON, L.C. (1977). Induction by 2-chloroethylphosphonic acid of viral-like lesions, associated proteins and systemic resistance in tobacco. *Virology* 80, 417 - 420.

59. VAN LOON, L.C. & VAN KAMMEN, A. (1970). Polyacrylamide disc electrophoresis of the soluble leaf proteins from *Nicotiana tabacum* var. 'Samsun' and 'Samsun NN'. II. Changes in protein constitution after infection with tobacco mosaic virus. *Virology* 40, 199 - 211.

60. VERMA, H.M. & AWASTHI, L.P. (1979). Antiviral activity of *Boerhaavia diffusa* root extract and the physical properties of the virus inhibitor. *Canadian Journal of Botany* 57, 926 - 932.

61. VERMA, H.M., AWASTHI, L.P. & SAXENA, K.C. (1979). Isolation of the virus inhibitor from the root extract of *Boerhaavia diffusa* inducing systemic resistance in plants. *Canadian Journal of Botany* 57, 1214 - 1217.

62. WHITE, R.F. (1979). Acetylsalicylic acid (aspirin) induces resistance to tobacco mosaic virus in tobacco. *Virology* 99, 410 - 412.

63. WYATT, S.D. & SHEPHERD, R.J. (1969). Isolation and characterization of virus inhibitor from *Phytolacca americana*. *Phytopathology* 59, 1787 - 1794.

THE EFFECT OF DEFENCE REACTIONS ON THE ENERGY BALANCE AND YIELD OF RESISTANT PLANTS

V. SMEDEGAARD-PETERSEN

Department of Plant Pathology
The Royal Veterinary and Agricultural University
Copenhagen, Denmark

INTRODUCTION

The basis for considering changes in the energy balance of resistant-reacting plants is the finding that incompatibility between hosts and pathogens is usually associated with marked in-increases in the biosynthetic activity of the resistant tissues. This lecture primarily deals with the energy balance of incompatible interactions between barley and the powdery mildew fungus, *Erysiphe graminis hordei*. However, other host-pathogen interactions and interactions between hosts and certain non-pathogens are also considered.

In incompatible combinations between barley and the powdery mildew fungus host plants react to inoculation with changes in biochemical activity including increased peroxidase and ethylene production (10), production of fungal inhibitors, phytoalexins (17, 18), synthesis of fluorescent, u-v absorbing, aromatic substances which have not been further identified (14), changes in the carotenoid content (8), and stimulated synthesis of nucleic acid (16). Furthermore, changes in the rate of respiration have been reported by a number of workers (15, 19, 21, 26, 27).

Although the role of such activities in resistance is disguised by the fact that they often occur in both compatible and incompatible interactions (7, 9, 12), there is good experimental evidence to suggest that at least some of them are involved in active defence reactions. Thus Ouchi *et al*. (18) demonstrated that barley plants inoculated with an avirulent race of powdery mildew became partly resistant to a race originally virulent on that plant. The resistance conferred by the primary infection with an incompatible race indicated the involvement of antifungal compounds, and Oku *et al.* (17) detected high phytoalexin activity 12 - 20 hours after inoculation in an incompatible barley-mildew combination. In compatible combinations no phytoalexins were found at this early stage of infection, although later antifungal activity also occurred in susceptible hosts.

CAUSAL RELATION BETWEEN INCOMPATIBLE HOST RESPONSE AND INCREASED RESPIRATION

An important aspect of the early stage of incompatible host-pathogen interactions is whether there exists a causal relationship between incompatible host reactions and changes in respiration. In other words, are active defence reactions associated with respiratory changes ?

Since many incompatible disease reactions, as previously mentioned, are associated with increased synthesis of compounds such as phytoalexins, nucleic acids, steroids and phenolics, there seems little doubt that these reactions also stimulate the host to increased respiration in order to furnish carbon units and energy for their synthesis.

To illustrate the potential role of respiration in the biosynthesis of compounds associated with incompatibility Daly (5) calculated the requirement for synthesis of rishitin in potato. He calculated, with certain reservations, that if glycolysis is the source of both pyruvate and ATP, a minimum of $4\frac{1}{2}$ molecules of glucose must be metabolized to furnish one molecule of rishitin.

It is well documented that increased respiration is a characteristic physiological feature of diseased plants. In contrast there is no extensive documentation of respiratory changes in tissue reacting incompatibly, and there are many inconsistencies in the results obtained by different authors working with the same host-pathogen interactions. Since the respiratory response in incompatible tissue is crucial from an energetic point of view, these contradictory results will be considered in a little more detail.

Comparing susceptible and resistant wheat cultivars infected with stem rust Samborski and Shaw (20) found that incompatible combinations showed an earlier and sharper rise in respiration than did compatible combinations. However, Heitefuss (9) did not find so clear-cut a correlation between the degree of compatibility and increase in respiration in combinations of wheat and stem rust, including those with the temperature-sensitive Sr6 gene. Neither were Antonelli and Daly (2) able to observe respiratory differences when near-isogenic wheat lines with Sr6 and sr6 alleles were inoculated with a specific race of *Puccinia graminis tritici*. The increases in respiration were parallel until the sixth day after inoculation; later the respiratory rate continued to increase only in the compatible combination. Bushnell (3) found a slower increase in respiration in resistant than in susceptible wheat tissues inoculated with wheat stem rust.

Millerd and Scott (15) compared respiration in barley cultivars with varying degrees of resistance against powdery mildew. One highly resistant cultivar reacted with an earlier rise in respiration than did the susceptible cultivar whereas another resistant cultivar even reacted to inoculation with a decrease in respiration. Comparing semi-resistant and susceptible varieties, there was no difference in oxygen uptake during the first 96 hours after inoculation.

Scott and Smillie (21) examined respiratory activities in leaves of two isogenic barley lines, one highly resistant and the other susceptible after infection with *Erysiphe graminis hordei*. In the incompatible combination there was only a slight increase compared to the controls. In the compatible combination there were no marked differences in infected leaves and non-infected controls until 96 hours after inoculation. Later the infected leaves showed a higher respiration.

Comparing respiration in resistant and susceptible barley plants inoculated with powdery mildew Paulech (19) found an enhanced respiratory rate in resistant plants at the time when the first appressoria had developed. The maximum rate was reached when two or three haustoria had developed at the infection site. In susceptible plants the rise in respiration started later and could first be detected when an average of two haustoria had developed at each infection site.

One reason for the inconsistencies may be the fact that most of the comparisons, except for wheat stem rust and the Sr6 gene (4, 9), have been made with cultivars of quite different or unknown genetic backgrounds in studies of disease physiology as have been discussed by Johnson (11) and Daly (4).

I shall now turn to the work with energy balance in resistant plants being done in our laboratory in Copenhagen. The object of this work has been to examine whether there exists in highly incompatible host-pathogen combinations an extra demand for respiratory host energy. As a basic model we used barley powdery mildew, and the question asked was : do those biosynthetical activities which obviously occur in inoculated, highly resistant plants consume energy to an extent that plant growth and yield are affected ?

In one series of experiments we examined the respiratory rate in incompatible and compatible combinations of the barley cultivar Sultan and two races of *Erysiphe graminis hordei*, one avirulent, the other virulent. The resistance of this cultivar is conditioned by a single gene derived from "Arabische" (29, 31). The avirulent race 15-0 did not cause visible symptoms on Sultan, whereas the virulent race 1-4 produced large sporulating colonies after 6 - 8 days. By using one barley cultivar and two races, differences in the genetical background of the host were eliminated.

Fig. 1 shows the time course of respiration in incompatible and compatible leaves of Sultan after inoculation with the two races. It is seen that resistant leaves react to inoculation with a very rapid, temporary increase in the rate of respiration starting as early as 12 hours after inoculation, reaching a maximum 80% above that of the uninoculated controls and then decreasing to approximately the normal level after three days. In susceptible leaves the increase starts three days after inoculation and remains at a high level until the beginning of senescence. It thus appears that the rate of respiration in susceptible, inoculated tissue increases gradually throughout symptom development.

Fig. 1 also shows the results from another typical respiration experiment with the barley cultivar Emir. Again, the resistant leaves react with a rapid, temporary increase in the uptake of oxygen starting about 12 hours after inoculation. The maximum rate is reached within 24 hours; later it decreases to the level of the uninoculated controls. In susceptible plants the increase in oxygen uptake typically starts closely to three days (72 hours) after inoculation.

The pronounced increase in the respiratory rate of incompatible tissue occurs simultaneously (12 - 24 hours after inoculation) with an increased production of fungitoxic substances, phytoalexins, which retard or stop fungal growth (14, 17, 18). The enhanced respiratory rate also coincides with enhanced RNA synthesis according to Oku *et al.* (16) who found that RNA synthesis in incompatible hosts increased soon after inoculation to a maximum rate after two days. Then it declined to the level of uninoculated control four days after inoculation. In compatible combinations RNA synthesis gradually increased from four days after inoculation.

The high correlation between increases in oxygen uptake and synthetic activity indicates that the enhanced respiratory rate in incompatible barley-powdery mildew interactions is part of energy-requiring biosynthetic processes, probably defence reactions against the pathogen.

The increase in respiration of incompatible hosts is significant 16 - 24 hours after inoculation, but some increase can be

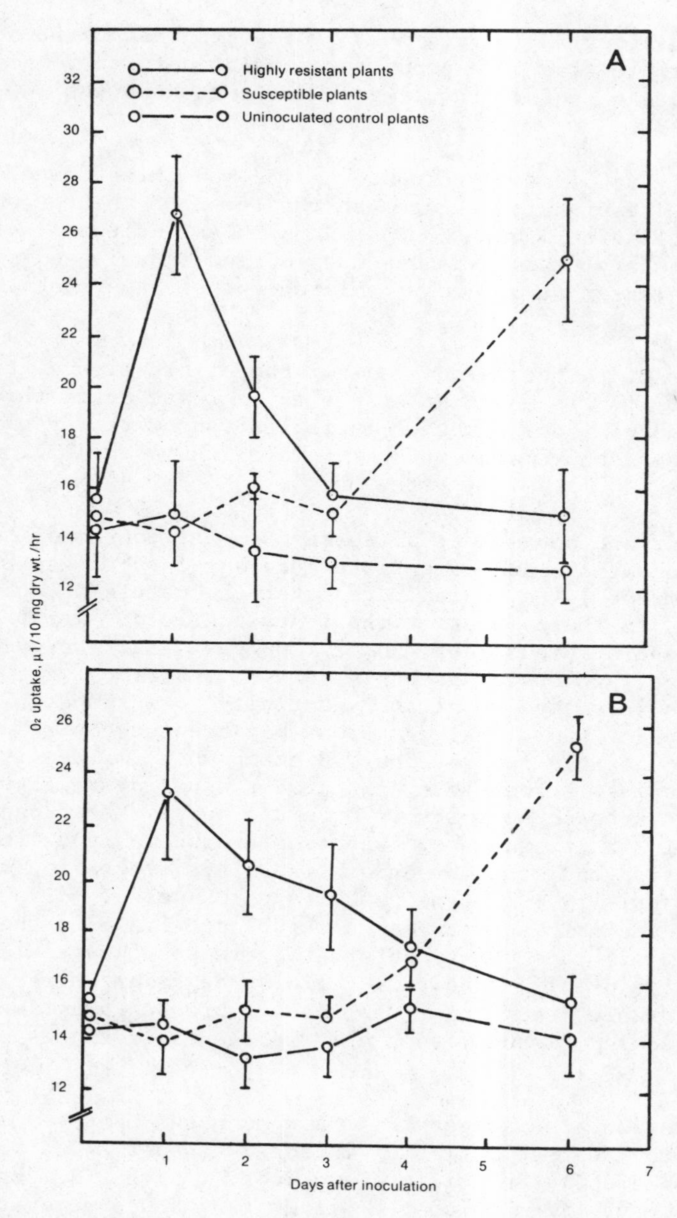

Figure 1 The effect of powdery mildew on the respiration of resistant and susceptible barley plants. A Plants of the cultivar Sultan inoculated with the avirulent race 15-0 (resistant reaction) or with the virulent race 1-4 (susceptible reaction). B Plants of the cultivar Emir inoculated with the avirulent race 13-0 (resistant reaction) or with the virulent race 1-4 (susceptible reaction)

detected even after 12 hours. This is important since it means that incompatible tissue responds to inoculation with increased oxygen uptake even before the fungus has penetrated the epidermal cell lumen (28).

The early respiratory response in incompatible tissue and the fact that the respiration returns to the level of that of uninoculated controls after three to four days rules out the possibility that the temporary rise is a secondary effect caused by disintegration of cellular structures in microscopical hypersensitive necroses.

In contrast to the responses in incompatible tissue respiration in compatible tissue never starts to increase until three days after inoculation. It then increases concomitantly with symptom development and remains at a high level until the beginning of senescence.

The clear and consistent difference in the course of respiration in incompatible and compatible interactions between barley and powdery mildew clearly indicates that the nature of enhanced oxygen uptake in these interactions is basically different. While the present and previous data (26, 27) suggest that energy-requiring biochemical processes, probably defence reactions, are responsible for the increased respiratory activity in incompatible interactions, other or additional reactions may account for respiratory changes in compatible interaction. Since powdery mildew is an obligate parasite, susceptible, infected tissue may react with an enhanced biosynthetic activity in order to supply the fungus with nutrients. In later stages of infection breakdown of cellular compartmentation and other kinds of tissue disintegration may lead to abnormal activity of enzymes such as peroxidases, polyphenol oxidases and ascorbic acid oxidases which contribute to the oxygen uptake. Daly (5) thus points out that increases in peroxidase reactions which usually accompany cellular degradation may account for a considerable fraction of the total increase in uptake of oxygen in diseased tissue, especially in the later stages of infection.

The question also arises as to how much the pathogen itself contributes to the respiration of infected tissues. The work of Allen and Goddard (1) and Millerd and Scott (15) has demonstrated that barley leaves from which the mildew fungus was removed respired at almost the same rate as did tissue carrying the fungus. In our experiments the marked differences in oxygen uptake in incompatible and compatible interactions during the first 72 hours after inoculation clearly indicate that the fungus does not contribute significantly during this period of infection.

ENERGY-REQUIRING DEFENCE PROCESSES REDUCE PLANT GROWTH, GRAIN YIELD AND GRAIN QUALITY IN RESISTANT BARLEY PLANTS INOCULATED WITH POWDERY MILDEW

In order to investigate whether the increased energy demand in inoculated, resistant plants is sufficient to influence plant growth, yield experiments were carried out in three growth chambers under identical and controlled environmental conditions (27). Each chamber of area 12 m^2 contained 32 containers each with 15 plants of the barley cultivar Sultan.

The plants in chamber I were uninoculated controls. The plants in chamber II were continuously inoculated with the avirulent race 15-0 which did not cause visible symptoms. Plants in chamber III were continuously inoculated with the virulent race 1-4 which caused large sporulating mildew infections on Sultan. Plants were inoculated by placing pots with infected plants of the susceptible cultivar, Proctor, in the chambers every week from the five leaf stage until just after heading.

Although resistant plants did not show visible disease symptoms after continuous inoculation with the avirulent race 15-0, the grain yield was significantly reduced by 7% and the kernel weight by 4%. The content of grain protein was reduced from 9.75 to 9.38% which was equivalent to a 11% reduction in the yield of grain protein per unit area (Tables 1 and 2). The straw yield of resistant plants was reduced by 3% and the straw length by 5.7 cm compared with uninoculated controls (Table 3).

Table 1 Yield of grain, kernel weight, and % protein in grain of resistant and susceptible plants of the barley cultivar Sultan inoculated with two races of powdery mildew

Host reaction	Yield of grain		Kernel weight		
	g per pot	% of control	mg	% of control	% protein in grain
Non-inoculated control	66.9 a	100	42.4 a	100	9.75 a
Resistant plants 1)	62.0 b	93	40.5 b	96	9.38 b
Susceptible plants 2)	49.2 c	74	37.6 c	89	9.25 c

Means followed by different letters (a, b, c) in the same column are significantly different at the 1% probability level

1) Plants inoculated with the avirulent race 15-0; no disease symptoms at heading

2) Plants inoculated with the virulent race 1-4;at heading 56% of total leaf area covered with mildew

Table 2 Yield of protein of resistant and susceptible plants of the barley cultivar Sultan inoculated with two races of powdery mildew

Host reaction	Yield of protein g per pot			Yield of protein % of control	
	grain	straw	total	grain	total
Non-inoculated control	6.52 a	2.59 a	9.11 a	100	100
Resistant plants 1)	5.82 b	2.48 b	8.30 b	89	91
Susceptible plants 2)	4.55 c	3.36 c	7.91 c	70	87

Means followed by different letters (a, b, c) in the same column are significantly different at the 1% probability level

1) Plants inoculated with the avirulent race 15-0; no disease symptoms at heading
2) Plants inoculated with the virulent race 1-4;at heading 56% of the total leaf area covered with mildew

Table 3 Yield of straw, length of straw, and % protein in straw of resistant and susceptible plants of the barley cultivar Sultan inoculated with two races of powdery mildew

Host reaction	Yield of straw		Length of	% protein
	g per pot	% of control	straw cm	in straw
Non-inoculated control	78.2 a	100	107.8 a	3.31 a
Resistant plants 1)	76.2 b	97	102.1 b	3.25 a
Susceptible plants 2)	68.9 c	88	102.9 c	4.88 b

Means followed by different letters (a, b, c) in the same column are significantly different at the 1% probability level

1) Plants inoculated with the avirulent race 15-0; no disease at heading
2) Plants inoculated with the virulent race 1-4; at heading 56% of the total leaf area covered with mildew

The susceptible plants in chamber III were fairly heavily infected, and at heading time 56% of the leaf area was covered with powdery mildew. The grain yield was reduced by 26%, and the kernel weight by 11%. The content of grain protein was reduced from 9.75 to 9.25% which was equivalent to a 30% reduction in the yield of grain protein per unit area (Tables 1 and 2). The straw yield of infected plants was reduced by 12% and the straw length by 4.9 cm compared with uninoculated control plants (Table 3). The protein content of infected straw was higher than in uninoculated controls.

The fact that highly mildew-resistant barley plants do not suffer visible damage by inoculation with powdery mildew thus does not mean that the plants are unaffected. The results indicate that resistant plants react to inoculation by energy-requiring defence reactions which cause a drain on the host energy available for useful work and ultimately leads to reduction in grain yield and grain quality.

RESISTANCE REACTIONS AGAINST OTHER PLANT PATHOGENS

The clear indications that mildew-resistant barley plants react to infection with increased respiration to obtain energy for biochemical defence reactions raise the question as to whether similar reactions occur in other diseases. We investigated this in net blotch of barley and barley leaf stripe (26).

Net blotch is caused by the fungus *Pyrenophora teres* (stat. conid. *Drechslera teres*). The effect of this pathogen and its isolated toxins on the respiration of barley has already been reported (24, 25).

Fig. 2 shows the results from respiratory studies with the barley cultivars CI 9647 and Wing, the former highly resistant, the latter fully susceptible to isolate N-197 of *P. teres*. It is seen that resistant plants react to inoculation with a sharp but temporary increase in oxygen uptake beginning a few hours after inoculation and reaching a maximum 80% above that of uninoculated control plants. It then declines sharply and proceeds at about the same level as that of the controls.

In susceptible plants respiration increases until the beginning of necrosis, then it decreases as the necrosis progresses. After 11 days infected leaves were wholly necrotic and there was no respiration.

Enzyme inhibitor studies indicate that the dramatic increase of respiration of susceptible barley leaves infected with the net

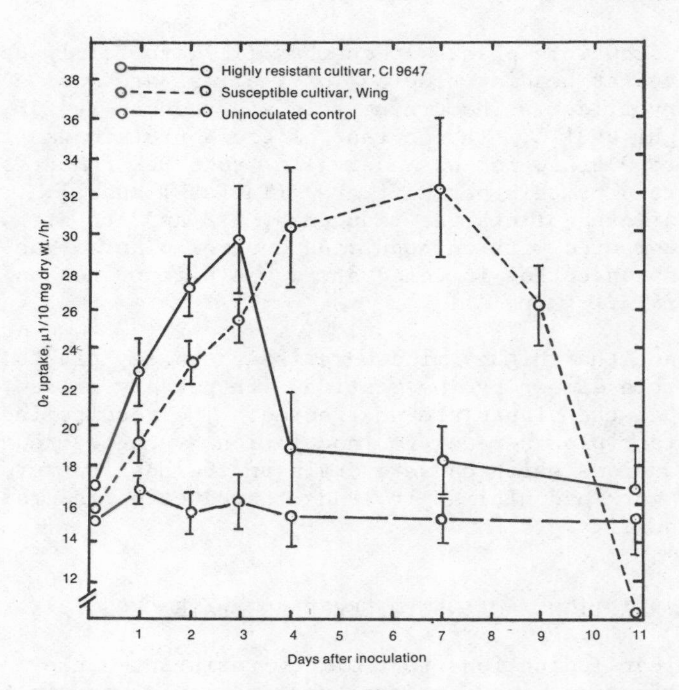

Figure 2 The effect of the net blotch pathogen *Pyrenophora teres* on the respiration of resistant and susceptible barley plants. The cultivar CI 9647 was resistant and the cultivar Wing susceptible to isolate N-197 used for inoculation

blotch pathogen is caused by uncoupling of oxidative phosphorylation from the electron transport chain (25). Similar studies with incompatible tissue would be most useful in order to evaluate whether the increased respiration in compatible and incompatible interactions results from different processes.

Barley leaf stripe is caused by the fungus *Pyrenophora graminea* (stat. conid. *Drechslera graminea*). In contrast to most other leaf-infecting fungi, the fungus is unable to cause disease by direct penetration of the leaf surface, and diseased plants always arise from infected seeds. After inoculation with conidia of *Pyrenophora graminea* leaves react with hypersensitive reactions as minute, dark lesions. The germ tubes form appressoria and penetrate the epidermis but the growth of infection hyphae soon ceases at a length *c.* 100 µm (23).

When conidia come into contact with barley leaves, the leaf tissue reacts almost immediately with a dramatic temporary increase in oxygen uptake (Fig. 3) in much the same way as do mildew-resistant leaves.

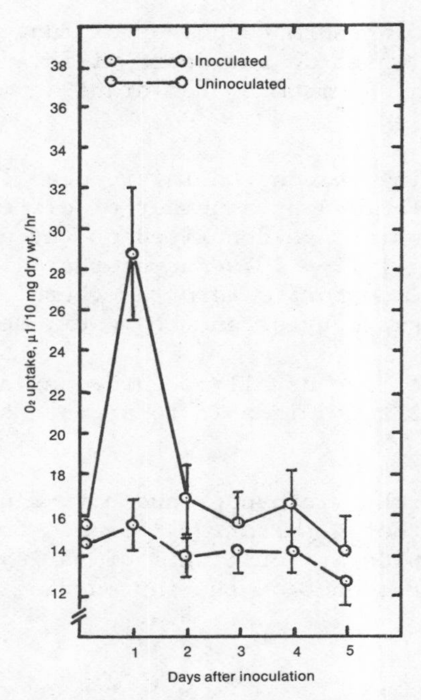

Figure 3 The effect of the barley leaf stripe pathogen *Pyrenophora graminea* on the respiration of inoculated barley leaves. Barley plants are resistant to leaf inoculation with this fungus (see text for further explanation)

The fact that resistance against three different fungi, including biotrophic and necrotrophic pathogens, involves energy-requiring processes suggests that this phenomenon may be general in resistant plants exposed to pathogens. Further studies with a wider spectrum of hosts and parasites are now required.

RESISTANCE REACTIONS AGAINST SAPROPHYTIC FUNGI

The finding that an avirulent race of powdery mildew can cause significant reductions in grain yield and grain quality without causing visible symptoms suggests some interesting practical aspects. Active resistance reactions are expressed only when the host comes into contact with a pathogen to which it is resistant. Under field conditions, therefore, high concentrations of avirulent spores of mildew or of other pathogenic fungi must be present to cause measurable yield reduction. Such high concentrations usually will not be present.

However, we must realize that the traditional concept of resistance is associated with a relatively small number of fungal pathogens such as mildews and rusts. But plants in a field crop daily are exposed to an enormous number of fungal spores and to other microorganisms which do not cause visible disease in plants because most plants are immune or resistant to most microorganisms in nature.

During the growing season the lower, dead leaves in a barley or wheat crop are colonized by a number of different saprophytic fungi the spores of which are deposited on the upper green leaves. Spore-trapping during 1977 - 1979 showed that *Alternaria alternata* (Fr.) Keissler and *Cladosporium herbarum* (Pers.) Link ex Gray frequently occur as saprophytes on the lower, dead leaves.

These two species are usually regarded as saprophytes although *Alternaria alternata* is able to promote senescence of green leaves (6, 22).

By the use of light microscopy and u.v.-fluorescent microscopy we have found that *Alternaria alternata* forms appressoria and penetrates the epidermal cell wall of barley leaves where to give an intense fluorescence of the surrounding cells (Fig. 4).

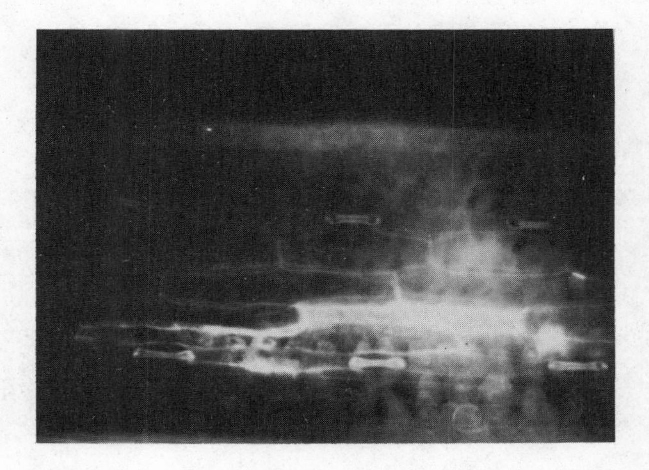

Figure 4 U.V. - fluorescence from epidermal cells of barley leaves two days after inoculation with *Alternaria alternata*

The strong fluorescence which extends into the mesophyll cells beneath the appressoria indicates the synthesis or accumulation of u.v.-absorbing substances. Similar u.v.-fluorescence appears prior to the development of hypersensitive necroses at the penetration sites in barley leaves inoculated with an avirulent race of powdery mildew (14, 30).

Measurements further demonstrated that barley leaves inoculated with spores of *Alternaria alternata* and *Cladosporium herbarum* reacted with a temporary, but marked increase in the rate of respiration with a maximum 24 - 48 hours after inoculation (Fig. 5).

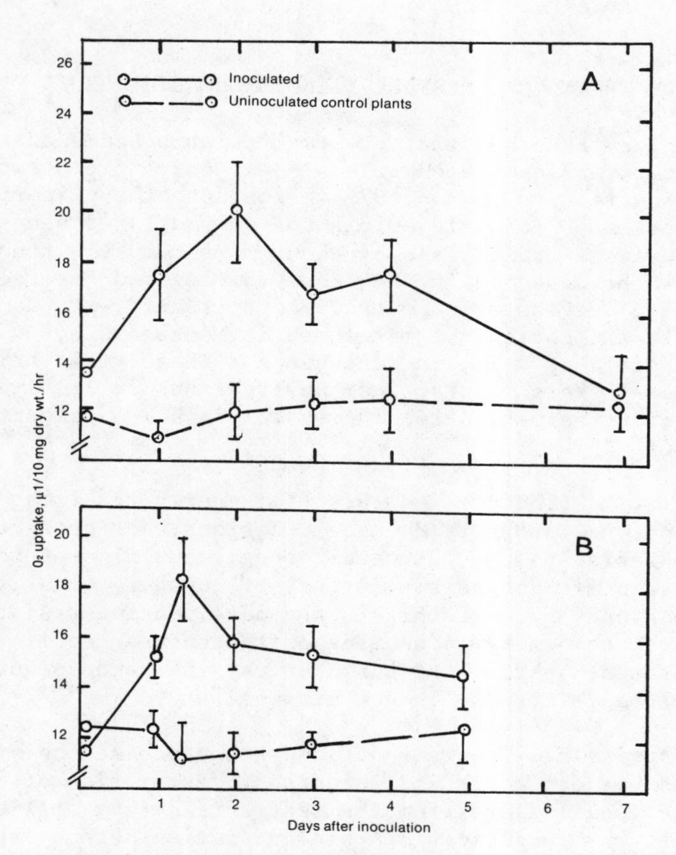

Figure 5 A The effect of *Alternaria alternata* on respiration of barley plants. B The effect of *Cladosporium herbarum* on respiration of barley plants. These two fungi are normally regarded as saprophytes

The similarities in host reactions incited by avirulent races of barley powdery mildew and *Alternaria alternata* suggest that this fungus should be regarded not as a saprophyte but as a pathogen to which barley is highly resistant. That these reactions consume host energy and thus deprive the plants of nutrients is supported by the observation that heavy inoculation of barley plants with *Alternaria alternata* promotes senescence as also recorded by Skidmore and Dickinson (22).

The results indicate that energy-requiring physiological reactions may be a general phenomenon in resistant plants, not only as a response to recognized pathogens, but also to certain fungi usually considered as saprophytes. The influence of saprophytes on plant growth could well be an intriguing and complex problem for future research.

CAN FUNGICIDE TREATMENT INCREASE YIELD IN HEALTHY CROPS ?

It is often recorded that treatment with broad-spectrum fungicides retards senescence (6). In extensive field experiments in the growing seasons 1978 - 1979 we found that application of the fungicide Maneb to barley plots not only delayed senescence but also increased grain yield by about 5% although virtually no disease could be detected in the crops (Smedegaard-Petersen unpublished data). Plants in control plots treated with manganese equivalent to the content of manganese in Maneb did not show any increase in yield or delay in senescence. These yield trials confirm several years of practical observations in Denmark showing that fungicide treatment often increases yield in apparently healthy crops.

The present findings indicate that energy-requiring physiological reactions, probably defence reactions, are characteristic not only of resistance to pathogenic fungi, but also of resistance to certain microorganisms in the phyllosphere usually regarded as saprophytes. If this is the case, active resistance reactions or other energy consuming host responses to microorganisms present on the leaf surfaces may influence plant growth and yield although disease is not visible.

Since active resistance reactions are expressed only by interactions between hosts and parasites, restriction or elimination of the fungal flora from the leaf surfaces by application of broad-spectrum fungicides may prevent active defence reactions and other physiological interactions between the host and the microorganisms. As a consequence more host energy becomes available for growth and plant production.

REFERENCES

1. ALLEN, P.J. & GODDARD, D.R. (1938). A respiratory study of powdery mildew of wheat. *Am. Jour. Bot.* 25, 613 - 621.

2. ANTONELLI, E. & DALY, J.M. (1966). Decarboxylation of indole acetic acid by near-isogenic lines of wheat resistant or susceptible to *Puccinia graminis* f. sp. *tritici*. *Phytopathology* 56, 610 - 618.

3. BUSHNELL, W.R. (1970). Patterns in the growth, oxygen uptake, and nitrogen content of single colonies of wheat stem rust on wheat leaves. *Phytopathology* 60, 92 - 99.

4. DALY, J.M. (1972). The use of near-isogenic lines in biochemical studies of the resistance of wheat to stem rust. *Phytopathology* 62, 392 - 400.

5. DALY, J.M. (1976). The carbon balance of diseased plants : Changes in respiration, photosynthesis and translocation. In : *Physiological Plant Pathology*, Eds. R. Heitefuss & P.H. Williams. p. 450 - 479. Springer Verlag.

6. DICKINSON, C.H. (1973). Effects of ethirimol and zineb on phylloplane microflora of barley. *Trans. Br. mycol. Soc.* 60, 423 - 431.

7. FRIC, F. (1976). Oxidative enzymes. In : *Physiological Plant Pathology*, Eds. R. Heitefuss & P.H. Williams. p. 617 - 631. Springer Verlag.

8. HASPEL-HORVATOVIC, E. & PAULECH, C. (1969). Changes of oxidized and reduced carotenoids, of the respiration and the photosynthesis of barley (*Hordeum sativum* L.) in connection with its resistance against powdery mildew. In : *Progress in Photosynthesis Research*, Ed. H. Metzner. p. 396 - 401. Tübingen 1969.

9. HEITEFUSS, R. (1965). Untersuchungen zur Physiologie des temperaturgesteuerten Verträglichkeitsgrades von Weizen und *Puccinia graminis tritici*. I. Veränderungen von Sauerstoffaufnahme und Phosphatstoffwechsel. *Phytopathol. Z.* 54, 379 - 400.

10. HISLOP, E.C. & STAHMANN, M.A. (1971). Peroxidase and ethylene production by barley leaves infected with *Erysiphe graminis* f. sp. *hordei*. *Physiol. Plant Pathol.* 1, 297 - 312.

11. JOHNSON, R. (1976). Development and use of some genetically controlled lines for studies of host-parasite interactions. In : *Biochemical aspects of plant-parasite relationships*, Eds. J. Friend and D.R. Threlfall, p. 25 - 41. Academic Press.

12. KUĆ, J. (1972). Phytoalexins. *Ann. Rev. Phytopath.* 10, 207 - 232.

13. KUĆ, J.A. (1976). Phytoalexins. In : *Physiological Plant Pathology*, Eds. R. Heitefuss and P.H. Williams, p. 632 - 652. Springer Verlag.

14. MAYAMA, S. & SHISHIYAMA, J. (1978). Localized accumulation of fluorescent and u.v.-absorbing compounds at penetration sites in barley leaves infected with *Erysiphe graminis hordei*. *Physiol. Plant Pathol.* 13, 347 - 354.

15. MILLERD, A. & SCOTT, K. (1956). Host pathogen relations in powdery mildew of barley. II. Changes in respiratory pattern. *Aust. Jour. Biol. Sci.* 9, 37 - 44.

16. OKU, H., OUCHI, S. & SATO, M. (1973). Nucleic acid metabolism in powdery mildew-infected barley leaves as associated with symptom development. *Rept. Tottori Mycol. Inst. Jpn.* 10, 511 - 516.

17. OKU, H., OUCHI, S., SHIRAISHI, T., KOMOTO, Y. & OKI, K. (1975). Phytoalexin activity in barley powdery mildew. *Ann. Phytopath. Soc. Japan* 41, 185 - 191.

18. OUCHI, S., OKU, H., HIBINO, C. & AKIYAMA, I. (1974). Induction of accessibility and resistance in leaves of barley by some races of *Erysiphe graminis*. *Phytopathol. Z.* 79, 24 - 34.

19. PAULECH, C. (1967). Einfluss des Pilzes *Erysiphe graminis* f.sp.*hordei* Marchal auf die Respiration anfälliger und resistenter Gerstensorten (*Hordeum sativum* L.). *Biologia* 22, 202 - 209.

20. SAMBORSKI, D.J. & SHAW, M. (1956). The physiology of host-parasite relations. II. The effect of *Puccinia graminis tritici* Eriks. and Henn. on the respiration of the first leaf of resistant and susceptible species of wheat. *Can. Jour. Bot.* 34, 601 - 619.

21. SCOTT, K.J. & SMILLIE, R.M. (1966). Metabolic regulation in diseased leaves. I. The respiratory rise in barley leaves infected with powdery mildew. *Plant Physiology*

41, 289 - 297.

22. SKIDMORE, A.M. & DICKINSON, C.H. (1973). Effect of phylloplane fungi on the senescence of excised barley leaves. *Trans. Br. mycol. Soc.* 60, 107 - 116.

23. SMEDEGAARD-PETERSEN, V. (1976). Pathogenesis and genetics of net-spot blotch and leaf stripe of barley caused by *Pyrenophora teres* and *Pyrenophora graminea,* 176 pp. *DSR Forlgag. Royal Vet. Agric. Univ., Copenhagen.*

24. SMEDEGAARD-PETERSEN, V. (1977). Respiratory changes of barley leaves infected with *Pyrenophora teres* or affected by isolated toxins of this fungus. *Physiol. Plant Pathol.* 10, 213 - 220.

25. SMEDEGAARD-PETERSEN, V. (1980). Enzyme-inhibitor studies of respiratory pathways in barley leaves infected with *Pyrenophora teres* and affected by its isolated toxins. *Royal Vet. Agric. Univ. Yearbook* 1980, 28 - 35.

26. SMEDEGAARD-PETERSEN, V. (1980). Increased demand for respiratory energy of barley leaves reacting hypersensitively against *Erysiphe graminis, Pyrenophora teres* and *Pyrenophora graminea.* *Phytopathol. Z.* 99, 54 - 62.

27. SMEDEGAARD-PETERSEN, V. & STØLEN, O. (1980). Resistance against barley powdery mildew associated with energy-consuming defence reactions which reduce yield and grain quality. *Royal Vet. Agric. Univ. Yearbook* 1980, 96 - 108.

28. STANBRIDGE, B., GAY, J.L. & WOOD, R.K.S. (1971). Gross and fine structural changes in *Erysiphe graminis* and barley before and during infection. In : *Ecology of leaf surface micro-organisms,* Eds. T.F. Preece and C.H. Dickinson. p. 367 - 379.

29. TORP, J., JENSEN, H.P. & JØRGENSEN, J. HELMS (1978). Powdery mildew resistance genes in 106 northwest European spring barley varieties. *Royal Vet. Agric. Univ. Yearbook* 1978, 75 - 102.

30. TOYODA, H., MAYAMA, S. & SHISHIYAMA, J. (1978). Fluorescent microscopic studies on the hypersensitive necrosis in powdery-mildewed barley leaves. *Phytopathol. Z.* 92, 125 - 131.

31. WOLFE, M.S. (1972). The genetics of barley mildew. *Rev. Pl. Path.* 51, 507 - 522.

CONTRIBUTED PAPERS

Each participant was invited to give a short paper at the Advanced Study Institute. Almost all did so. So far as was possible these papers were given in association with the Lectures to which they were most closely related. The contents of these papers are summarized below. The addresses of the authors are given in the list of Participants at the end of the book.

S.A. ARCHER - THE ROLE OF CELL WALLS IN ACTIVE DEFENSE

The walls of higher plant cells have long been considered as relatively inert structures within which the metabolically active protoplast is encased. By virtue of their considerable mechanical strength cell walls are an important non-specific barrier to colonization by micro-organisms and constitute a major component of passive resistance. Superficially therefore cell walls are not a promising topic for a symposium on active defence mechanisms. However this attitude, which originates from pathologists' preoccupation with cell wall degradation in certain soft rot and related syndromes, neglects much recent evidence for major and very rapid changes to cell walls as a post-infectional phenomenon.

Just as the role of cell walls in passive resistance has probably been over-estimated, so their involvement in active defence has until very recently been much neglected. Indeed the unthickened cell walls of primary tissues are a surprisingly poor barrier to penetration. Witness the apparent ease with which many necrotrophic pathogens can colonize the intercellular region of plants and the extent to which biotrophic pathogens penetrate cell walls to form haustoria.

Three main ways are envisaged in which cell walls can participate in active resistance. Firstly, walls may be involved in the recognition of pathogens and so trigger the plant's resistance mechanisms. Secondly, cell walls may themselves be the medium through which resistance operates, via an induced change in structure or chemistry. Thirdly, the chemical composition of the wall, by virtue of its intimate exposure to the pathogen, might control the synthesis of the pathogen's degradative enzymes. In practice there is little support for this last hypothesis, and it is now known that the secretion of cell wall-degrading enzymes is controlled by the availability of sugar monomers rather than by the configuration of carbohydrate polymers.

With some species of plant pathogenic bacteria there is a good correlation between the ability of strains to bind to host cell walls following intercellular infiltration,and the elicitation of hypersensitivity. In tobacco incompatible but not virulent isolates of *Pseudomonas solanacearum* become attached to mesophyll cell walls by an active process involving swelling of the wall matrix and the deposition of fibrillar and granular material which binds to and envelops the pathogen. Bacterial surface polymers are implicated in the binding, but the molecular mechanisms responsible for translating this into hypersensitivity are unknown.

The structural and chemical modifications to cell walls that occur upon infection are frequently ascribed a role in resistance. Localized deposits at infection sites, referred to as papillae, and consisting of mixtures of callose, lignin, silica and suberin may be formed within hours of challenge by a micro-organism but are as frequently associated with successful as with unsuccessful invasion, so their role in resistance is ambiguous.

D.F. BATEMAN, C.J. BAKER AND S.L. McCORMICK - DEGRADATION OF CUTIN MEMBRANES BY CUTIN ESTERASE

Cutin esterase in the absence of other hydrolytic enzymes can cause the disruption of intact isolated cutin membranes prepared from grape berries. An isozyme of cutin esterase was purified from 6-day old culture filtrates of *Fusarium solani* f. sp. *pisi* grown on 0.5% apple cutin. Cutin membranes were prepared from the skins of green grapes (var. Calmeria) by a 2 h wash in methanol, followed by a 2 h wash in chloroform, and then overnight incubation in acetone. The dewaxed skins, placed between two rubber gaskets, were then treated with a 0.5% pectinase and 1.5% cellulase mixture for 24 to 48 h at pH 4.5 to yield intact cutin membranes c. 5μ in thickness. Membranes were then placed in diffusion cells and the effects of cutin esterase action on movement of molecules, including enzymes, across membranes and on the strength of the membranes were determined over time.

Intact control membranes did not permit the passage of ^{14}C-glucose, but membranes exposed to 1 unit of cutinase (1 unit of enzyme released 1 μmole of acetate/min from *p*-nitrophenyl acetate at pH 8.5) permitted ^{14}C-glucose and higher molecular weight solutes, including enzymes, to pass through the membranes. Cutinase itself (MW 15,000) reached equilibrium between the half cells of diffusion chambers within c. 48 h. Studies with a series of purified fungal cell degrading enzymes that included xylanase (MW 15,000), endo-pectate lyase (MW 30,000) and α-L-arabinofuranosidase (MW 60,000) revealed that these enzymes passed through cutin membranes that had been exposed to cutin esterase for a period of 72 h. Further experiments showed that membranes exposed to cutin esterase ruptured when exposed to a hydrostatic force of 0 to 0.2 neutons/cm^2 whereas control membranes, exposed to inactive cutinase, required a minimum force of 1.1 neutons/cm^2 to cause rupture.

The results reported here, coupled with published results of fungal penetration phenomena as revealed by the electron microscope, plus the known ability of pathogenic fungi to produce cutin esterase indicates that this enzyme may play a significant role in the ingress of fungi into higher plants. Also, since cutinase action permits the movement of cell wall degrading enzymes such as endo-pectate lyase across cutin membranes this enzyme may contribute to the early phases of pathogenesis prior to fungal ingress.

J.A. CALLOW - THE STRUCTURE AND ACTIVITY OF A GLYCOPEPTIDE ELICITOR OR TOXIN FROM *FULVIA FULVA*

Tomato cultivars containing the Cf_2 and Cf_4 genes for resistance to *Fulvia fulva* (syn. *Cladosporium fulvum*) react to avirulent races of the pathogen by a form of hypersensitive response accompanied by necrosis and phytoalexin synthesis. The paper described the isolation and partial characterization of a polydisperse glycoprotein from filtrates of cultures of *F. fulva* with the properties of a non-specific elicitor of resistance.

The major high molecular weight fraction of culture filtrates is a polydisperse, phosphorylated heteroglycopeptide of mean MW 70,000 (1). On DEAE-cellulose three components were distinguished of varying protein/carbohydrate ratio and phosphorylation. Each component was further fractionated by affinity chromatography on con A-Sepharose. Acid, alkali and enzyme treatments indicate a family of polymers based on a common structure of a peptide of MW 5 - 10,000, oligomannose side chains, 1-4 residues long, attached to the peptide, and a core polysaccharide containing mannose, glucose and galactose which is phosphorylated and bears galactofuranose side chains. This basic structure is essentially common to races 0,4 and 1, 2, 3 of the fungus.

The glycopeptides interact with tomato cells inducing rapid changes in membrane permeability to electrolytes, but without race or cultivar specificity (2). Over longer periods they also elicit phytoalexin synthesis, necrosis and "callose" formation, responses associated with resistance to the fungus *in vivo*. These activities are a function of the glycan moiety since periodate oxidation abolishes activity. An important role for the oligomannose side chains is indicated since digestion with exomannosidase (but not glucosidases or galactosidases) also abolishes activity.

The activities described suggest a function for the glycopeptides as general or non-specific resistance elicitors. However, an alternative function may be suggested. *F. fulva* is a biotrophic parasite and grows intercellularly without haustoria, and initially without apparent tissue damage. The fungus presumably relies on nutrients within the free space of the tissue. The glycopeptides described here may therefore function more as toxins than as elicitors *in vivo*, since their effects on membrane permeability will potentially augment the supply of nutrients to the parasite. The action of the glycopeptides in eliciting phytoalexins and necrosis may represent secondary, artifactual activities resulting from the use of unnaturally high concentrations of the glycopeptides in laboratory experiments.

1. Dow, J.M. & Callow, J.A. (1979a). *J. Gen. Microbiol.* 113, 57 - 66.

2. Dow, J.M. & Callow, J.A. (1979b). *Physiol. Plant Pathol.* 15, 27 - 34.

F. CERVONE AND R.H.A. COUTTS - ABSORPTION OF POLYGALACTURONASE FROM *COLLETOTRICHUM LINDEMUTHIANUM* BY FRENCH BEAN TISSUE AND PROTOPLASTS

Almost all plant pathogens produce enzymes that degrade one or more of the polysaccharides of plant cell walls. Of these enzymes, polygalacturonase (PG) and pectic lyase (PL) are the most important in pathogenesis but until recently there was little evidence that they were determinants in specificity. Our results support the hypothesis that PG may have a role in specificity of plant diseases and that a site of action for PG may exist in the plant cell membrane.

Polygalacturonase from *Colletotrichum lindemuthianum* was absorbed by tissue from bean plants but not by tissue from non-host plants. Polygalacturonases from fungi which were not pathogen to bean were little affected by bean tissues. Absorption of enzyme by bean tissue was partially nullified by certain sugars or by treating tissue with heat, chloroform or sodium periodate. The permeability of bean cells was increased by PG from *C. lindemuthianum* but not by other polygalacturonases. By using both compatible and incompatible combinations of French bean cultivars and *C. lindemuthianum* races, it was found that enzyme absorption is much faster in incompatible than in compatible systems.

To determine the site of enzyme absorption isolated bean protoplasts were incubated with *C. lindemuthianum* PG in a plasmolyticum in which they remained intact. Enzyme activity in the supernatant after gentle centrifugation greatly decreased after 3 h incubation. Activity was recovered from the protoplasts by adding enough buffer to rupture them. In a similar experiment without plasmolyticum no enzyme absorption was detected.

These results provide evidence that the function of PG is not only to macerate plant tissue and kill plant cells. If enzyme absorption before maceration is a critical step in pathogenesis, PG also play a role in recognizing the tissue to be attacked. A binding site for PG may well be present in the plant cell membranes where binding proteins for specific sugars are present (many PGs are glycoproteins). A PG localized on the membrane would be well placed to affect both cell wall and protoplast.

D.D. CLARKE - THE ACCUMULATION OF CINNAMIC ACID AMIDES IN THE CELL WALLS OF POTATO TISSUE AS AN EARLY RESPONSE TO FUNGAL ATTACK

The resistance of potato tuber tissue to avirulent isolates of *Phytophthora infestans* is based on a hypersensitive reaction. One of the components of this reaction is the accumulation of brown lignin-like polymers in the walls of the penetrated and surrounding cells. These polymers make the walls resistant to enzymic degradation and thus may play a role in resistance by inhibiting wall penetration and so confine the fungus to the lesion.

We have found that one of the earliest reactions of potato tuber tissue to inoculation by avirulent isolates of *P. infestans* leads to the accumulation of a group of highly fluorescent compounds. They first accumulate on starch grains within the cells around the inoculation site, but after a few hours the fluorescence disappears from the starch grains as accumulation begins within the cell walls. The fluorescence persists within the walls for about 12 hours before finally disappearing at the time of hypersensitive cell death when brown pigments accumulate within the walls.

The same pattern and time scale of accumulation occurs in response to inoculation with virulent isolates of *P. infestans*, or with isolates of *Fusarium solani* var. *coeruleum*, or *Phoma exigua* var. *foveata* except that the walls remain fluorescent much longer, only losing their fluorescence when the much delayed cell necrosis occurs.

A number of the fluorescent compounds have been characterised and shown to be amides of cinnamic or *p*-coumaric acid with tyramine or octopamine. Similar amides of caffeic and ferulic acids are probably also produced. Thus the compounds are related to a whole family of esters or glycosides of oxygenated cinnamic acids and coumarins which accumulate in potato tissue on infection. However, unlike these latter compounds which accumulate within the vacuoles of the cells, the amides move out of the cells and accumulate within the walls. When they first move into the walls they are freely extractable with methanol, but they soon become bound and then can only be removed with hot NaOH, indicating their involvement in lignification-like reactions. However their role in these reactions is not clear. When mixtures of the compounds are fed to healthy tissue the fluorescence disappears, but its disappearance is not associated with the accumulation of brown pigments within the walls. Thus the browning reactions which occur in tissues undergoing a hypersensitive reaction or which eventually occur in infected tissue are probably not developed directly from the amides and the role of the amides in the infection process is still under investigation.

Dr D.D. Clarke acknowledges gratefully the contributions of Dr P. Mohamed and Dr N.J. McCorkindale, Chemistry Department, Glasgow University in the isolation and identification of the amides.

R.H.A. COUTTS - THE USE OF PLANT PROTOPLASTS IN STUDIES OF ACTIVE DEFENCE AGAINST PLANT VIRUS INFECTION

Following the development of isolated leaf mesophyll protoplast systems in several host/virus combinations many of the basic questions regarding virus infection and replication have had to be posed again. Most of the physical defence barriers to virus infection and replication are removed by conversion to protoplasts and the host range of some viruses has been extended. Protoplast systems allow the virus to be presented to the cells in multivarious ways, under differing conditions of concentration, electrical charge and promoters of infection. Whole leaf systems by physical considerations are not suitable for such investigations because infection sites are transient and the cell-wall barrier is a particularly good physical defence. The juxtapositioning of leaf cells *in vivo* is necessary for a hypersensitive response and the separation of the infected cells from one another by plasmolysis early after infection suppresses the local lesion reaction in leaves, by isolating protoplasts *in vivo*. Similarly, protoplasts derived from hypersensitive reacting hosts show no such reaction when inoculated as protoplasts *in vitro* with a virus that causes necrosis *in vivo*. Indeed the phenotypic expression of genes defining necrotic reactions may require cell to cell contact. With very high resistance and susceptibility representing the extremes of plant reactions towards virus infection, the placement of hypersensitivity tolerance and immunity in such schemes is difficult. Biochemical explanations for total resistance, subliminal infection, immunity and hypersensitivity are still to be found, although their transfer to new cultivars has been achieved in some plant species. Indeed the physical characteristics of leaves may still be of great importance in any resistance situation and the ability of viruses to move from cell to cell and cause systemic infection may be a feature of the viral genome itself. Studies on these phenomena particularly genetically defined resistance in isolated protoplasts have intrigued virologists. The maintenance of immunity and resistance in isolated plant protoplasts has been studied in several host/virus combinations, but no clear picture emerges, reactions often depending on the infecting virus strain. The only true maintenance of resistance at the protoplast level involves GCR-237 tomato leaf protoplasts and TMV-L. With Arlington cowpea leaf protoplasts and CPMV-SB, and China cucumber leaf protoplasts and CMV-W, resistance is expressed only partially, whereas whole plants are totally resistant. While no biochemical parameters for any of the above reactions *in vitro* are illustrated or defined, the artificial nature of isolated protoplasts casts a doubt on the validity of using such systems as compared to whole plants.

R.H.A. COUTTS AND E.E. WAGIH - ALTERATIONS IN THE SOLUBLE PROTEINS OF COWPEA LEAVES AND CUCUMBER COTYLEDONS FOLLOWING TOBACCO NECROSIS VIRUS (TNV) INFECTION AND THEIR IMPLICATED RÔLES IN ACTIVE DEFENCE MECHANISMS

Both cowpea primary leaves (*Vigna sinensis* cv. Blackeye No. 5) and cucumber cotyledons (*Cucumis sativus* cv. Ashley) respond hypersensitively to infection with TNV with the development of spreading local lesions, three and six days respectively after inoculation. Concurrently reproducible alterations occur in the soluble protein fractions of these tissues as shown by discontinuous-polyacrylamide gel electrophoresis. In cowpea two apparently novel fractions α and β are associated with 'shoulder' proteins, while in cucumber, one fraction, termed γ has been observed. The γ protein has a molecular weight (MW) 24,500 d, whereas the α and β fractions have a common polypeptide MW 10,000 d and the β fraction contains two additional polypeptides, MW 25,500 d, 36,000 d in smaller amounts. The 'novel' proteins in both cowpea and cucumber are detectable only in necrotic tissue and in uninoculated half leaf tissue adjacent to necrotic half leaves. The function of these proteins is obscure, but the α and β fractions are not isozymes of either polyphenoloxidase (PPO) or ribonuclease (RNAse) but the 'shoulder' proteins may be associated with peroxidase (PO) activity when necrosis is complete. The γ protein is not associated with either PO or PPO isozymes, but an RNAse isozyme of host-origin with the same R_f value is reproducibly observed by disc-polynucleotide gel electrophoresis.

Following the homologous challenge of individual cowpea leaves and cucumber cotyledons induced resistance has only been demonstrated as a reduction in lesion number in half leaves adjacent to necrotic tissue. In upper leaves of cucumber, a reduction in lesion number and size has been observed after challenge following the prior inoculation of both cotyledons. In neither system has the presence of the novel proteins been absolutely correlated with induced resistance, particularly in the later example where the γ protein is absent from partially resistant leaves.

The 'novel' proteins are either amplified or induced through the host genome. Interestingly, similar proteins in both cowpea and cucumber tissue appear in the soluble fraction following a period of mannitol-induced osmotic stress. Such a response may be similar to the stress reaction of plants following localization of virus and/or associated necrosis. The 'novel' proteins have no apparent direct rôles in resistance, but indirect ones cannot be excluded. The proteins may well be involved in as yet uninvestigated enzymatic, hormonal or structural changes in both inoculated and other leaves.

G. DÉFAGO - GENETICAL ANALYSIS OF TOMATIN RESISTANCE AND PATHOGENICITY OF *FUSARIUM SOLANI* MUTANTS

Tomatin inhibits the growth of *Fusarium solani in vitro*; it does not diffuse from intact tomato cells. To prove that tomatin is a resistance factor i.e. that the fungus causes enough tomatin to leach from the host cell to stop its growth, we have produced *F. solani* mutants resistant to tomatin and analysed the genetics of their resistance to tomatin, of their pathogenicity for green tomatoes containing tomatin, and of their content of sterols which may be a mechanism of resistance to tomatin.

Six mutants grew at 43 per cent of the controls in the presence of 800 ppm tomatin (wild strain, 10 ppm). One mutant obtained after treatment with nitrosoguanidine (NTG) and one after treatment with UV light were crossed with *Nectria haematococca*, the sexual form of *F. solani*. The number of ascospores per ascus was reduced to mainly 4. The analysis of 421 F_1 monocultures gives a 80 per cent probability of a dihybrid segregation of tomatin and mating type. The analysis of some crossing recombinants gives a probability of 99.6 per cent that the mutations obtained after treatment with NTG or UV are on the same locus. The mutants colonized the flesh of green tomatoes; the wild strain does not. Mutants and the wild strain grew similarly in the flesh of red tomatoes not containing tomatin.

In 64 F_1 strains tested, resistance to tomatin and increased pathogenicity for green tomatoes were linked. In the mutants and in 48 F_1 strains which were studied resistance to tomatin was linked with a decrease in sterol content; no qualitative changes were observed. The mutants and the F_1 strain resistant to tomatin are slightly more resistant to digitonin and to solanins (two other saponins) and to nystatin, an antibiotic acting on membrane sterols, but not to triarimol, a fungicide acting on sterole biosynthesis. Two mutants from another *F. solani* wild strain have a higher sterol content and are resistant to tomatin; they are resistant to triforine and highly resistant to solanins. Their pathogenicity for potato tubers is independent of the amount of solanins in the tuber; this is not the case for the wild strain.

Resistance to tomatin, increase in pathogenicity for green tomatoes and reduction in sterol content cannot be separated genetically. This means that tomatin and probably solanins are liberated by *F. solani* in the very early stages of infection and contribute effectively to stop further growth. Furthermore, changes in sterol content can be considered as a mechanism of resistance to tomatin and solanins.

A.E. DESJARDINS, L.M. ROSS, M.W. SPELLMAN AND P. ALBERSHEIM - PROTECTION OF SOYBEANS AGAINST *PHYTOPHTHORA MEGASPERMA* VAR. *SOJAE* BY MOLECULES ISOLATED FROM FUNGAL CULTURE FILTRATES

Phytophthora megasperma var. *sojae* (PMS) is the causal organism of soybean root and stem rot, a disease of great economic significance in the United States and Canada. PMS exists as a number of physiologic races which demonstrate a race specific relationship in their infection of host soybean cultivars. The hypothesis that this race specificity is mediated by carbohydrate-containing molecules secreted by the fungus is being investigated in this laboratory. Molecules purified from culture filtrates of incompatible races of PMS were previously shown to protect soybeans from infection by compatible races. However, subsequent progress in isolation and characterization of the active moieties has been slow because of poor reproducibility of the biological assay. Our major efforts have therefore been focused, with partial success, on identification and control of the factors contributing to variation in the assay. We have shown that plant vigor is extremely important for successful protection experiments and we have manipulated many variables in search of optimal growth and assay conditions. Although variability in individual experiments continues, the accumulated data indicate that preparations from incompatible races are relatively active, while those from compatible races are relatively inactive as protection factors. The active material is fractionated from culture fluid by precipitation with ammonium sulfate followed by chromatography on columns of phenyl-substituted agarose and diethylaminoethyl-cellulose. Material obtained from four races of PMS consists mainly of glycoproteins with an average molecular weight of 150,000 daltons by molecular sieving chromatography. Upon polyacrylamide slab gel electrophoresis under denaturing conditions, more than twenty polypeptide bands can be resolved. The preparations contain from 15 to 20 per cent carbohydrate, predominantly mannose, glucose and glucosamine. We are continuing the purification and characterization of the protection factors by chemical, enzymatic and serological methods. Preliminary studies of the biological mode of action of the glycoprotein preparations indicate that they do not inhibit PMS mycelial growth *in vitro*. They are also relatively poor elicitors of the phytoalexin glyceollin in the soybean cotyledon assay and the elicitation observed is not race specific. With a more reliable bioassay and the chemical techniques available in this laboratory, we hope to isolate and identify the protection factors from the complex glycoprotein preparations and to elucidate their role in soybean resistance and susceptibility to PMS. Supported by the U.S. Department of Energy (EY-76-S-02-1426), the National Institutes of Health (1F32GMO 7104-01) and the Rockefeller Foundation (RF 78035).

B.J. DEVERALL - ELICITATION OF NECROSIS BY RUST FUNGI

Infection of wheat bearing the *Lr20* allele by strains of leaf rust *Puccinia recondita* f. sp. *graminis*, avirulent with respect to this allele, results in an extensive hypersensitive response at 20°C. Microscopic inspection of whole mounts of leaves stained and cleared with trypan blue/lactophenol and chloral hydrate respectively revealed no difference between avirulent and virulent infection sites during the day following inoculation. During and after the second day, necrosis of cells occurred widely around apparently unaffected cells penetrated by haustoria of avirulent strains. Extensive mycelial colonies of avirulent and virulent strains could be established in leaves held after inoculation at 30°C where the *Lr20* allele is inoperative. Transfer of these leaves to 20°C resulted in extensive necrosis only around the avirulent colonies. The hypothesis that the necrosis was caused by a product of avirulent strains was tested by means of leaf transplant experiments, as reported in *Physiological Plant Pathology* (1978) 12, 311 - 319.

The results of the transplant experiments implied that avirulent strains of leaf rust produced a diffusible substance able to elicit necrosis only in wheat bearing the *Lr20* allele. Attempts have been made to detect *Lr20*-specific toxicity in different types of aqueous extract from avirulent strains. These extracts comprised leachates and macerates of germinated spores and diffusates from exposed mesophylls of leaves bearing mycelia established at 30°C. Extracts and water controls were tested as droplets in cavity slides overlaid with exposed mesophylls of segments of uninfected leaves. The first three experiments suggested the detection of *Lr20*-specific toxicity in extracts of germinated spores but all subsequent experiments have revealed a low level of non-specific activity in extracts of avirulent and virulent strains. Work is continuing with refinements to the bioassay procedures and attempts to obtain extracts from infection structures *in vitro* and established mycelia in leaves.

Infection of soybean leaves by the rust *Phakopsora pachyrhizi* occurs by direct penetration of cuticle and epidermal cells. Rapid necrosis occurs in the penetrated cells and a changed stain retention can be detected in underlying palisade mesophyll cells between which infection hyphae pass. As reported in *Transactions of the British Mycological Society* (1977) 69, 411 - 415 and (1980) 74, 329 - 333, millipore-filtered fluids, in which aseptically produced uredospores have germinated, caused browning and phytoalexin production in seed cavities of soybean pods. A more rapid response of a resistant soybean cultivar than of a susceptible cultivar increases interest in a possible selective elicitor of necrosis in the germination fluids, which is being studied further.

D.G. ELGERSMA - ACCUMULATION OF RISHITIN IN SUSCEPTIBLE AND RESISTANT TOMATO PLANTS AFTER INOCULATION WITH *VERTICILLIUM ALBO-ATRUM*

The phytoalexin rishitin has been isolated from potato, tomato and tobacco plants. In a study of mechanisms of resistance of tomato plants to *Verticillium albo-atrum* we were interested in the relation of the accumulation of rishitin to resistance as well as in the number of propagules present in the infected tissue. Six-week-old tomato plants of the susceptible cultivar 'Maascross' and the resistant cultivar 'Multicross' were stem-inoculated with a conidial suspension of 10^7 conidia per ml. Two, 3, 4, 7 and 11 days after inoculation stem pieces cut at 0.5 to 5.5 cm and at 5.5 to 10.5 cm above the site of inoculation were extracted and concentrations of rishitin determined by gas chromatography. During the first 4 days after inoculation the accumulation of rishitin in the lower stem pieces of susceptible and resistant cultivars was similar. At 7 and 11 days after inoculation concentrations were lower in resistant than in susceptible plants. Rishitin accumulation was less in the upper than in the lower stem pieces. In the upper stem pieces there were, however, no differences in concentrations of rishitin between susceptible and resistant plants at 2 and 3 days after inoculation and there were lower accumulations in resistant plants at 4, 7 and 11 days after inoculation. Rishitin was never detected in control plants treated with water instead of a conidial suspension.

Growth of the pathogen in stem pieces similar to those used for rishitin extraction was determined by homogenizing and plating out. One day after inoculation, distribution patterns of the inoculum in susceptible and resistant plants were similar. The numbers of propagules isolated from stem pieces of resistant and susceptible plants during the first 4 days after inoculation were similar. At 7 days after inoculation, however, the number of propagules appeared to be significantly lower in resistant than in susceptible plants. To investigate if concentration of conidia in inocula affects rishitin accumulation we inoculated plants with conidial suspensions of 10^5, 10^6, 10^7 and 10^8 conidia/ml. Three days later accumulation of rishitin was stimulated with increased concentration of conidia but no differences between susceptible and resistant plants were observed. It is concluded that rishitin accumulation is not a primary factor in resistance.

M-T. ESQUERRÉ-TUGAYÉ, D. MAZAU, D. ROBY AND A. TOPPAN -
INTERACTION OF ELICITORS WITH PLANT TISSUES AND PROTOPLASTS INDUCES THE SYNTHESIS OF ETHYLENE AND OF HYDROXYPROLINE-RICH GLYCOPROTEINS

Enrichment of plant cell walls by a hydroxyproline-rich glycoprotein was first demonstrated a few years ago in melon seedlings infected with *Colletotrichum lagenarium* and later found to occur in several plants infected with different pathogens, either fungi or bacteria (*C.R. Acad. Sci. Paris* 276, 525 - 528,1973, and unpublished data). Amounts of this glycoprotein are directly assessed by automatic analysis of the hydroxyproline-oligoarabinosides which characterize the molecule ($Hyp-Ara_n$,with n = 1 to 4).

Early triggering or inhibiting of biosynthesis by treatment with ethylene or free hydroxyproline indicated that the accumulation of cell wall glycoprotein is a response closely associated with the defence of plants (*Plant Physiol.* 64, 320 - 326,1979). In melon seedlings inoculated with *C. lagenarium*, this response is induced, through ethylene, by cell surface interaction between host and pathogen. The role of ethylene as a signal which triggers the host response has first been demonstrated in infected tissues. These tissues produce large amounts of ethylene which can be lowered by using specific inhibitors of ethylene synthesis such as canaline or aminoethoxy vinyl glycine. In the presence of these inhibitors and of ^{14}C-proline, as a precursor of cell wall hydroxyproline, both ethylene and cell wall glycoprotein synthesis are inhibited. The role of cell surface interaction in triggering both ethylene and cell wall hydroxyproline-glycoprotein synthesis is supported by experiments using fungal glycopeptides. As was formerly demonstrated, the cell wall and the culture filtrate of *C. lagenarium* contain glycopeptides which are able to induce, in healthy plants, the increased ethylene production which normally occurs in infected plants (3rd International Congress of Plant Pathology, Munich, Abstract p. 229, 1978).Interaction of these molecules with organs or protoplasts of the host shows that the host cell wall seems necessary for stimulation of ethylene biosynthesis to occur. Experiments recently performed with ^{14}C-proline and healthy tissues indicate that the fungal glycopeptides are able to stimulate both the synthesis of ethylene and of cell wall glycoproteins. Only microgram quantities of glycopeptides are required to elicit significantly both responses. Since these responses are associated with the defence of plants, these glycopeptides may be called elicitors of cell wall glycoproteins.

This is the first report on the involvement of elicitors in the triggering, through ethylene, of hydroxyproline-rich glycoprotein biosynthesis.

J. FRIEND - PHENOLICS AND RESISTANCE - SOME SPECULATIONS

Active resistance of potato tubers to several fungal pathogens is associated with the deposition of insoluble phenolic materials which ought to be termed lignin-like to differentiate them from "classical" lignin. Among probable structures are :

i) Cinnamic acids esterified to hydroxyl groups of cell wall polysaccharides

ii) Polymeric oxidation products of cinnamic acids covalently bound to cell wall carbohydrate or protein

iii) Cell wall polysaccharides cross-linked by smaller phenolic oxidation products such as by 5,5-diferulate or β, β'-diferulate

In the case of some potato cultivars containing major genes for resistance to *Phytophthora infestans* the rate of appearance of the lignin-like material is correlated with increases in phenylalanine ammonia lyase in incompatible combinations.

In tuber discs of three cultivars showing different degrees of field resistance to a complex race of *P. infestans* the rates of PAL increase are similar, as they are in the early stages of infection of discs of either susceptible or resistant tubers with *Fusarium solani* var. *coeruleum*. In these latter two host-parasite systems, as well as in that of tubers inoculated with *Phoma exigua* var. *foveata* (P.T. Gans, 1978, *Ann. Appl. Biol.* 89, 307), there are increases in chlorogenic acid levels in tuber tissue after inoculation. In each case there is a greater increase in the chlorogenic acid level in the uninoculated control tuber tissue. Tuber discs of major gene resistant varieties show a drop in level of chlorogenic acid on inoculation. One explanation could be that, in the inoculated tissue, carbon is being diverted from biosynthesis of chlorogenic acid to that of insoluble phenolics. A second possibility is that chlorogenic acid is being turned over very rapidly and under the appropriate conditions it could act as a reservoir of caffeoyl units which can be further metabolized to the insoluble phenolics which accumulate in the particular host-parasite system. This second possibility seems highly likely in the potato - *F. solani* interaction where there is a correlation between tuber resistance and the content of both chlorogenic acid and of hydroxycinnamoyl-CoA: quinate hydroxycinnamoyl transferase (CQT). Although this enzyme is usually assumed to be involved in chlorogenic acid biosynthesis, it is reversible and could be, in appropriate circumstances, controlling the metabolism of chlorogenic acid to caffeoyl CoA which would be further metabolized to insoluble phenolics. It will be interesting to examine the activity of this enzyme in tuber tissue infected by several pathogens.

B. FRITIG AND R. MASSALA - AMINOOXYACETATE, A COMPETITIVE INHIBITOR OF PHENYLALANINE AMMONIA-LYASE, MODIFIES THE HYPERSENSITIVE RESISTANCE OF TOBACCO TO TOBACCO MOSAIC VIRUS

An increased production of metabolites of the phenylpropanoid pathway occurs in many plants during infection with various pathogens, particularly in incompatible host-pathogen interactions. Necrotic local lesions appear on tobacco leaves reacting hypersensitively to infection by tobacco mosaic virus (TMV) and virus spread is restricted to the vicinity of these lesions. The resistant cells surrounding the lesions show pronounced increases in activity of enzymes involved in the phenylpropanoid pathway followed by an accumulation of soluble phenols and of lignin and lignin-like polymers. How accumulation of these metabolites, host cell necrosis and virus localization are interrelated is not clear. We have shown previously that increases in activity of phenylpropanoid enzymes are well correlated with the efficiency of the virus localizing mechanism. Furthermore, inhibitors of protein synthesis have been reported to increase lesion size and partially suppress the localizing mechanism. But they also suppress the increases in activity of the phenylpropanoid enzymes since these increases have been shown to arise from activated *de novo* synthesis of the enzymes. Therefore, the use of metabolic inhibitors cannot help to examine whether activation of the phenylpropanoid pathway is involved in virus localization or whether these two processes are unrelated.

We propose here a different biochemical approach using a competitive inhibitor of a key enzyme of the pathway, phenylalanine ammonia-lyase (PAL). Tobacco leaves were supplied with α-aminooxyacetate (AOA), a competitive inhibitor of deamination of phenylalanine *in vitro*. The rationale was that this treatment would specifically reduce the flux of phenylpropanoid derivatives without suppressing the stimulus itself, i.e. without drastically changing primary metabolism, as is observed with inhibitors of transcription and translation. AOA supplied to infected leaves did not change the number of TMV-induced local lesions, but it increased 2- to 4-fold the size of the lesions for the three tobacco-TMV combinations examined. It was effective even when supplied at low concentrations (10 to 50 μM) and up to just 6 hours before appearance of the lesions. It weakened the mechanism of virus localization as evidenced by the presence of virus at the lesion edges. Virus content of these large lesions, however, did not parallel the increases in their areas. AOA reduced the flux of phenylpropanoid derivatives without suppressing activated *de novo* synthesis of the enzymes. These results suggest that the enhanced production of phenylalanine-derived metabolites is directly involved in hypersensitive resistance to viral infection.

A. FUCHS - SOME ASPECTS OF FUNGAL DEGRADATION OF AND SENSITIVITY TO PHYTOALEXINS

In recent years, phytoalexin research has been especially focussed on elicitation, and the chemical nature of the elicitors involved. Much less attention is being paid to regulatory mechanisms governing cellular concentration and activity of phytoalexins. It often seems taken for granted that phytoalexins once being formed 'do what they are supposed to do' without being subject to the normal regulatory mechanisms accepted for other cellular activities. For instance, it often seems hardly to be realized that phytoalexins show metabolic turnover, and thus give rise to degradation products with altered intrinsic antifungal activity or to non-fungitoxic metabolites. Further, and despite remarks to be cautious in using *in vitro* experimental data to explain *in vivo* situations, fungal sensitivity to phytoalexins is often treated as if independent of cellular conditions, such as pH, concentration of glucose and other 'catabolites'. Therefore, instead of representing a '*status quo*', the infected plant cell is probably also in a 'steady state' so far as phytoalexins are concerned.

This notion guides our experimental work on fungal degradation of and sensitivity to isoflavonoid phytoalexins in legumes. One of the aspects being studied concerns the biological implications of the stereochemistry of these phytoalexins. Differences in substitution pattern in the A and D rings of the pterocarpan skeleton in, for instance, inermin, medicarpin, phaseollin and pisatin, do not seem to interfere strongly with ability of a fungus to carry out a given type of degradation. On the other hand, substitution at C-6a as well as absolute configuration are among the key determinants for degradation of pterocarpans by a given fungal species. Low or non-sensitivity can be either an intrinsic feature of the fungal species involved or a consequence of rapid breakdown of the phytoalexin concerned. However, the ability to degrade a phytoalexin does not in itself imply non-sensitivity to the parent compound or even to its metabolites. Whereas *Fusarium oxysporum* f. sp. *lycopersici* and *F. oxysporum* f. sp. *pisi* are equally sensitive to pisatin, *in vivo* the pea pathogen apparently circumvents its toxic action through rapid degradation. Unlike substrate specificity of fungi with respect to phytoalexin breakdown, sensitivity of fungi to phytoalexins does not seem to depend on the absolute configuration; for instance, the (+) and (-) enantiomers of inermin inhibited the growth of *Monilinia fructicola* to the same extent.

In order to screen rapidly fungi for phytoalexin-degrading abilities and to examine the effect of 'environmental' factors, a micromethod has been developed which allows quantitative study in 3-ml aliquots of medium, with the phytoalexins being added in a ^{14}C-labelled form.

A. GERA AND G. LOEBENSTEIN - USE OF PLANT PROTOPLASTS FOR STUDYING RESISTANCE MECHANISMS AGAINST VIRUSES

The approach of using plant protoplast for studying resistance against viruses was hampered by the finding of Japanese researchers that in isolated protoplasts from resistant plants, virus multiplies to the same extent as in protoplasts from susceptible tobacco cultivars. This discrepancy between low virus multiplication in intact resistant tobacco and the high titer reached in protoplasts was found by us to be due to the presence of the plant hormone 2,4-dichlorophenoxyacetic acid (2,4-D) in the protoplast incubation medium. 2,4-D markedly enhanced TMV multiplication in protoplasts from resistant cultivars while in susceptible cultivars 2,4-D reduced virus multiplication.

An inhibitor of virus multiplication was found to be produced by isolated plant protoplasts from a resistant tobacco variety and released into the liquid medium. This substance(s) inhibits virus multiplication both in protoplasts from resistant and susceptible plants when applied 5, 12 and 18 hours after inoculation of the test protoplast. It is not produced in protoplasts from a susceptible cultivar.

R.N. GOODMAN - A ROLE FOR BACTERIAL EXTRACELLULAR POLYSACCHARIDE (EPS) IN DISEASE RESISTANCE ?

Evidence is accumulating that indicates that bacterial extracellular polysaccharide (EPS) is a virulence factor rather than one which impinges directly on resistance. Experiments in my laboratory with both virulent (V) and avirulent (AV) strains of *Erwinia amylovora* revealed that the EPS-deficient AV strain is agglutinated *in vivo* by host-elaborated (by xylem parenchyma) agglutinins.

An agglutinating factor (AF) with specificity for the AV and V strains of *E. amylovora* has been isolated from apple seed, stem and leaf tissue. AF agglutinates the AV strains at much lower concentrations than V isolates. It is plausible that abundant EPS on the surface of V strains of *E. amylovora* masks the mechanism that triggers the release of host elaborated agglutinins. It is also possible that endogenous plant agglutinins are precluded from reaching receptors on the surface of cells of V strains by their mantle of EPS.

The AF we have recently isolated is a small protein with a molecular weight of $\sim$ 13,000 daltons. Its agglutinating activity is specific for *E. amylovora* as other Erwiniae are recalcitrant to AF as are species from other genera including *Pseudomonas, Xanthomonas, Corynebacterium, Agrobacterium* and *Micrococcus*.

The agglutinating activity of AF is 16 times greater against AV than V strains which may reflect, for example, a larger number of exposed receptor sites for AF on the cell surfaces of AV strains. Yet another explanation lies in the discovery that AF is precipitated at very low concentrations by EPS. Hence it seems probable that EPS which is present in copious amounts in infected apple tissue competes with the receptor sites on the bacterial cell surface for AF. Thus large amounts of non-adhering EPS (loose capsule) in the millieux of bacterially invaded host xylem vessels could complex with AF being released from neighbouring host xylem parenchyma cells, permitting, thereby, the continued proliferation of virulent *E. amylovora* cells. In effect EPS *neutralizes* the immune response.

D.S. INGRAM, R.I.S. BRETTELL AND E. THOMAS - PRODUCTION OF TOXIN RESISTANT MAIZE PLANTS THROUGH TISSUE CULTURE

First the potential for application of the following tissue culture procedures in the production of novel disease resistant plants was briefly discussed: (a) the induction and selection of mutants among cultured cells and protoplasts; (b) the regeneration of plants from the somatic cells of species that are genetic mosaics; (c) the modification of the genomes of cultured cells and protoplasts with exogenous nucleic acids; (d) somatic hybridization through protoplast fusion and organelle transfer; (e) the avoidance of incompatibility mechanisms in interspecific crosses through *in vitro* fertilization or embryo culture. It was concluded that tissue culture procedures will have an increasingly important role to play in the breeding of disease resistant plants.

Next, the following series of experiments which has led to the production of maize plants resistant to the T-toxin of *Drechslera maydis* race T was outlined. (a) The demonstration of sensitivity to T-toxin in cells and mitochondria of tissue cultures from Texas male sterile (Tms) maize and of resistance in cells and mitochondria of cultures from normal male fertile (N) maize. (b) The selection of T-toxin resistant sectors (without treatment with mutagens) from tissue cultures initiated from immature embryos of Tms maize and grown on a medium containing a just-lethal concentration of toxin. (c) The demonstration that the novel resistance was located in the mitochondria of the cultured cells. (d) The regeneration of 58 toxin resistant (but male fertile) plants from the selected cultures and of 31 resistant/fertile plants, 19 sensitive/sterile plants and 1 sensitive/fertile plant from control cultures never exposed to toxin. (e) The demonstration that the progeny of the regenerants showed the phenotype of the female parent with respect to toxin resistance and male fertility.

It was concluded that the data are consistent with the heritable changes observed being the result of an altered mitochondrial genome and that the experiments outlined clearly demonstrate that tissue culture methods may be used to produce novel disease resistant plants.

P. LANGCAKE - THE DICHLOROCYCLOPROPANES : ACTIVATION OF HOST RESISTANCE MECHANISMS IN RICE BLAST DISEASE ?

WL 28325 (2,2-dichloro-3,3-dimethyl cyclopropane carboxylic acid) is representative of a group of compounds, the dichlorocyclopropanes, that give systemic protection of rice plants against the rice blast disease caused by *Pyricularia oryzae*. It is highly specific for rice blast, giving very little control of *P. oryzae* on alternative hosts (ryegrass or barley) or of other diseases of rice e.g. sheath blight, *Pellicularia sasakii*. The preferred route of application is through roots.

In studies of the mode of action of this compound, only weak effects against the pathogen were found *in vitro* and these depended on the composition of the growth medium. The compound did not inhibit pigment formation by *P. oryzae* on agar media as does tricyclazole (Tokousbalides and Sisler, *Pestic. Biochem. Physiol.* 8, 26 (1978), nor was there any evidence for its conversion to a fungitoxic metabolite in rice plants. It seemed likely therefore either that the compound effects host plant physiology or that it interferes with the pathogenicity of *P. oryzae*. In either case, this would lead to the expression of a resistant type of response by the host.

Histological studies by light and electron microscopy confirmed that such resistant type responses do occur when treated plants are challenged with *P. oryzae*. Macroscopically, these appeared as small brown hypersensitive flecks. Microscopically, they were seen as gross degeneration of penetrated cells and of cells immediately surrounding the infection site. These cells became filled with brown melanin-like material and degeneration of the fungus was seen at the same time. The enhanced accumulation of melanin-like material in treated, infected plants has been measured quantitatively and it may be associated with the increased levels of peroxidase that are detectable following treatment with WL 28325.

Infection of treated plants also induced the formation of phytoalexin-like compounds that have been identified as the momilactones A and B, diterpenes previously isolated in Japan from rice husks. They have potent activity against *P. oryzae*. The localization, timing and concentrations at which the momilactones are produced in relation to infection of treated plants suggest that they contribute significantly to the inhibition of fungal growth.

The precise mechanism of action of WL 28325 has not been determined as yet. It has not been possible to distinguish critically between effects of the compound on the pathogenic mechanisms of the fungus and effects on resistance mechanisms of the host. However, there is evidence that WL 28325 alters the physiology of the host.

P. LANGCAKE - DISEASE RESISTANCE IN *VITIS* SPP. AND THE PRODUCTION OF STRESS METABOLITES

The cultivated grapevine (*Vitis vinefera* L.) produces a range of stress metabolites, the viniferins, that are derived from resveratrol (3,4',5-trihydroxy stilbene). The most important of these, ε- and α-viniferins, are toxic to a range of fungi *in vitro*, while resveratrol itself is of low fungitoxicity. The general characteristics of the production of these compounds are analogous to those of the production of the isoflavanoids in leguminous plants. Thus, they have similar biosynthetic origins, they can be elicited by a range of biotic (e.g. pathogenic fungi) and abiotic (e.g. UV light) stimuli and they appear to have chemotaxonomic significance. Studies reported here show that the viniferins contribute to the ability of *V. vinifera* and *V. riparia* to resist attack by the pathogens *Botrytis cinerea* and *Plasmopara viticola*.

Commercial cultivars of grapevine are susceptible to downy mildew (*P. viticola*). To study the role of stress metabolites, therefore, comparisons were made with *V. riparia* which is resistant to *P. viticola*. *V. riparia* was found to produce a range of stress metabolites similar to that found in *V. vinifera*. Microscopical studies of infection in leaves showed that resistance is not expressed until after the formation of the first haustorium, and was evident by 18 - 20 hours after inoculation. This coincided with the rapid accumulation of resveratrol, ε- and α-viniferins. These were localised in and immediately around infection sites. It was estimated that ε- and α-viniferins accumulated to concentrations approaching 2 mg/g fresh weight within 4 days of inoculation. It is likely that these concentrations are sufficiently high to inhibit growth of the pathogen. In grapevines, stress metabolite concentrations remained very low.

Stress metabolite production and the resistance of grapevine to *B. cinerea* were also studied. The susceptibility of successive leaves on shoots of *V. vinifera* was compared with the levels of stress metabolites produced during infection. Susceptibility decreased with leaf age. The main stress metabolite, α-viniferin, was produced both in the lesion tissue and in apparently healthy tissue 1 - 2 mm in advance of the lesion. In both regions, there was a highly significant inverse correlation between susceptibility and the logarithm of the concentration of α-viniferin. In the least susceptible leaves, levels of α-viniferin approached 0.6 mg/g fresh weight, several times in excess of the ED_{90} (155 μg/ml) against *B. cinerea in vitro*. In the most susceptible leaves, concentrations

(< 10 μg/g) of α-viniferin were insufficient to inhibit growth of the pathogen.

Leaves of *V. riparia* are considerably more resistant to *B. cinerea* regardless of age. Stress metabolite concentrations in lesions caused by *B. cinerea* on *V. riparia* were much greater than those in lesions on *V. vinifera*. In *V. riparia*, ε-viniferin was the predominant stress metabolite although α-viniferin was also detected.

G. LAZAROVITS, P. STÖSSEL AND E.W.B. WARD - SPECIFICITY AND GLYCEOLLIN PRODUCTION IN THE REACTION OF SOYBEAN HYPOCOTYLS TO *PHYTOPHTHORA MEGASPERMA* VAR. *SOJAE*

Six day old soybean (*Glycine max* L.) cultivars were tested for resistance or susceptibility to races of *Phytophthora megasperma* var. *sojae* (Pms) by placing droplets of zoospore suspensions on hypocotyls of etiolated seedlings. Within 24 hours resistant hypersensitive interactions produced brown necrotic lesions, whereas susceptible tissues were soft, water soaked and rotted. Concentrations of the phytoalexin glyceollin were consistently higher in resistant than in susceptible reactions.

Hypocotyls increased in resistance to both compatible and incompatible races from the top (youngest part) to the bottom (oldest part). Typical susceptibility to compatible races was limited to the youngest part of the hypocotyl; at the middle and bottom necrosis and glyceollin were similar to those found in incompatible interactions. The interaction with incompatible races became even more incompatible towards the bottom where there was a reduction in both necrosis and glyceollin concentration.

This change in host response corresponded to a difference in tissue age of 1 - 2 days. With the cultivar Altona and Pms races 4 (incompatible) and 6 (compatible) appressoria formation or frequency of penetrations were similar regardless of race or inoculation site.

Incompatibility in all race and site combinations was overcome temporarily by heat treatment of the hypocotyl or by wiping the hypocotyl surface with organic solvents prior to inoculation. Duration of the heat treatment needed to overcome resistance was proportional to the incompatibility of the race-site combination, suggesting that differences were quantitative. Partial elimination of resistance was accompanied frequently by increases in necrosis and glyceollin production, presumably due to more extensive tissue colonization. Glyceollin production was closely correlated with necrosis. Tissue at the bottom of the hypocotyl was more sensitive to glucan elicitor and to localized freezing injury as measured by glyceollin production and necrosis.

The results suggest that race-specific and age-related resistance are basically similar, that differences between resistance and susceptibility are quantitative and that the balance between the two generally favours resistance. Only in the immature tissues at the top of the hypocotyl is the pathogen able to tip the balance toward compatibility.

M. LEGRAND, P. GEOFFROY AND J. COLLENDAVELLOO - ENZYME ACTIVITIES AND HYPERSENSITIVE RESISTANCE OF TOBACCO TO TMV; DENSITY LABELLING STUDIES OF THE MECHANISM OF REGULATION

Infection of plants by viruses may lead to hypersensitive reactions characterized by the necrotic lesions and localization of the virus. Phenylpropanoid metabolism is strongly activated in hypersensitive tobacco leaves infected by tobacco mosaic virus (TMV) and it has been shown that the activities of enzymes involved in this pathway increase during infection. In this study the molecular mechanism of the regulation of the activities of phenylalanine ammonia-lyase (PAL) and o-diphenol-O-methyltransferases (OMTs) after infection has been investigated by a density labelling method with deuterium supplied as heavy water (D_2O). Labelled and unlabelled enzymes were distinguished by their buoyant densities on CsCl, RbCl or KBr. In contrast to radiochemical and immunological methods, density labelling does not require extensive purification of the enzymes since these are detected on the gradient by their activity. Comparison of the activity profiles enables us to estimate both the half-life of the active enzyme (increase in bandwidth at half peak height) and its specific labelling (shift in density from the native enzyme).

An increase in enzyme activity may arise from different mechanisms of regulation (1) an increased rate of synthesis of the active enzyme (2) a decreased rate of degradation or (3) an activation of an inactive form of the enzyme. An increased rate of synthesis is the only mechanism leading to higher specific labelling of the stimulated enzyme compared with that of the unstimulated enzyme. This was observed for PAL and OMT activities in infected tobacco leaves together with no significant changes in the size of the pool of amino acids. Therefore, it was concluded that there was a higher rate of *de novo* synthesis of these enzymes during infection. Furthermore, it appeared that OMT activity of tobacco arises from three enzymes which have different substrate specificities and are increased differently after infection. The major enzyme of healthy leaves methylates very efficiently phenylpropanoid type substrates to lignin precursors. The two other enzymes which preferentially increase during infection have a broader specificity. They methylate efficiently some unusual diphenols produced by infection to methoxylated products that readily polymerise. Thus, the increase of enzyme activities and the changes in the pattern of methylating enzymes might be related to an enhanced synthesis of lignin or lignin-like polymers involved in virus localization.

J.W. MANSFIELD AND T.M. O'NEILL - MECHANISMS OF RESISTANCE OF NARCISSUS BULBS TO *BOTRYTIS* SPP.

Narcissus smoulder disease is caused by *Botrytis narcissicola*. Other species of *Botrytis* typically are unable to infect the daffodil plant in the field, although *B. cinerea* may colonize senescing tissues. Conidia of *Botrytis* spp., including *B. narcissicola*, inoculated on to fleshy bulb scales or leaves in droplets of sterile water typically fail to colonize the tissues producing at most, only limited lesions within which growth of invading hyphae is restricted. The addition to inoculum droplets of exogenous nutrients, for example pollen grains, or the use of mycelial inocula allows *B. narcissicola* but not other species to form spreading lesions.

Although germination of conidia in water on bulb scales usually approached 100 per cent, many failed to produce infection hyphae. Various types of reaction material (papillae) fluorescing yellow under u.v. radiation formed within living cells at sites where penetration appeared incomplete. The most common reactions were localized granular deposits and thickened lateral walls. Histochemical tests showed that deposits, thickened walls and often apparently normal walls around reaction sites were lignified. Incorporation of ethanol insoluble phenolic polymers into walls at sites of attempted penetration was confirmed by microautoradiography using ^{14}C labelled cinnamate. Lignification was also observed in response to inoculation with mycelium of non-pathogenic species but not in spreading lesions produced by *B. narcissicola*.

No pre-formed fungitoxic compounds were detected in bulb scales but phytoalexins accumulated in tissue bearing limited lesions. Inhibitors were detected in microscope slide bioassays with *Botrytis* spp. and also by *Cladosporium herbarum* thin layer chromatography (t.l.c.) plate assay. Milligram quantities of the phytoalexins were isolated from diethyl ether extracts of *B. cinerea* infected tissue by combinations of gel filtration (Sephadex LH20), t.l.c. and high performance liquid chromatography. The phytoalexins were separated into two groups by gel filtration. Members of the first group to elute from the column lacked characteristic u.v. absorbance and were not identified. The second group were phenolics and the three principal inhibitors were identified as 7-hydroxyflavan, 7,4'-dihydroxyflavan and 7,4'-dihydroxy-8-methylflavan. Flavans unsubstituted in the pyran ring are rare natural products and have not previously been reported to possess fungitoxic activity. The fungitoxicity of the flavans, unidentified phytoalexins and crude extracts containing the inhibitors were tested against sporelings of *B. cinerea*, *B. fabae*, *B. narcissicola* and *B. tulipae*. Isolates of *B. narcissicola* were as sensitive as non-pathogens to the inhibitors. Changes in phytoalexin concentrations occurring during the formation of limited and spreading lesions were monitored using t.l.c. plate bioassays. The inhibitors accumulated rapidly in tissue bearing limited lesions but were present only in trace amounts in and around spreading lesions. Successful colonization by *B. narcissicola* appeared to depend on the suppression of active mechanisms of resistance. The mechanisms by which suppression might be achieved and structure/activity relationships among the flavanoids were discussed.

H. MARAITE - INDUCED RESISTANCE IN MUSKMELON WILT : FACTS AND LIMITATIONS

Various experiments were designed to clarify if decrease of wilt in muskmelons induced by simultaneous or prior inoculation with less virulent strains of the pathogen or incompatible races of *F. oxysporum* is due to competition with the pathogen or to the induction of a resistance mechanisms in the plant.

Wilt development was delayed when the roots of muskmelon seedlings were dipped in suspensions (10^8/ml) of conidia of a weakly virulent mutant of the pathogen *Fusarium oxysporum* f. sp. *melonis*, of *F. oxysporum* f. sp. *lycopersici*, or of a saprophytic strain of *F. oxysporum* (inducers), 42 hours before injection of a suspension of spores of a virulent strain of the pathogen (challenger) into the soil. Wilt decreased with increase of the interval between induction and challenge and with decrease of the concentration of the inoculum of the challenger below 10^7 conidia/ml. A similar delay in wilt was observed when suspensions of conidia of the inducer were injected into the soil or the sand, or pulverized on the roots of plants grown under near aseptic conditions on rolls of germination paper. The efficacy of the preinoculation treatment was improved by a booster treatment. Inhibition of the inducer by a benomyl treatment, followed by the inoculation with a benomyl resistant strain of the pathogen, did not affect the decrease of wilt. When roots and hypocotyls of seedlings were split before transplanting into two separate pots and the roots of one side were treated with the inducer, a delay in colonization by the challenge inoculated at the other side was observed compared to untreated plants. This delay in disease development was, however, much less important than the one observed when inducer and callenger were inoculated at the same side. Cell free extracts of the saprophytic strain induced in soil cultures the same decrease of wilt as did living conidia. They were, however, ineffective when applied to roots of plants grown on germination paper or to stem punctures, 24 hours before inoculation of the same site by the pathogen. In the latter system complete protection was observed with living conidia.

It is concluded that induced resistance is probably a major

component in the observed decrease of wilt. However, no fungitoxic compounds were detected in ethanol or water extracts of treated roots. The decrease is only partially systemic and transitory, probably because of the continuous formation of unprotected roots. This is a major limitation for the practical use of the induced protection against Fusarium wilt.

G.P. MARTELLI - ULTRASTRUCTURAL ASPECTS OF POSSIBLE DEFENCE REACTIONS IN VIRUS-INFECTED PLANT CELLS

Among the hypotheses put forward to explain why the escape of virus from hypersensitive local lesions is impeded, the formation of physical barriers blocking intercellular connections in advance of infection has been advocated.

Cylindrical (or pinwheel) inclusions of potyviruses which, according to some authors, may prevent virus translocation by plugging plasmodesta, are coded by the viral genome and therefore constitute very unlikely expressions of the host defence mechanism.

Structural modifications of the cell wall and the cell wall-plasma membrane interface such as paramural bodies, callose deposits and cell wall outgrowths, also may be possible barriers to virus spread.

Paramural bodies are usually associated with callose deposits and cell wall protrusions in the building of which they are thought to take part. These structures occur in localized and systemic infections induced by members of many but not all taxonomic groups of plant viruses. An involvement in virus confinement is difficult to reconcile with their rare occurrence in hypersensitive or chlorotic kesions elicited by many viruses, as opposed to their relatively high frequency in systemically invaded tissues where virus movement is not impeded.

Deposits of callose and other wall-related substances (glycoproteins, lignin, suberin) enjoy a widespread reputation as agents restricting virus movement. However, in non-hypersensitive lesions accumulations of callose seem to occur less frequently than in hypersensitive lesions, irrespective of whether the inducing virus becomes systemic or not. Furthermore, callose is also found in hosts which become systemically infected and accumulates prominently and in hosts where systemic spread of the virus is not preceded by local lesion formation. To constitute an effective physical barrier against virus spread, callose should not only block in advance of infection all plasmodesmata of cells adjoining infected areas, but should counteract possible secondary formation of plasmodesmata that could allow virus to escape. Even so, a successful barrier of callose, or any other wall-related substance, does not explain the establishment of the zone resistant to super-infection surrounding the lesion.

Similar concepts are applicable to cell wall outgrowths i.e. the finger-like projections of the wall extending into the cytoplasm, associated primarily with infection by members of some virus groups having isometric particles. Their formation may be stimulated by structures (individual virions, virus-containing tubules) with a diameter above 45 nm when attempting to traverse plasmodesmata. Cell wall protrusions may indeed represent expressions of a localized reaction to viral infection but they appear unable to confine efficiently the inducing viruses which are very successful systemic invaders of the plants they infect.

A.J. MAULE - PROTOPLAST TECHNOLOGY WITH PARTICULAR REFERENCE TO CUCUMBER

Recent developments in the whole field of plant pathology indicates some reluctance on the part of research workers to use protoplast technology to aid their investigations. This seems to stem from the belief that the methodology is practically too complex so this talk was designed to illuminate some key aspects and remove some of the mystique from the subject. Conditions for the isolation and treatment of protoplasts will vary from species to species so only general points can be made.

Optimum conditions for growth of plants will determine the ease with which tissues can be manipulated, the time taken for enzymic digestion and the stability of the resulting protoplast suspension, and will be species and tissue specific. Under appropriate conditions almost any tissue will yield protoplasts.

For tissue digestion a wide range of commercial pectinase and cellulase enzymes can be used in sequence or together. The resulting protoplast population will be less than 100 per cent viable. Separation of viable protoplasts from other material can be achieved efficiently on discontinuous carbohydrate and Ficoll gradients. Assessment of protoplast viability before and after treatments is important in all protoplast experiments and a survey of different methods for cucumber protoplasts has led us to use routinely the exclusion dye phenosafranine which gives the best correlation with a vital metabolic function.

Protoplasts can be used experimentally to observe the action of substances on the plasma membrane (e.g. phytoalexins or their elicitors), or in the cytoplasm. Virus infection of protoplasts has shown us that uptake into the cytoplasm can be stimulated by polycations and polyethylene glycol and, more recently, by liposomes. This last technique also has the potential for introducing many small and high molecular weight compounds quantitatively into protoplasts.

Effects may be seen in individual protoplasts as changes in viability or as the formation of specific cytopathic structures, or in populations, as alterations in particular enzymes or substrates. It is important, however, that the qualitative character of the effect should not be influenced by the means of assessment, and to this end the effect should be measured in as many ways as are practically possible. The nature of protoplast isolations is such that they tend to show variation between batches and it is therefore important that each experiment is designed to be complete, requiring no quantitative reference to another.

A.J. MAULE, M.I. BOULTON AND K.R. WOOD - USE OF PROTOPLASTS IN VIRUS RESISTANCE STUDIES

The use of protoplasts has allowed mechanisms of resistance to virus disease to be divided into those dependent upon cell to cell interactions and those which are expressed in single cells. Available evidence based mainly on the use of metabolic inhibitors, suggests that the term 'active defence' is applicable to the former but not the latter.

Resistance requiring cell to cell contact for its expression is seen in many host/virus combinations. Hence, protoplasts from resistant plants show no restriction to virus multiplication and no symptoms. In contract, resistance which functions in single cells appears to be comparatively uncommon.

Cucumber mosaic virus (CMV) infections of the resistant cucumber cultivar, China (Kyoto) are essentially asymptomatic, with systemic spread of low levels of replicated virus occurring. Previous published work has shown that this resistance is partially sensitive to Actinomycin D, and was interpreted as an active defence phenomenon.

Inoculation of cucumber protoplasts with CMV nucleoprotein in comparative experiments using China (Kyoto) and a susceptible cultivar Ashley, showed that the expression of resistance occurs in single cells. The possibility that this observation resulted from alterations in the protoplasts arising from the artificial procedures used in protoplast isolation and inoculation, was discounted after looking for differences in several physical and metabolic criteria between the two varieties. No differences could be detected in protoplast surface charge, cell viability, total RNA synthesis and the ability to replicate two other RNA viruses.

Following protoplast infection with CMV-RNA the resistant response is still expressed, indicating that the block to efficient virus replication in China (Kyoto) lay beyond uncoating of the virus, i.e. at some stage in transcription or translation.

CMV has a multipartite genome of which the three largest RNA species are essential for re-infection of the host. The proportionate synthesis of these key RNA molecules in resistant and susceptible protoplasts has been under investigation. Treatment of protoplasts with UV light and Actinomycin D to suppress host RNA synthesis, ^3H-labelling of virus RNA and their subsequent separation by PAGE and detection by fluorography showed the proportions of the three RNAs to be different between the two cultivars. Whether this difference is a resistance-specific phenomenon or just normal variation between hosts, is currently under study.

The insensitivity of resistance in China (Kyoto) protoplasts to Actinomycin D is in contrast to the reported situation in whole plants and may indicate multiple resistance to CMV in this cucumber variety with active and passive mechanisms operating in whole plants and only passive mechanisms in single cells. Resistance in China (Kyoto) is believed to depend on more than one gene.

J.D. PAXTON - DEGRADATION OF GLYCEOLLIN BY *PHYTOPHTHORA MEGASPERMA* VAR. *SOJAE*

Degradation of phytoalexins by plants and their pathogens is now under active investigation. Phytoalexins and many other "secondary" products of plant metabolism are now recognized as being rapidly metabolized and not as simply dead end products. The ability to degrade the host phytoalexin has been shown for several pathogens of plants and this appears to be a prerequisite for virulence in some pathogens.

The concentration of the phytoalexin glyceollin increases for 2 days in soybean plants inoculated with various pathogens and then decreases. Yoshikawa and co-workers have shown that soybean plants metabolized glyceollin but the mycelium of *Phytophthora megasperma* var. *sojae* presumably was not able to metabolize glyceollin within 24 hours. In the susceptible soybean plant glyceollin concentrations exceed levels that are inhibitory to the growth of *P. megasperma* var. *sojae in vitro*. The attack by *P. megasperma* var. *sojae* of young soybean plants represents one of the few host-parasite systems that lacks an active defence mechanism in the susceptible plant. Once infection has started it generally proceeds until the entire plant is rotted.

This indicates that in Phytophthora root rot of soybean glyceollin is rapidly metabolized and that *P. megasperma* var. *sojae* probably has some mechanism for detoxifying it. With this information in mind I studied the degradation of glyceollin by *P. megasperma* var. *sojae* in soybean broth.

An isolate of *P. megasperma* var. *sojae* race 1 was grown on soybean broth (3 gm of soybeans/100 ml of water) for 5 days. A small piece of this mycelium was transferred to a sterile tube containing 1 ml of soybean broth containing 0.25, 50 or 100 μg of glyceollin (The ED_{50} of glyceollin against *P. megasperma* var. *sojae* is 50 μg/ml). After seven days the broth was extracted with 1 ml of chloroform and absorbance at 285 nm read on a spectrophotometer. A clean spectrum for glyceollin was obtained and in the tubes containing 100 μg/ml of glyceollin 99 per cent of the glyceollin was recovered (the fungus did not grow). In the tubes containing 0, 25 or 50 μg/ml progressively less growth occurred but after 7 days no glyceollin was found in the chloroform extracts. These results were confirmed by thin layer chromatography on silica gel plates in hexane : ethyl acetate : methanol (60 : 40 : 2, v/v/v) and an additional spot at a lower R_f was consistently seen in tubes where glyceollin disappeared.

This suggests that *P. megasperma* var. *sojae* race 1 can metabolize glyceollin to a more polar product.

J.D. PAXTON - PHYTOALEXINS (Reproduced with permission after slight amendment from *Plant Disease* (1980) 64, 734)

At the NATO Advanced Study Institute on "Active Defence Mechanisms in Plants" a group of plant pathologists assembled to revise the definition of the term phytoalexin. A similar attempt at the 1978 International Congress of Plant Pathology in München, Germany failed and it was considered important to create a new working definition of phytoalexin reflecting some of the advances in our field since Müller and Börger first proposed the term phytoalexin (1).

In attendance at this meeting were Drs. P. Albersheim, J. Bailey, D. Bateman, J. Callow, D. Clarke, G. Defago, B. Deverall, D. Elgersma, M. Esquerre-Tugaye, J. Friend, B. Fritig, A. Fuchs, M. Heath, R. Heitefuss, D. Ingram, N. Keen, J. Kuć, P. Langcake, G. Lazarovits, J. Mansfield, J. Pacton, W. Rathmell, F. Schönbeck, P. de Wit, and R. Wood, representing a wide range of phytoalexin researchers. The working definition arrived at by consensus is : Phytoalexins are low molecular weight, antimicrobial compounds that are both synthesized by and accumulated in plants after explsure to microorganisms.

This definition may be contrasted with J. Hardy's direct translation of Müller and Börger's (1) definition of a phytoalexin.

1. The premature death of the parasite on the tubers of the resistant W varieties is *not* due to any toxic "principle" already present in the tuber before the infection nor to the absence of any substance necessary for the normal development of the fungus, but to *a change in the state of the host cells which come into contact with the parasite.* This change of state results in a "paralysis" or the premature death of the fungus (cf. Meyer (29); Müller, Meyer and Klinkowski

(35)). The principle inhibiting the development of the fungus is formed or activated only in the course of this change of state which we have termed the "defensive reaction".

2. The defensive reaction is linked with the living state of the host cell. This does not mean, however, that a tissue which is parasitized by a virulent *Phytophthora* strain, but is still alive, may not at the same time be capable of responding to the attack of an avirulent strain with the changes of state characteristic of the defensive reaction.

3. The inhibiting principle must be of a material nature. It is formed or activated in the reacting host cell and may be regarded as the end product of a "necrobiosis" released by the parasite.

4. This not yet isolated and therefore still hypothetical "defence substance" is "non-specific". It has an inhibiting action on other parasitic fungi of the potato tuber as well as *Phytophthora infestans*. Saprophytic fungi are also inhibited in their development by this substance. However, the various parasitic species differ in their sensitivity towards this "phytoalexin".

5. The decisive factor for the fate of the parasite and hence also for the "immune" behaviour of the host is discovered only in the sensitivity of the host cells to certain material influences emanating from the *Phytophthora* fungus: the greater the sensitivity, the higher the resistance. The reaction product is accordingly not specific, but only the genotypically determined *readiness of the host cell to react*, which is manifested in the speed with which the hypothetical defensive substance is formed.

6. The defensive reaction is confined to the tissue colonized by the fungus and its immediate neighbourhood. There is no immunization embracing the whole individual.

7. In the resistant varieties we find an "immunization" of the portions of tissue invaded by the parasite, in the susceptible varieties the opposite is the case: the host cells invaded by a virulent *Phytophthora* strain, but still alive, also become "sensitive" to fungi which are incapable of attacking an intact potato tuber after association with the parasite for some time. Here, too we note gradual differences in the capacity of the individual fungus to colonize the tissue attacked by *phytophthora*.

8. What is inherited is only the capacity to "acquire" the resistance at the place of infection and only here, but not the resistant "state" in itself. This state must first be "acquired", and this happens only after the plant has come into contact with the pathogenic agent. This serves to release the "mechanism" which transforms the portions of tissue attacked by the parasite from the "indifferent"

to the "resistant" state.

Müller in 1956 revised his definition of phytoalexin to "Phytoalexins are defined as antibiotics which are the results of an interaction of two different metabolic systems, the host and the parasite, and which inhibit the growth of microorganisms pathogenic to plants". (2).

It is hoped that the revised definition presented here will help clarify current interpretation of what constitutes a phytoalexin. It is also hoped that continuing research in this area will clarify the role of phytoalexins in plant disease resistance.

1) Müller, K.O. and Börger, H. (1940). Experimentelle Untersuchungen über die Phytophthora-Resistenz der Kartoffel. *Arb. Biol. Reichsanstalt. Lan-u. Forstwirtsch., Berlin,* 23, 189 - 231.

2) Müller, K.O. (1956). Einige einfache Versuche zum Nachweis von Phytoalexinen. *Phytopathol. Z.* 27, 237 - 254.

W.S. PIERPOINT - SYNTHESIS OF PATHOGENESIS-RELATED (PR) PROTEINS IN DETACHED AND UNDETACHED LEAVES OF TOBACCO

The infection of tobacco leaves with viruses that give localised, necrotic lesions, also induces the synthesis of a family of proteins which were previously present, at the most, in very small amounts. These appear not only in infected leaves, but also in adjacent uninfected leaves, suggesting that they have a role in the resistance to further virus infection that is induced in these leaves. We have followed the appearance of these proteins in infected and in uninfected leaves of *Nicotiana tabacum* (var. Xanthi nc), and also their synthesis in detached leaves which were subjected to chemical stimuli.

The proteins were recognised by their mobility during PAGE. Leaves were extracted at a low pH, and the proteins, after separation by exclusion chromatography, were digested with trypsin and chymotrypsin, and electrophoresed in 15 per cent acrylamide gels. This procedure selectively extracts the PR-proteins, and allows their separation from normal plant proteins with which they may be confused. Spectroscopic scanning of the stained gells allows a semi-quantitative estimation of the proteins, especially PR-1a and PR-1b.

The proteins first become detectable in tobacco leaves, 3 days after inoculation with tobacco mosaic virus, and increase until by the tenth day they comprise about 0.1 mg/gm fresh weight, or about 1 per cent of the total leaf protein. The time-course of their appearance differs from that of trypsin-inhibitors which are also formed after inoculation. PR-proteins first appeared in uninoculated leaves 4 days after their appearance in adjacent inoculated leaves, and increased to about a fifth of the amount in inoculated leaves. The proteins also developed in uninoculated leaves of plants from which inoculated leaves had been removed, provided that removal occurred later than 4 days after inoculation.

The possibility that a chemical message moves from inoculated to uninoculated leaves to initiate the synthesis of PR-proteins was investigated in preliminary experiments using detached leaves. Such leaves produce large amounts of PR-proteins under the influence of compounds such as acetyl salicylic acid, which are known to induce their synthesis in intact plants. Extracts made from infected leaves according to the protocol that extracts PIIF, the wound-hormone described by Ryan, were ineffective. Concentrated extracts of the low molecular-weight compounds from both healthy and TMV-infected tobacco leaves, although appreciably phytotoxic, induce PR-proteins in detached leaves. However, this may not be the effect of a specific chemical in the extracts, but the result of general stress that the extracts cause : mannitol solutions, which probably stress the leaves osmotically also induce the synthesis of some PR-proteins.

Mr N.P. Robinson and Dr J.F. Antoniw collaborated in this work.

W.G. RATHMELL - ACTIVE DEFENCE MECHANISMS OF PLANTS IN RELATION TO THE PROTECTION OF CROPS FROM PATHOGENS WITH CHEMICALS

The main problems associated with the use of chemicals to control plant diseases are that control may be insufficient, that current application methods are inefficient in bringing the chemical into contact with the pathogen, that chemicals may have undesirable effects upon non-target organisms and that strains of pathogens resistant to chemicals may arise. Two areas of study of active defence mechanisms in plants suggest types of molecule, namely phytoalexins and elicitors, that might be put to practical use for disease control and overcome some of these problems.

Results from our laboratory and elsewhere show however that, for example, pterocarpanoid phytoalexins have no antifungal activity when sprayed at application rates below 100 $\mu g\ ml^{-1}$ in bioassays *in vivo* against a wide range of diseases. Synthetic fungicides are active at application rates much lower than this (1 - 50 $\mu g\ ml^{-1}$) in equivalent tests. There is at present no reason to suppose that toxophores based upon pterocarpanoid phytoalexins can be used to overcome any of the problems encountered in chemical disease control.

Elicitors have been found effective and highly active in reducing the susceptibility of plants to fungi in laboratory experiments. There are, however, no reports of successful use in the field. Many studies suggest that the use of elicitors would divert plant metabolism from yield production to damaging or largely gratuitous active defence, although there may be exceptions to this.

Data generated in our laboratory suggest that synthetic compounds that can induce the resistance of plants to pathogens are surprisingly common. A highly significant correlation has been observed between the ability of a compound to cause some form of phytotoxicity when applied to soil in which rice or wheat seedlings are growing and to induce resistance to a subsequent inoculation of the plant's foliage with *Pyricularia oryzae* or *Puccinia recondita*. Between two and five per cent of compounds tested caused phytotoxicity and induced resistance in the treated plants and the linkage between the two effects was highly significant statistically. Any substance which causes a change in the metabolism of a plant to which it is applied, whether or not there is visible damage, may interfere with the plant's susceptibility to a pathogen. Interpreted in this way, our results obviously have important implications for experiments in which substances of plant or pathogen origin, such as elicitors, are studied for their ability to induce resistance.

Despite the fact that large numbers of chemical compounds may be capable of inducing the resistance of plants to pathogens, there are no certain examples of chemicals with this mode of action finding practical use. The potential advantages of such 'pathogenicity agents' over conventional fungicides or bactericides in overcoming the problems listed at the outset remain theoretical.

P. RICCI - INDUCED RESISTANCE IN CARNATION INFECTED BY *PHYTOPHTHORA PARASITICA*

Non-rooted carnation cuttings could be made resistant to a further inoculation with *P. parasitica* by dipping them for some 20 hours in water containing an elicitor from the same fungus.

The elicitor was first found in zoospore germination fluids and shown to be probably a carbohydrate of molecular weight over 10,000. The resistance eliciting activity was then detected in culture filtrates and in the soluble fraction released by heat from purified cell walls of the fungus.

Induced resistance was localized to a few mm at the treated base of the cutting. It was associated with the accumulation of two nitrogen occurring phytoalexins (the structures not yet completely known), and of fluorescent, phenolamid-like compounds, and a modification of cell-walls in a continuous layer a few cells deep benath the surface of the base.

Carnation callus cultures were grown on a modified Murashige and Skoog medium containg 1 mg NAA and 0.8 mg N6-benzyladenine per litre. When half-dipped in the "cell-wall elicitor" for 24 hours, callus fragments also became resistant; placed in a zoospore suspension (10^5 zoospores for each fragment), fragments remained uninfected while water-treated controls were fully colonized by hyphae with many sporangia. By itself, the elicitor immediately or after incubation with callus, did not cause lysis or inhibit germination of zoospores. Therefore the absence of fungal development after elicitation was due to resistance induced in the callus tissue. This resistance was accompanied by accumulation of the same phytoalexins and similar modifications of cell-walls.

As with cuttings, induced resistance expressed itself quantitatively. By diluting the elicitor or by increasing the concentration of the challenge inoculum the response could be shifted from total protection to more or less delayed invasion. But invasion itself was either complete or did not occur; serial sections of cuttings or callus in which resistance had been induced never showed the fungus inhibited within a restricted lesion.

After several months of sub-culture on the same medium, the callus tissue has shown a decrease in its response to the cell-wall elicitor. We have already demonstrated large variations in the susceptibility and response to elicitors of cuttings depending on the season at which they are taken from the mother plant, the most susceptible being the least responsive.

Thus it appears that the intensity of the multiple response to the fungal elicitor leading to resistance is regulated by the physiological state of the host tissue. This may be one way through which environmental factors predispose the plant to the disease.

R. ROHRINGER, J. CHONG, W.K. KIM AND D.E. HARDER - CYTOCHEMISTRY OF THE HOST/PARASITE INTERFACE AND THE USE OF GOLD MARKERS IN CEREAL/RUST INTERACTIONS

Previous histological work (1) indicated that incompatibility between wheat (*Triticum aestivum* L.) containing the *Sr6* gene for resistance and stem rust (*Puccinia graminis* f. sp. *tritici*) containing the *P6* gene for avirulence may be determined at the interface between host cells and newly formed haustoria. It is hypothesized that a structural component of the invaginated host plasmalemma interacts with a structural component of the haustorial neck wall, representing the product of the gene for resistance and the product of the gene for avirulence, respectively. We examined with cytochemical methods (2) and electron microscopy the composition of the structures involved in this interaction and in the oat (*Avena sativa* L.) crown rust (*Puccinia coronata* f. sp. *avenae*) system, and we propose an approach with which this hypothesis may be tested experimentally.

Our results indicate that the neck wall and body wall of haustoria contains protein, lipid, Con A receptors, and Thièry-positive substances, possibly β-(1→6) glucan. Wheat germ lectin (WGL) receptors, probably chitin, were found in hyphal walls and in the body wall of old haustoria, but not in that of young haustoria or in haustorial neck walls. The extra-haustorial matrix contained protein, Thièry-positive substances, Con A receptors and probably cellulose, but no lipid or WGL receptors. The neck bands of the haustoria of *P. coronata* f. sp. *avenae* contained Si (α-band) or Fe and P (β-band). The single neck band present in *P. graminis* f. sp. *tritici* resembled the β-band of *P. coronata* f. sp. *avenae*.

The method (3) employing gold markers for visualizing Con A and WGL receptors may be useful in the search for the putative products of the genes for resistance and avirulence. If a product can be extracted from hyphal walls and conjugated with gold, and if it binds to the product of the gene for resistance, it may be possible to detect it by observing binding of the gold marker to the host plasmamembrane. Alternatively, the inverse approach may be used by isolating constituents of the host pasma membrane containing the putative product of the gene for resistance, conjugating these extractives with gold, and observing if the gold marker has an affinity for the haustorial neck wall. In either case, the appropriate combinations of host and parasite must be used to determine if binding, when observed, is gene-specific.

1. SAMBORSKI, D.J., KIM, W.K., ROHRINGER, R., HOWES, H.K. & BAKER, R.J. (1977). Histological studies on host-cell necrosis conditioned by the *Sr6* gene for resistance in wheat to stem rust. *Can. J. Bot.* 55, 1445 - 1452.

2. CHONG, J. & HARDER, D.E.(1980). Ultrastructure of haustorium development in *Puccinia coronata avenae*. I. Cytochemistry and electron probe X-ray analysis of the haustorial neck ring. *Can. J. Bot.* 58, 2496 - 2505.

3. HORISBERGER, M. & ROSSET, J. (1977). Colloidal gold, a useful marker for transmission and scanning electron microscopy. *J. Histochem. Cytochem.* 25, 295 - 305.

H. ROSS - DEFENCE AGAINST VIRUSES IN THE POTATO

More than 20 viruses attack the potato but only Potato Virus A (PVA), PVM, PVS, PVX, PVY and PLRV are sufficiently destructive to justify breeding for resistance. The potato and related wild Solanum have developed non-necrotic and necrotic types of resistance. In the non-necrotic type (quantitative or infection resistance) the virus is confined to the site of entry or to the inoculated leaf or establishes itself only in a few plants or remains at low concentrations. This resistance, controlled mainly by polygenes, occurs generally and for all potato viruses. However, in potato it has not progressed to the state in which the virus remains localized or there is zero reaction. The highest degree of infection resistance is found in wild species. Necrotic reactions can be ranked according to their severity, from systemic hypersensitivity to localized hypersensitivity and extreme resistance. The necrotic reactions are governed by major genes but minor genes can vary the reaction in a manner that breeders often regret. There are useful major genes for necrotic response to all potato viruses except PLRV. A weak type of necrotic reaction is very common; infection starts with necroses but the virus moves out and a systemic mosaic develops. When the reaction is stronger the mosaic is mixed with necrotic streaks and finally the plant may die. This reaction can be useful when combined with a high degree of infection resistance. Then, as in cultivars Apta and Carla, infected plants disappear or tubers do not sprout, leaving the potato field clean. Local hypersensitivity, as local lesions, is a more intense type of hypersensitivity. The virus does not spread systemically and the plants remain virutally free from virus. It is not the mechanical barrier of dead cells around the lesions that prevents the virus from becoming systenic. Local lesion formation is connected with induced systemic resistance. Highly expressive major genes in potato cultivars and primitive cultivars localize the viruses PVA, PVS, PVX and PVY to the point of entry. This means that in crosses resistant x susceptible 50 per cent of seedlings carry the gene and will never show infection in the field. If in these cases inoculation is by graft on to a virus source, the hypersensitive reaction is expressed as a top necrosis. If inoculation occurs in the field by contact with diseased leaves or contaminated machines, virus lesions remain invisible and the plants are virtually free from infection. However that is not all. There is another strong defence mechanism called extreme resistance. An alternative term "immunity" is wrong in my opinion. Plants with the gene Rx for extreme resistance to PVX show no sign of infection when sap or graft inoculated. Grafted plants with the gene Ry for extreme resistance to both PVA and PVY show small star-like white flecks while lower leaves show large necrotic patches. The main shoot ceases to grow and side buds grow out to symptomless branches. Virus cannot usually be recovered. Although extreme resistance is related to localized hypersensitivity it must be separated because the reactions are governed by alleles of the same locus. The allele for extreme resistance is dominant to that for localized hypersensitivity. The zero reactions to PVY and the gene Ry were detected in *S. stoloniferum* in the Max-Planck-Institute. Today six cultivars carrying Ry have been bred in the German Federal Republic, one in the Netherlands and one in Poland. The gene Ry is non-strain-specific. Because of the interest in non-strain-specific genes among phytopathologists and breeders it should be stated that 11 non-strain-specific genes for virus resistance in potato are known against 8 strain-specific genes. Such a high number of non-strain-specific genes is not uncommon among genes for virus resistance. This may be due to the smallness of the virus genome with its restricted mutation potential compared with that of other pathogens, e.g. fungi.

The most urgent problem in breeding for virus resistance in potato is increasing the hitherto unsatisfactory level of infection resistance to PLRV. Selection among protoplasts may provide a solution. Loebenstein has now developed a method to detect differences in virus replication in nn- and NN-tobacco protoplasts.

K. RUDOLPH AND F. EL-BANOBY - MECHANISMS OF RESISTANCE IN BEAN LEAVES TOWARDS BACTERIA

Many bacterial leaf spots start as watersoaked spots. This indicates that the intercellular spaces, normally air-filled, become filled with water, thus allowing the bacteria to multiply. We found that extracellular polysaccharides (EPS) from bacterial cultures produced persistent water-soaked spots in the absence of bacteria. This effect was host-specific at species and cultivar level. EPS from *Pseudomonas phaseolicola* race 2 but not from race 1 produced persistent water-soaking in cv. 'Red Mexican', whereas EPS from both races produced persistent water-soaking in cv. 'Red Kidney'. Similar specific reactions were found for EPS preparations from *Pseudomonas lachrymans, P. pisi, P. glycinea, Xanthomonas malvacearum* and *X. translucens* f. sp. *cerealis*.

When purified EPS from *P. phaseolicola* was infiltrated into the mesophyll of susceptible and resistant bean leaves, electron microscopy revealed that infiltrated EPS was degraded in the resistant tissue within 12 hours. In contrast, the intercellular spaces of the halo-blight susceptible cv. 'Red Kidney' were still completely filled with EPS 3 days after infiltration, and the water-soaked leaf area persisted.

To prove results, *in vitro* studies were also done. The so called intercellular fluid (IF) was obtained by a method developed by Söding and Klement. Leaves were infiltrated with phosphate buffer under vacuum. Then the leaves were centrifuged over a sieve so that the infiltrated water was forced out after rinsing the intercellular spaces. This fluid was named 'so called' IF because in reality there is nothing like an intercellular fluid in the mesophyll of leaves. The IF from 18 bean cultivars was mixed with EPS from *P. phaseolicola*, and the liberation of free sugars was determined. All resistant cultivar IF degraded EPS, whereas IF of the susceptible cultivars caused little or no degradation. Non-degraded EPS was precipitated with ethanol, dialysed, lyophilized, dissolved in water to the usual concentration and infiltrated into bean leaves. The recovered EPS which had been incubated with IF from resistant cultivars was nearly inactive. However, the EPS which had been incubated with IF from susceptible cultivars produced persistent water-soaking. It was concluded, therefore, that in resistant bean leaves the non-persistence of water-soaking caused by bacterial EPS is due to enzymatic degradation and inactivation of EPS.

F. SCHÖNBECK - NON-SPECIFIC RESISTANCE AND SUSCEPTIBILITY INDUCED BY NON-PATHOGENIC ORGANISMS

Disease resistance and susceptibility of plants are not static but elastic features that can vary independently of genetic alteractions. Two models of this are as follows.

VA-Mycorrhiza This can increase and decrease resistance of plants to pathogens. Diseases caused by root invading fungi or nematodes appear to be diminished in most cases in mycorrhizal roots. Diseases of aerial plant parts caused by fungi or viruses are more severe in mycorrhizal plants. Increased or decreased resistance, caused by mycorrhizal infection, appears, other than in virus diseases, to depend more on the kind of plant part infected by the pathogen than on the pathogen itself. The effect of VA-mycorrhiza must be interpreted as an effect acting on the pathogen through the host.

Saprophytic bacteria Culture filtrates of 106 isolates of saprophytic bacteria were applied to leaves of French beans; 2 or 3 days later the plants were inoculated with uredospores of *Uromyces phaseoli*. Fifteen per cent of the tested bacteria decreased or increased rust infection by more than 50 per cent. Two of the most active bacteria (one Gram-negative, one Gram-positive) were selected for more detailed studies of the resistant response. Resistance inducing compounds were found not only in culture filtrates but could also be extracted from bacterial cells with alcoholic solvents. The compounds are not large molecules and most likely are not polysaccharides or proteins.

The period necessary to initiate the resistant response was at least 2 days. The effect lasted for 10 days when the inducer was applied only once; it could be extended by repeated application. The induced resistance spread within the plant, but neither crossed the mid-vein nor moved down the stem. It was effective against obligate biotrophic parasitic fungi, but not against perthotrophic pathogens.

Spore germination and appressorium formation were not inhibited on plant leaves treated with the inducers. Penetration by the infection peg was, however, markedly reduced in the case of powdery mildews. The spread of mycelium within leaves and the formation of haustoria were also diminished remarkably. In bean cultivars reacting with necrosis to the invading rust fungus, the number of necrotic spots decreased when an inducer was applied before the challenge inoculation.

When unprotected plants were inoculated with rust spores from protected plants successful infection was reduced by 50 per cent compared with spores from unprotected leaves. This loss of infectivity persisted in following generations. The germ-tubes of the spores from protected plants were shorter and much more branched than were those of normal spores.

L.M. SHANNON AND C.N. HANKINS - ENZYMIC PROPERTIES OF PHYTOHEMAGGLUTININS

A phytohemagglutinin from mung bean seeds was purified and found to possess a strong enzymatic activity (α-galactosidase). The lectin and enzymatic activities co-purified by methods that would resolve components with significantly different size or charge properties. The purified extract contained a single tetrameric glycoprotein (MW 160,000) composed of identical (or nearly identical) subunits (MW 45,000). The lectin and enzyme possessed indistinguishable carbohydrate (substrate or inhibitor) specificities, both bind to red blood cells; the binding of both to red blood cells is reversible by the same spectrum of sugars, both are equally heat-labile in the presence or absence of sugars, and both are equally sensitive to a sulfhydryl reagent. These results provide compelling evidence that both the lectin and enzymatic activities reside on a single protein species. Under appropriate conditions this protein is capable of enzymatically altering those erythrocyte receptors with which it interacts, resulting in a dissolution of cell aggregates. The disaggregated erythrocytes are permanently altered and are no longer agglutinable by mung bean hemagglutinin although they remain agglutinable by many other legume lectins. Clot dissolving activity is readily observed, even with crude extracts, as a disappearance (dissolution) of erythrocyte aggregates at a rate which is directly proportional to the concentration of extract used. Recently we detected clot dissolving activity in several other legume species, including two which contained previously well characterized non-clot dissolving hemagglutinins. These proteins are closely related to each other and to the mung bean lectin, thus suggesting that each of these plants may contain a homologue from a specific class of enzymic phytohemagglutinin.

G.S. SIDHU - ACTIVATION OR SUPPRESSION OF DISEASE REACTION GENE IN DISEASE COMPLEXES

Various types of true and incidental parasites establish disease complexes when present together on the same host plant. In such situations one parasite may induce (a) resistance or (b) susceptibility against the other parasite. They influence each other's presence either directly through intra- and interspecific interactions and/or by modifying the biochemistry of their common host. All biochemical changes are eventually controlled by the respective genomes of the interacting organisms. Therefore, the genetic basis of the modified disease responses can be studied. Two experiments on such interactions were conducted. (i) *Induced susceptibility* A virulent culture of root-knot nematode (*Meloidogyne incognita*) induced susceptibility in a resistant tomato host which became susceptible to the wilt pathogen *Fusarium oxysporum* f. sp. *lycopersici*. When an F_2 tomato population was exposed to each of these pathogens *singly*, the progeny segregated into four reaction classes 9/16 R-R, 3/16 R-S, 3/16 S-R, 1/16 S-S corresponding in turn to resistance to both, to one, to the other, or to neither of the two pathogens. This indicated two independently segregating dominant R-genes, one effective against the nematode and the other against the fungus. Nevertheless, when the same F_2 population was inoculated with both the pathogens *sequentially*, the four reaction classes were modified to three,9/16 R-R, 3/16 R-S, 4/16 S-S, indicating a recessive epistasis. This epistatic ratio (9:3:4) depended on the suppression of the allele for fungal resistance by the recessive gene for nematode susceptibility in the class 3/16 R-S. It is interesting to note that a ratio 9:7 (R:S) would be obtained if the F_2 progeny, otherwise inoculated with both the pathogens, was evaluated only against the fungus. This is a pseudogenetic ratio and is misleading in breeding for disease resistance. (ii) *Induced resistance* A tomato plant genetically resistant and susceptible to *Fusarium* and *Verticillium* respectively became relatively resistant to *Verticillium* wilt when inoculated with both the pathogens *sequentially*. The F_2 tomato progeny when tested *singly* gave a dihybrid ratio 9/16 R-R, 3/16 R-S, 3/16 S-R, 1/16 S-S (resistant to both, to one, to the other, to neither of the two pathogens). However, when tested *sequentially* a ratio 12/16 R-R, 3/16 S-R, 1/16 S-S (corresponding to resistance to both, to *Verticillium*, to none) appeared which is a characteristic of dominant epistasis. This epistatic (12:3:1) ratio occurred because of the activation of the *Verticillium* susceptible gene by the allele for *Fusarium* resistance in the reaction class 3/16 R-S (resistant to *Fusarium* but susceptible to *Verticillium*). It should be noticed that the ratio 12:3:1 is a pseudogenetic ratio which appears under appropriate conditions of testing and could be misleading in breeding for resistance. Activation or suppression of genes by one parasite against another seems to be a fairly common characteristic in host-parasite systems.

S. SPIEGEL AND A. STEIN - CHANGES IN POLYRIBOSOMES FOLLOWING INJECTION OF A RESISTANCE INDUCING SYNTHETIC POLYANION INTO TOBACCO LEAVES

Injection of ethylene maleic anhydride copolymer (EMA) into leaves of Samsun NN tobacco induced resistance to tobacco mosaic virus (TMV), decreasing both lesion number and lesion size. Development of resistance following application of the inducer was gradual and was sensitive to actinomycin D. This suggested that the transcription mechanism of the cell is involved. Enhanced incorporation of ^{3}H-uracil into the ribosomal fraction of the treated leaves and a significant increase in specific radioactivity (cpm/O.D. 260) became evident at an early stage of the induction period followed by a gradual decrease. This indicated that the translation apparatus of the cell is involved. Protein synthesis is associated with a high proportion of ribosomes in a polymeric form. Therefore we looked at polyribosome profiles after injection of EMA into the leaf tissue.

A time-course study of polyribosome profiles of the polyanion- or water-injected leaves indicated an increase in the polyribosome content in EMA-injected leaves at an early stage of the induction period, followed by a gradual decrease. This phenomenon was specific for Samsun NN tobacco plants which were induced by EMA and was not found when EMA was injected into Samsun tobacco plants which are uninducible by EMA. Also, injection of the polyanionic compound vinyl methyl ether/0.5 methylester of maleic anhydride (VME/MAes), which does not induce resistance, into Samsun NN tobacco leaves did not cause this effect. A comparison of the dissociation rates of polyribosomes from EMA-injected and control leaves in the presence of puromycin indicated that the material observed in a sucrose gradient in the polysome area is true polysomal material with similar rates of dissociation to monosomes and subunits. Monosomes from induced-resistant tissue of Samsun NN were more stable to dissociation in high-KCl media than were monosomes from control or polyanion-treated Samsun plants indicating that active protein synthesis occurs.

The RNase activity of the post-ribosomal supernatant of EMA-injected and control leaves was tested on polyribosomes from unexpanded tobacco leaves and from pea hypocotyls, both having large polysomes. The activity of RNase from both sources was similar for both polyribosome preparations.

It is suggested that the increase in polyribosome content during the induction period reflects increased activity of the translation system and may be linked to the process leading to resistance.

R.C. STAPLES AND B.-F. HUANG - CONTROL OF APPRESSORIUM FORMATION IN THE BEAN RUST FUNGUS

Germ tube development by the bean rust fungus, *Uromyces phaseoli*, is guided by at least three contact stimuli, those for orientation, appressorium formation, and downward growth of the infection peg. Production of haustorial mother cells requires in addition contact of the infection hyphae with the mesophyll wall.

One of the earliest biosynthetic events to occur in bean rust uredospores induced to form infection structures is the synthesis of nuclear DNA. Studies of DNA synthesis were carried out by allowing the spores to germinate in the presence of radioactive adenosine. We found that uredospores begin the synthesis of nuclear DNA some time after the second hour of germination, about the time when the nuclei begin to divide. Until then, DNA synthesis is entirely confined to the mitochondria, and replication of nuclear DNA does not occur.

If viable nuclei are required for formation of appressoria, irradiation of germinating uredospores with gamma rays should inhibit differentiation but not germ tube elongation. By this method, we found that functional nuclei are required for differentiation of infection structures but not germ tube elongation. Cordycepin also inhibits nuclear activity; however, only vesicle formation is prevented not appearance of appressoria. While this drug inhibits nuclear division, it does not inhibit synthesis of DNA, and we concluded that while nuclear division need not precede formation of appressoria, replication of DNA is required. In agreement, Actinomycin D inhibits synthesis and appearance of appressoria.

During studies of gene activity by uredospores forming appressoria, we examined the synthesis of proteins by bean rust uredospores. Of the many proteins synthesized, two groups of proteins are formed only by spores destined to develop infection structures. Clearly, uredospores synthesize a wide range of proteins when induced to germinate and differentiate. Spores committed to develop infection structures shift their synthetic program, and a wide range of proteins are produced which are different from those present in germinating spores.

We have known for some time that uredospores contain a complete system for synthesis of proteins. In an effort to determine whether mRNA is synthesized or activated by spores from endogenous reserves, we assayed the content of polyadenylic acid (polyA). The analyses show that while uredospores contain reasonable levels of polyA (1.2 μg/mg RNA), synthesis of additional polyA did not begin until after six hours when the vesicle develops.

The shift in protein synthesis must be controlled at the level of translation from mRNA stored in the spore. Gene activation apparently does not occur until about the time of the second round of nuclear division after nuclear migration into the vesicle.

I.F.H.J. SUGIYAMA AND V.J. HIGGINS - VARIATIONS WITH TIME IN BIOLOGICAL ACTIVITY OF *CLADOSPORIUM FULVUM* GLYCOPEPTIDES RELEASED *IN VITRO*

The *Cladosporium fulvum*-tomato interaction is a model system for the study of resistance and susceptibility mechanisms within a well documented gene-for-gene host-parasite relationship. Recent studies have concentrated on the role of extracellular glycopeptides in determining cultivar specificity. These glycopeptides have been shown to cause necrosis, membrane permeability changes and callose deposition, and to elicit phytoalexin production in tomato leaves and fruit. To date, no cultivar specificity has been demonstrated for these molecules. The present study involved reassessment of the *in vitro* production and release of *C. fulvum* glycopeptides as a preliminary to determining their possible roles in eliciting host responses *in situ*.

C. fulvum Race 1 was cultured on a synthetic liquid medium and culture filtrates sampled at 72 hour intervals up to 21 days post-inoculation. Culture filtrates were assayed for carbohydrate and protein content and biological activity (necrosis and electrolyte leakage bioassays) before and after a simple purification procedure involving gel filtration through Sephadex G-25 (V_0 fractions) or exhaustive dialysis. Fungal growth (dry weight) was maximal 9 days post-inoculation but extracellular carbohydrate and protein of crude filtrates increased linearly throughout the sampling period. Necrosis-inducing activity of crude filtrates assayed at their original concentrations peaked 6 days post-inoculation and thence remained high. However, assay at lower concentrations (below 50 μg glucose equivalents/ml) resulted in a biphasic activity profile with reduced activity 9 and 12 days post-inoculation. Electrolyte leakage assays at 10 μg glucose equivalents/ml showed similar activity profiles. These trends were accentuated in dialysates and Sephadex G-25 V_o fractions similarly assayed. Preliminary studies indicate that active molecules from the two activity peaks differ markedly in chemical composition. Dialysates showed a decrease in protein and high molecular weight carbohydrate corresponding to samples of low biological activity. This suggests that degradation of the initial active molecules occurs after staling commences. None of the ctive preparations showed cultivar specificity. The data underline the importance of sampling times in *in vitro* studies of this kind. It seems likely that previous workers using 3 week batch filtrates may have enriched their samples with active molecules released post-staling. Intuitively, these are poor candidates as initial determinants in specificity. However, glycopeptides released *in vitro* during staling may be released *in situ* due to stress caused by the resistant response or by sub-optimal conditions for infection. Research is underway to determine if active *C. fulvum* glycopeptides are released *in situ* and if these, or other molecules on the hyphal tip surface initiate the host response.

J.T. TIPPETT - BARRIER ZONES TO VASCULAR PATHOGENS

Barrier zone formation is a non-specific resistance mechanism in trees to vascular pathogens; their formation favours the recovery and survival of infected trees. The barrier zones formed in response to vascular pathogens were atypical xylem rings of axial parenchyma. These protective tissues were common in trees which survived for some years after infection by vascular pathogens. The barrier zones once formed protect the vascular cambium as they limit outward radial spread of pathogens and phytotoxins; they wall off invaded tissues internally.

The prevention of infection plays only a relatively minor role in the resistance of many trees to vascular pathogens as trees are constantly wounded and parts are constantly shed which allows direct entry of fungi into the xylem. Trees once invaded by vascular pathogens initiate defence mechanisms, such as tyloses and gum production, to limit vertical spread of the pathogens, but mechanisms restricting outward radial spread are also important in disease resistance. Barrier zone formation is such a mechanism.

Barrier zones have been observed in trees recovering from various vascular diseases. The anatomy of the zones has been studied in the following examples; Dutch elm disease, *Ulmus americana* L. and *Ceratocystis ulmi* (Buism.); sugar maple sap streak, *Acer saccharum* Marsh. and *Ceratocystis coerulescens* (Munch) Bakshi; verticillium wilt, *Acer platanoides* L. and *Verticillium albo-atrum* Reinke and Berth; mimosa wilt, *Albizia julibrissin* Durazz and *Fusarium oxysporum* f. sp. *perniciosum* (Hepting) Toole. Circular patterns of sapwood discoloration were common in branch and stem disks collected from trees exhibiting some resistance to these diseases. Pathogen invasion was often limited to single annual growth increments and barrier zones circumscribed such necrotic rings of sapwood. In all examples, barrier zone parenchyma was derived as a result of septation of fusiform initials. The cells retained their nuclei and cell membranes, and accumulated starch and, subsequently, polyphenols. The bands of xylem parenchyma constituting the barrier zones were 5 to 20 cell layers wide.

The protective nature of the barrier zones may be related to their ability to accumulate polyphenols. Another property of the zones is that the parenchyma forms a continuous symplast beneath the cambium, hence restricting passive diffusion of phytotoxic materials from areas of necrotic sapwood towards the cambium.

Other anatomists and pathologists have realized that cambial activity may be affected by pathogens, but that the tissues produced after injury or infection may have protective functions and be involved in resistance has not been discussed. Recognition of nonspecific resistance mechanisms in perennial plants will aid in understanding the basis of the varying susceptibility of some tree species to pathogens.

E.C. TJAMOS - SYSTEMIC INDUCTION OF PROTECTION TO VERTICILLIUM WILT IN CUCUMBERS

High molecular weight (HMW) preparations of *Verticillium albo-atrum* culture filtrates applied to the leaves of wilt susceptible cucumber plants can trigger a resistance mechanism(s) and induce systemic protection against the same pathogen.

Differences between treated and untreated plants were seen as limited symptom development expressed as mean percentages of wilted, chlorotic or desiccated leaves over the total number of expanded leaves, and as restricted vessel colonization calculated as mean percentages of colonized proto- and metaxylem vessels in sections cut below the cotyledonary node 30 days after root inoculation.

This systemic protection was initially obtained by spraying cucumber plants once 4 days before inoculation. It was later shown that more than one applications either before or/and after inoculation were far more effective. This booster effect was also evident when a high inoculum concentration was used to secure the best possible infection. Further studies which investigated whether the factor(s) involved are thermolabile, demonstrated that both autoclaved and non-autoclaved HMW preparations can induce protection. A crude cell wall extract preparation obtained from *V. albo-atrum* conidia did not induce protection. Treatment of the HMW component of culture filtrates with protease or ribonuclease before application to cucumber plants did not affect protection. Plants sprayed with autoclaved, non-autoclaved or enzyme treated preparations developed similar disease patterns which differed statistically from those of the controls. Also, plants sprayed with a ribonuclease preparation were protected. This striking effect was obtained in a second trial in which ribonuclease and HMW culture filtrate preparations were compared.

The existence of a potential resistance mechanism in Verticillium susceptible cucumber plants was clearly shown. The mechanism can be challenged by the HMW component of *V. albo-atrum* culture filtrates but not by cell wall extracts. The possible involvement of other polysaccharides cannot be excluded. The inducer(s) of systemic protection are heat resistant and their activity is retained after treatment with protease or ribonuclease. The protection was made more evident by using a constant inoculum level (10^7 conidia ml^{-1}) and increasing either the concentration of the HMW preparations (5.5, 8.4 and 33 mg protein 100 ml^{-1}) or the times of application prior to,or prior to and/or after root inoculation (1 - 4 times). Increase of the conidial concentration 5 x 10^7 conidial ml^{-1}) suppressed the protection unless a booster was applied after inoculation. Finally, the mechanism(s) can be also triggered by a ribonuclease preparation (1 mg in 60 ml of a sucrose sodium nitrate liquid medium).

J. TORP - THE BARLEY/POWDERY MILDEW INTERACTION : CHARACTERIZATION OF FUNGAL CULTURES BY TWO-DIMENSIONAL GEL ELECTROPHORESIS

Proteins from conidia of the barley powdery mildew fungus (*Erysiphe graminis* f. sp. *hordei*) were examined by two-dimensional polyacrylamide gel electrophoresis. Preliminary results indicate that the protein patterns obtained provide an efficient way to characterize and identify individual cultures.

Conidia were collected from heavily sporulating, young barley plants which had been inoculated two weeks earlier with a genetically pure culture (a clone) of the fungus. The conidia were sucked into, and retained by an aqueous detergent, by means of a small mouthpiece connected to a source of vacuum. Various vital stains showed that more than 90 per cent of the conidia were actively metabolizing one hour or more after being collected. However, to avoid autohydrolysis, the suspension of conidia was immediately frozen in liquid N_2, and freeze-dried. After drying, the conidia may be stored frozen in a desiccator for some months without any apparent alterations of the proteins. The dried conidia contained 4.5 per cent nitrogen.

The most efficient extraction of the proteins was obtained by sonicating the conidia in a weak, neutral buffer with urea and β-mercaptoethanol. The samples were then incubated for one hour at $4^{o}C$, centrifuged, and the supernatant applied to the gel. This procedure solubilized 50 - 60 per cent of the nitrogen present in the conidia.

The proteins were first separated according to their isoelectric point by electrofocussing in a 6 per cent polyacrylamide slab gel containing 6M urea and 2 per cent ampholines. The ampholines used gave an S-shaped pH-gradient in the range of pH 3.5 to 8.5. The samples were applied in pre-cast wells in the gel. In the second dimension, the proteins were separated by SDS-polyacrylamide gel electrophoresis using a discontinuous system with a 5 per cent stacking gel and a 12 per cent running gel. By staining the gels to equilibrium with Coomassie Blue, *c*. 200 spots could be identified. The spots were fairly evenly distributed over the gel which shows that there was no conspicuous relation between molecular weight and isoelectric point of the proteins extracted.

So far, it has been possible to distinguish the different cultures of the fungus by means of their protein patterns. Any two cultures differed in 2 - 6 spots, whereas the majority of the spots appeared in identical positions and intensities. Most of the variation was found in the pi-values.

The available observations indicate that the patterns are insensitive to fluctuations in environment during fungal growth. The two-dimensional protein separation technique therefore appears to be a useful way of 'fingerprinting' different genotypes of the fungus, and may prove to be valuable in ontogenic, biochemical and genetic experiments.

This work was supported by a grant from the Danish Agricultural and Veterinary Research Council.

A.M. TRONSMO - TEMPERATURE AS A PREDISPOSING FACTOR IN RESISTANCE TOWARDS LOW TEMPERATURE FUNGI ON GRASSES

Snow moulds on grasses caused by *Fusarium nivale*, *Typhula incarnata*, *Typhula ishikariensis* or *Sclerotinia borealis* are among the most important plant diseases in northern countries.

Damage caused during the winter is influenced by autumn temperatures. A warm autumn followed by a sudden snowfall may result in severe damage. In contrast, snow following a 2 - 3 week period at a few degrees above freezing and sufficient light intensity gives much less disease. This difference can be explained by hardening of plants exposed to low temperatures. Hardening is known to be essential for plants to tolerate freezing but its effect on disease resistance has been little studied.

Artificially hardened plants have been shown to be significantly more resistant towards snow moulds than unhardened plants, and there is a significant correlation between freezing tolerance and disease resistance (Arsvoll, 1977).

During hardening the low temperature stresses the plants. Biological stress is defined as "an environmental factor capable of inducing a potentially injurious strain in living organisms", and hardening can be defined as "an exposure to sublethal stress which results in resistance to an otherwise lethal stress factor".

Low temperature stress will predispose plants; their resistance should increase or decrease depending on the level of this stress.

The predisposing effect of temperature was studied with artificially hardened (3 weeks of + 1°C and light) and unhardened *Phleum pratense* plants. Prior to inoculation with *F. nivale* or *T. ishikariensis* plants were exposed to stress at -4 or -8°C. Unhardened, unstressed plant suffered 50 per cent damage, unhardened, stressed plants suffered 97 per cent damage, whereas hardened plants suffered 30 per cent damage. The results indicate that when temperature predisposes plants, susceptibility will increase or decrease, dependent on the level of this stress.

The possibility that availability of water restricts the pathogen was also investigated. The water potentials of plants during hardening leaves on intact plants and crown tissues showed a decrease from -4.5 to -15 bars and from -9.0 to -13.5 bars respectively.

The ability of *F. nivale* and *T. ishikariensis* to grow at different water potentials was studied on PD-broth amended with KCl or PEG. Growth of both fungi was reduced by 50 per cent when the water potential was decreased from -7 to -20 bars.

Water potential can partly but not wholly explain increased resistance in hardened plants.

Arsvoll, K. 1977. *Meld. Norg. LandbrHøgsk.* 56 (28), 1 - 14.

D. HENDRINA WIERINGA-BRANTS - THE RÔLE OF THE EPIDERMIS IN VIRUS-INDUCED LOCAL LESIONS ON COWPEA AND TOBACCO LEAVES AND POSSIBLE INFLUENCE ON ISOLATED PROTOPLASTS

Removal of the lower epidermis from virus-inoculated cowpea or tobacco leaves resulted in reduction of lesion numbers when a cowpea strain of TMV or a D-strain of TNV was used. The time-dependent reduction was greatly influenced by shading the plants 24 hours prior to inoculation. The contact period between epidermis and mesophyll required for lesion formation differed markedly in shaded and non-shaded plants. For cowpea with TMV this period was 1.5 and 8 hours; in cowpea with TNV, 1 and 6 hours; in Xanthi nc tobacco and TMV 1 and 8.5 hours; in tobacco and TNV, 1 and 6 hours respectively. So shading hypersensitive reacting plants prior to inoculation seems to affect drastically the time needed for infectious virus entities to enter mesophyll tissue from infected epidermal cells. Epidermal strips from inoculated leaves were kept for 5 days, homogenized by ultrasonic treatment and tested for virus activity. TMV and TNV multiplied in isolated epidermal strips of tobacco but not in those of cowpea. Epidermis-stripped leaves of cowpea and tobacco were inoculated with TMV or TNV. Cowpea mesophyll could not be infected in this way, nor with TNV-RNA. However, inoculation of tobacco mesophyll with TMV or TNV did provoke lesion formation. Clearly there are differences in the course of virus infection between cowpea and tobacco when TMV or TNV is used.

To study these differences more extensively, virus multiplication was examined in mesophyll protoplasts isolated from cowpea and tobacco leaves after treatment with pectolyase T 23. In cowpea protoplasts synthesis of TMV started 8 hours later than in tobacco protoplasts and the final virus concentration 3 days after inoculation was less. Similar results were obtained for TNV in cowpea and tobacco protoplasts. It is possible that necrosis is not expressed in mesophyll protoplasts because of the absence of epidermal cells in protoplast cultures. When inoculated epidermal strips are added to infected protoplast cultures a much higher percentage of protoplasts die than in the control cultures. Doke observed that potato protoplasts can react "hypersensitively" after treatment with mycelial cell wall components by "ghost"-formation; this also occurred in cowpea protoplasts infected with virus. It is not yet clear if this "hypersensitive"reaction of virus-infected protoplasts is correlated with necrosis of leaf cells caused by virus infection.

P.J.G.M. DE WIT - PARTIAL CHARACTERIZATION AND SPECIFICITY OF GLYCOPROTEIN ELICITORS PRESENT IN FILTRATES OF CULTURES AND CELL WALLS OF *CLADOSPORIUM FULVUM*

Cladosporium fulvum, causal agent of tomato leaf mould produced high molecular glycoprotein elicitors in shake cultures that induced rishitin accumulation in tomato fruit tissue. Glycoprotein elicitors could also be isolated from cell walls of *C. fulvum*.

Elicitors from culture filtrates (CFE) and cell walls (CWE) were not race or cultivar-specific. Elicitors were even not host-specific, because they induced the accumulation of pisatin in pea pods and glyceollin in soybean cotyledons. The elicitors did not induce rishitin and medicarpin in potato tuber discs and in jack bean cotyledons,respectively.

Elicitors were sensitive to $NaIO_4$, α-mannosidase, pronase, proteinase K and NaOH, confirming that they are glycoproteins.

The chemical composition of CFE depended on the age of the culture and the growth medium. CFE of 3 week cultures varied less in chemical composition and was more active than CFE of 1 week cultures. The glucose content of CFE of fast growing cultures was higher than that of slowly growing cultures. The chemical composition of CWE was constant and independent of the age of the culture and the growth medium. CWE was about five times as active as CFE. It is likely that CFE is derived from cell walls and that its variability is caused by breakdown by enzymes in culture filtrates.

CWE was further fractionated with hexadecyltrimethylammonium bromide in borate buffer and yielded a glycoprotein with a high mannose and galactose content. This glycoprotein was further purified on Sepharose-4B and Sephadex G-200; it bound completely to Concanavalin A Sepharose-4B. In sequential purification mannose (or galactose) content of the glycoprotein was positively correlated with rishitin and necrosis-inducing activity.

The purified elicitor was homogeneous but rather polydisperse with a mannose : galactose ratio of 5 : 4 for 6 races of *C. fulvum*;

the glucose content was slightly variable. Methylation analysis showed that some mannose and nearly all galactose residues were terminal. About 90 per cent of the galactose residues were released by weak acid which is explained by the presence of galactofuranosyl residues. The main chain of the polysaccharide in the glycoprotein consisted of 1 → 2 linked mannose residues.

The main amino acids in the protein part of the glycoprotein were serine, threonine, proline, glutamic and aspartic acid. Serine and threonine were involved in linkages between polysaccharide and protein.

The purified glycoprotein elicitor had an extremely high rishitin and necrosis-inducing activity.

M.A. YILMAZ, N. KASKA, A. ÇINAR AND O. GEZEREL - REDUCTION OF VIRUS DISEASE EFFECTS ON TOMATO BY BARRIERS IN ÇUKUROVA REGION

Virus diseases are widely distributed in southern Turkey. During a survey in glasshouses along the Mediterranean belt, we did not find any tomato plants free of virus. Even tomatoes grown from resistant seeds introduced by well known firms were infected. Viruses detected in the Çukurova region were TMV, CMV, PYV, PVX, TMSV and TYLCV. TYLCV (Tomato Yellow Leaf Curl Virus), the most destructive, is whitefly-borne. Losses in yield caused by TYLCV were reduced by using cheesecloth as a barrier. Seedlings of Linda and Super Marmande cultivars were transplanted in a glasshouse at the 4 - 10 leaf stage with 12 plots of 10 seedlings for each cultivar. Six plots were covered with 2 m high fine mesh cheesecloth cages after transplanting; the other 6 plots were left open. Yield, soluble solids, vitamin C content of fruit, stem weight, stem height, root weight, root lengths, macro and micro element uptake were determined at the end of the growing period.

Tomatoes grown in open plots showed systemic mosaic yellowing and upward curling type of leaf symptoms 2 - 4 weeks after transplanting. Similar symptoms appeared in caged plants only one month before final harvesting. This shows that transmission of the virus to the plants by whiteflies was prevented by cheesecloth during early growth. Late infection by TYLCV did not have much effect on yield or other factors. Fruit size in TYLCV infected plants was significantly less than in healthy plants. The increase in mean fruit weight of plants grown in cages was approximately 2-fold for both cultivars. The mean yield was 2.8 fold higher for Linda and 2.7 fold for Super Marmande cultivars in caged plants compared to those in open plots. The vitamin C content of fruit of healthy and TYLCV infected plants were similar for both cultivars. The plants in open plots were stunted and showed considerable decreases in soluble solids, stem weight, stem length, and root length. Macro and micro element uptake of healthy and TYLCV infected plants were similar.

This research was supported by Turkish Scientific and Technical Research.

PARTICIPANTS

ALBERSHEIM, Professor P. Department of Chemistry, University of Colorado, Boulder, Colorado 80309, U.S.A.

ARCHER, Dr. S.A. Botany Department, Imperial College, Prince Consort Road, London SW7 2BB, U.K.

BAILEY, Dr. J.A. Long Ashton Research Station, Long Ashton, Bristol, BS18 9AF, U.K.

BATEMAN, Professor D.F. North Carolina State University, Box 5847, Room 100 Patterson Hall, Raleigh, N.C. 27650, U.S.A.

BEM, Dr. F. Benaki Phytopathological Institute, Delta 8, Kiphissia, Athens, Greece.

CALLOW, Dr. J.A. Department of Plant Sciences, The University of Leeds, Baines Wing, Leeds LS2 9JT, U.K.

CERVONE, Dr. F. Istituto Dell'Orto Botanico, Universita di Roma, Largo Cristina di Svezia 24, 00165 Roma, Italy.

CLARKE, Dr. D.D. Botany Department, Glasgow University, Glasgow, Scotland, U.K.

COOPER, Dr. R.A. Department of Biological Sciences, Bath University, Bath, U.K.

COUTTS, Dr. R.H.A. Botany Department, Imperial College, Prince Consort Road, London SW7 2BB, U.K.

COUVARAKI, Miss C. Athens College of Agricultural Sciences, Votanikos, Athens 301, Greece.

DÉFAGO, Dr. Miss G. Eidg. Technische Hochschule, Institut für Phytomedizin, Universitätstrasse 2, 8092 Zürich, Switzerland.

DESJARDINS, Dr. Miss A. Department of Chemistry, University of Colorado, Boulder, Colorado 80309, U.S.A.

DEVERALL, Professor B.J. Department of Plant Pathology, University of Sydney, Sydney 2006, N.S.W., Australia.

ELGERSMA, Dr. D.M. Phytopathological Laboratory, "Willie Commelin Scholten", Javalaan 20, Baarn, The Netherlands.

ELLINGBOE, Professor A.H. Genetics Program and Department of Botany and Plant Pathology, Michigan State University, East Lansing, Michigan 48824, U.S.A.

ESQUERRÉ-TUGAYÉ, Dr. Miss M.T. Université Paul Sabatier, Centre de Physiologie Végétale, 118 route de Narbonne, 31062 Toulouse Cedex, France.

FRIEND, Professor J. Department of Plant Biology, The University, Hull, HU6 7RX, U.K.

FRITIG, Dr. B. Institut de Biologie Moléculaire et Cellulaire, Laboratoire de Virologie, 15 Rue Descartes, 67000 Strasbourg, France.

FUCHS, Dr. A. Department of Phytopathology, Agricultural University Wageningen, 9 Binnenhaven, Wageningen, The Netherlands.

FULTON, Professor R.W. Department of Plant Pathology, University of Wisconsin, 1630 Linden Drive, Madison, Wisconsin 53706, U.S.A.

GEORGOPOULOS, Professor S.G. Athens College of Agricultural Sciences Votanikos, Athens 301, Greece.

GERA, Dr. A. Agricultural Research Organization, The Volcani Center, Institute of Plant Protection, P.O.B. 6, Bet Dagan 50-200, Israel.

GIANINAZZI, Dr. S. Station d'Amélioration des Plantes, I.N.R.A., B.V. 1540, 21034 Dijon Cedex, France.

GIBBS, Dr. A.J. School of Biological Sciences, Australian National University, P.O. Box 475, Canberra 2601, Australia.

GOODMAN, Professor R.N. Department of Plant Pathology, University of Missouri, 108 Waters Hall, Columbia, Missouri 65211, U.S.A.

GRANITI, Professor A. Universita degli di Bari, Istituto di Patologia Vegetale, Via Giovanni Amendola 165/A, 70126 Bari, Italy.

GULSOY, Mr. H.E. Erenköy Plant Protection Institute, Erenköy, Istanbul.

HARRISON, Dr. B.D. Scottish Crop Research Institute, Invergowrie, Dundee, Scotland, U.K.

HEATH, Dr. Mrs.M.C. Department of Botany, University of Toronto, Toronto, Ontario M5S 1A1, Canada.

HEITEFUSS, Professor R. Institut für Pflanzenpathologie und Pflanzenschutz der Georg-August-Universität, Grisebachstrasse 6, 3400 Göttingen-Weende, Federal Republic of Germany.

INGRAM, Dr. D.S. University of Cambridge, Botany School, Downing Street, Cambridge CB2 3EA, U.K.

KEEN, Professor N.T. Department of Plant Pathology, University of California, Riverside, California 92521, U.S.A.

KUĆ, Professor J. Department of Plant Pathology, University of Kentucky, Lexington, Kentucky 40546, U.S.A.

LANGCAKE, Dr. P. Shell Biosciences Laboratory, Sittingbourne Research Centre, Sittingbourne, Kent, U.K.

LAZAROVITS, Dr. G. Agriculture Canada, Research Institute, University Sub-Post Office, London, Ontario, Canada.

LEGRAND, Dr. M. Institut de Biologie Moléculaire et Cellulaire, Laboratoire de Virologie, 15 Rue Descartes, 67000 Strasbourg, France.

LOEBENSTEIN, Professor G. Agricultural Research Organization, The Volcani Center, Institute of Plant Protection, P.O.B. 6, Bet Dagan 50-200, Israel.

MANSFIELD, Dr. J.W. Wye College, Wye, Ashford, Kent, U.K.

MARAITE, Dr. H. Laboratoire de Phytopathologie, Université Catholique de Louvain, Place Croix du Sud 3, B-1348 Louvain-la-Neuve, Belgium.

MARTELLI, Professor G. Universita degli di Bari, Istituto di Patologia Vegetale, Via Giovanni Amendola 165/A, 70126 Bari, Italy.

MATTA, Professor A. Istituto di Patologia Vegetale dell' Universita degli Studi di Torino, 10126 Torino, Via Pietro Giuria 15, Italy.

MAULE, Dr. A.J. Department of Microbiology, University of Birmingham, Edgbaston, Birmingham B15 2TT, U.K.

PAXTON, Dr. J.D. Department of Plant Pathology, University of Illinois at Urbana-Champaign, B-519 Turner Hall, Urbana, Illinois 61801, U.S.A.

PIERPOINT, Dr. W.S. Rothamsted Experimental Station, Harpenden, Hertfordshire, U.K.

PSALLIDAS, Dr. P.G. Benaki Phytopathological Institute, Delta 8, Kiphissia, Athens, Greece.

RATHMELL, Dr. W.G. I.C.I. Plant Protection Division, Jealott's Hill Research Station, Bracknell, Berkshire, U.K.

RICCI, Dr. P. I.N.R.A. Station de Pathologie Végétale, BP 78, F-06602 Antibes, France.

ROHRINGER, Dr. R. Agriculture Canada, Research Station, 195 Dafoe Road, Winnipeg, Manitoba R3T 2M9, Canada.

ROSS, Professor H. University of Cologne, Dompfaffenweg 33, 5000 Köln 30, Federal Republic of Germany.

RUDOLPH, Dr. K. Institut für Pflanzenpathologie und Pflanzenschutz der Georg-August-Universität, Grisebachstrasse 6, 3400 Göttingen-Weende, Federal Republic of Germany.

SCHÖNBECK, Professor F. Institut für Pflanzenkrankheiten und Pflanzenschutz, Universität Hannover, Herrenhauser Str. 2, 3000 Hannover 21, Federal Republic of Germany.

SEQUEIRA, Professor L. Department of Plant Pathology, University of Wisconsin, 1630 Linden Drive, Madison, Wisconsin 53706, U.S.A.

SHANNON, Professor L.M. Department of Biochemistry, University of California, Riverside, California 92521, U.S.A.

SIDHU, Dr. G.S. Department of Biological Sciences, Simon Fraser University, Burnaby, B.C., Canada.

SMEDEGAARD-PETERSEN, Dr. V. The Royal Veterinary and Agricultural University, Department of Plant Pathology, Thorvaldsenavej 40, DK-1871, Copenhagen, Denmark.

SPIEGEL, Dr. Miss S. Agricultural Research Organization, The Volcani Center, Institute of Plant Protection, P.O.B. 6, Bet Dagan 50-200, Israel.

STAPLES, Professor R.A. Boyce Thompson Institute, Tower Road, Ithaca, N.Y. 14853, U.S.A.

STAUB, Dr. T. Ciba-Geigy Limited, Division of Agricultural Chemicals, 4002 Basel, Switzerland.

SUGIYAMA, Mr. J. Department of Botany, University of Toronto, 6, Queen's Park Crescent, Toronto, Ontario M5S 1A1, Canada.

TIPPETT, Dr. Miss J. C.S.I.R.O., Division of Forest Research, P.O. Box 144, Kelmscott, Western Australia.

TJAMOS, Dr. E.C. Benaki Phytopathological Institute, Delta 8, Kiphissia, Athens, Greece.

TORP, Dr. J. Agricultural Department, Risø National Institute, DK-4000, Roskilde, Denmark.

TOUZÉ, Professor A. Université Paul Sabatier, Centre de Physiologie Végétale, 118 route de Narbonne, 31062 Toulouse Cedex, France.

TRONSMO, Dr. Mrs.A.M. Agricultural University, 1432 Ås-NLH, Norway.

VAN LOON, Professor L.S. Department of Plant Physiology, The Agricultural University, Arboretumlaan 4, 6703 BD, Wageningen, The Netherlands.

WIERINGA-BRANTS, Professor Mrs.D.H. Phytopathological Laboratory, "Willie Commelin Scholten", Javalaan 20, Baarn, The Netherlands.

de WIT, Ir. P.J.G.M. Department of Phytopathology, Agricultural University, 9 Binnenhaven, Wageningen, The Netherlands.

WOLF, Dr. G. Institut für Pflanzenpathologie der Universität, D-3400 Gottingen, Grisebachstrasse 6, German Federal Republic.

WOOD, Professor R.K.S. Botany Department, Imperial College, Prince Consort Road, London SW7 2BB, U.K.

YILMAZ, Professor M.A. C.U. Ziraat Fakültesi, Bitki Koruma Bolümü, Adana, Turkey.

ZIOGAS, Mr. B.N. Athens College of Agricultural Sciences, Votanikos, Athens 301, Greece.

INDEX